Biomedical Applications of DNA Recombinant Technology

Biomedical Applications of DNA Recombinant Technology

D. Yashwant Dugyala

RANDOM

RANDOM PUBLICATIONS

NEW DELHI (INDIA)

Biomedical Applications of DNA Recombinant Technology

ISBN 978-93-5111-604-2

Published in 2015 in India by

RANDOM PUBLICATIONS

4376-A/4B, Gali Murari Lal, Ansari Road
New Delhi-110 002
Phone : +9111-43580356, 011-23289044, 011-43142548
e-mail: sales@randompublications.com,
info@randompublications.com, randomexports@gmail.com

Reprinted 2026

Type Setting by : Friends Media, Delhi-110089

Preface

Recombinant DNA molecules are sometimes called chimeric DNA, because they are usually made of material from two different species, like the mythical chimera. R-DNA technology uses palindromic sequences and leads to the production of sticky and blunt ends.

The DNA sequences used in the construction of recombinant DNA molecules can originate from any species. For example, plant DNA may be joined to bacterial DNA, or human DNA may be joined with fungal DNA. In addition, DNA sequences that do not occur anywhere in nature may be created by the chemical synthesis of DNA, and incorporated into recombinant molecules. Using recombinant DNA technology and synthetic DNA, literally any DNA sequence may be created and introduced into any of a very wide range of living organisms.

Proteins that can result from the expression of recombinant DNA within living cells are termed recombinant proteins. When recombinant DNA encoding a protein is introduced into a host organism, the recombinant protein is not necessarily produced. Expression of foreign proteins requires the use of specialized expression vectors and often necessitates significant restructuring of the foreign coding sequence.

Recombinant DNA differs from genetic recombination in that the former results from artificial methods in the test tube, while the latter is a normal biological process that results in the remixing of existing DNA sequences in essentially all organisms.

The principal element in creating recombinant DNA constructs is the incorporation of the foreign DNA segment to be cloned (the so-called insert) into the vector DNA. The insert to be cloned is often a segment of the genomic DNA of an organism to be investigated, or a cDNA (complementary DNA) segment synthesised via reverse transcription from RNA molecules of the organism of interest. The genomic DNA fragments to be cloned are often produced by the fragmentation of the genome of the investigated organism by using restriction endonuclease enzymes, and the subsequent isolation of these restriction fragments. Restriction endonuclease-catalysed fragmentation is also often necessary during the cloning of cDNA products.

The book has been designed for under-graduates, post-graduates and technicians who wish to know and use the principles and techniques of Recombinant DNA Technology.

I would like to thank my team for standing beside me throughout my career and writing this book. My special thanks go to "Random Publications" who have published the book.

– D. Yashwant Dugyala

Contents

1

Recombinant DNA Technology

INTRODUCTION

Using Recombinant DNA technology, we can isolate and clone single copy of a gene or a DNA segment into an indefinite number of copies, all identical. These new combinations of genetic materials or Recombinant DNA '(rDNA)' molecules are introduced into host cells, where they propagate and multiply. The technique or methodology is called Recombinant DNA technology or "Genetic engineering". The first recombinant DNA molecules were generated by Paul Berg, Herbert Boyer, Annie Chang, and Stanley Cohen in 1973.

To obtain rDNA steps involved are:

- The DNA fragment containing the gene sequence to be cloned also known as ('insert') is isolated.
- Insertion of these DNA fragments into a host cell using a "vector" (carrier DNA molecule).
- The rDNA molecules are generated when the vector self replicates in the host cell.
- Transfer of the rDNA molecules into an appropriate host cell.
- Selection of the host cells carrying the rDNA molecule using a marker.
- Replication of the cells carrying rDNA molecules to get a genetically identical cells population or clone.

ENZYMES USED IN RECOMBINANT DNA TECHNOLOGY

RESTRICTION ENZYMES

For cloning of DNA, the DNA is cut at specific sites, which are recognised and cleaved by specific enzymes. These enzymes are known as restriction enzymes. These restriction enzymes recognise short sequences of double stranded DNA as targets for cleavage. Different enzymes recognise different but specific sequences, each ranging from 4-8 base pairs. The enzymes are named by a three letter (or four letter) abbreviation that identifies their origin *e.g.* AluI is derived from *Arthrobacter luteus*, *EcoRI* is derived from *E.Coli,* HpaI is derived from Haemophilus parainfluenzae. Besides cleavage, modification in the form of methylation is also brought about by some enzymes called modification enzymes sometimes also called "methylases". Different species of bacteria contain their own sets of restriction enzymes and corresponding methylases. Depending on the different modes of action, the restriction enzymes have been classified into three main classes- type I, type II, type III. Out of these, type II restriction enzymes are used in recombinant technology as they can be used in vitro to recognise and cleave with in specific DNA sequences usually consisting of 4-8 nucleotides.

DNA Ligases

These enzymes form phosphodiester bonds between the adjacent molecules and, covalently links two individual fragments of double stranded DNA. The most commonly used enzyme for cloning experiment is T4 DNA ligase.

Alkaline Phosphatase

The enzyme alkaline phosphatase (AP) removes the phosphate group from the 5' end of a DNA molecule leaving a free 5' hydroxyl group hence it is used to prevent unwanted self-ligation of vector DNA molecules in cloning procedures. This enzyme is isolated from bacteria (BAP) or calf intestine (CAP).

Cloning Vectors for Recombinant DNA

Vectors are the vehicles used to carry a foreign DNA sequence into a given host cell.

A vector should have:

- Origin of replication,
- A selectable marker to select the host cells containing the vector from among those which do not have the vector,
- One unique restriction endonuclease recognition site to enable foreign dna to be inserted into the vector in order to generate a recombinant dna molecule,
- It should be relatively small in size.

Plasmids as Vectors

Plasmids are defined as autonomous elements, whose genomes exist in the cell as extrachromosomal units. They are self replicating, circular duplex DNA molecules present in number of copies in a bacterial cell, yeast cell or in organelles in eukaryotic cells. These naturally occurring plasmids have been modified to serve as vectors in the laboratory.

pBR322 Vectors

One of the most commonly used cloning vector in gene cloning procedures is pBR322 (named after Bolivar and Rodriguez who constructed this) derived from E. coli plasmid ColE1. It is 4,362 bp DNA with genes for resistance against two antibiotics- tetracycline and ampicillin. It was constructed after several alterations and modifications in earlier cloning vectors.

pUC Vectors

The plasmids belonging to pUC series of vectors are 2,700 bp long with ampicillin resistance gene, the origin of replication derived from pBR322 and the lac Z gene derived from E.coli. When the DNA fragments are cloned in this region of pUC, the lac gene is inactivated.

Yeast Plasmid Vectors

Special vectors used to introduce DNA segments in yeast cells or a eukaryotic cell is being used to develop genetically engineered yeast cells. *e.g.* YIp or yeast integrative plasmids which allows transformation by crossing over and have no origin of replication. YEp or yeast episomal plasmids carry 2 micron DNA sequence including the origin of replication and *rep* gene. These vectors have been widely used to study yeast genome.

Shuttle Vectors

The vectors containing two types of origin of replication which helps them to exist in both eukaryotic cell as well as E.coli are known as shuttle vectors. *e.g.* Yep vector.

Retriever Vectors

A class of vectors which are used to retrieve specific genes from the normal chromosome of an organism like yeast through recombination. They are very useful in isolation of genes from yeast for molecular experiments like sequencing.

Vectors Based on Bacteriophages

Bacteriophages are viruses that infect bacterial cells by injecting their DNA into these cells.

They are used as vectors because they have a linear DNA molecule, which generate two fragments after breakage. These are later joined with foreign DNA to generate chimeric phage particle. The injected DNA is selectively replicated and expressed in the host cell resulting in a number of phages which burst out of the cell (lytic pathway) and further infect the neighbouring cells. *e.g.* M13, Lambda (λ).

Cosmids as Vectors

Cosmids are plasmid particles into which specific DNA sequences like cos sites of phage lambda are inserted and they are constructed by combining certain feature of plasmids and the 'cos' sites of phage lambda. The advantage of the using cosmids for cloning is that its efficiency is high to produce a complete genomic library of 10(6)-10(7) clones from only 1 microgrm of insert DNA.

Phagemids as Vectors

Phagemids are prepared artificially by combining features of phages with plasmids. *e.g.* pBluescript II KS is derived from pUC19 and is 2961 bp long.

BAC Vectors

BACs or Bacterial Artificial Chromosomes are vectors based on the natural, extra-chromosomal plasmid of E.coli- the fertility or F-plasmid, and are being used in genome sequencing projects. A BAC vector contains genes for replication and maintenance of the F-factor, a selectable marker and cloning sites and can accommodate up to 300-350 kb of foreign DNA.

Plant and Animal Viruses as Vectors

A number of plant and animal viruses can also be used as vectors for introducing foreign genes into cells and/or for gene amplification in host cells. Plant viruses like Gemini viruses, cauliflower mosaic virus or CaMV and tobacco mosaic virus/TMV) are three group of viruses that have been used as vectors for cloning of DNA segments. CaMV has a double stranded DNA molecule, 8 kbp in size which infects the members of Cruciferae family. Geminiviruses are a group of single stranded DNA plant viruses infecting cassava, maize and other cereals.

Animal virus vectors also deliver the foreign genes into the cultured cells which get integrated into the host genome. The expression of foreign genes can also be amplified using the promoters of the virus gene. The cloned genes can be used in gene therapy, for the synthesis of important proteins etc. A vector based on Simian Virus 40 (SV 40) was used in the first cloning experiment involving mammalian cells in 1979. Retroviruses like murine and avian retroviruses are very useful vectors as they are capable of infecting a large variety of cell types and can clone large genes. Herpes virus is a non-integrating large sized virus (150 kb) which is another useful vector which can be used for expression of large intact genes.

Artificial Chromosome (YAC and MAC) Vectors

YACs or Yeast Artificial Chromosomes are used as vectors to clone DNA fragments of more than 1 Mb in size. These long molecules of DNA can be cloned in yeast by ligating them to vector sequences that allow their propagation as linear artificial chromosomes. These long DNA molecules can be generated and allow construction of comprehensive libraries in microbial hosts. A lot of work is going on to create mammalian artificial chromosomes (MACs) following the isolation of mammalian telomeres and centromere. However YACs have a low cloning efficiency (1000 clones/microgm) DNA as against 106-107 clones/ microgm DNA for cosmids) and also it is difficult to recover large amount of pure insert DNA from individual clones.

WORKING PROCESS OF ENZYMES

ACTIVE SITES AND CATALYTIC MECHANISMS

Open any textbook of biochemistry and you will be presented with an overwhelming number of figures depicting the crystal structures of a multitude of enzymes. Of course, the three-dimensional structures of enzymes are crucial to their functions. Many enzymes are just regular protein molecules, composed of nothing else than the 20 standard amino acids. If you mix of all these amino acids in free form at, say, 10 mM each, this mixture will not have any significant catalytic activity. It therefore is the precise arrangement of the amino acids in the enzyme molecule that brings about the function.

As the á-amino and á-carboxyl groups of the amino acids are hooked up to each other in the polypeptide chain, they usually do not directly contribute to the catalytic effect of the enzyme. Instead, it is the side chains that are directly engaged win the reaction. A very good example of this is chymotrypsin. Chymotrypsin is one of the major proteases in the human digestive tract, where its job is to knock down large protein molecules into small peptides that are then further processed by peptidases. How does chymotrypsin do that? The enzyme-catalysed reaction is similar to alkaline hydrolysis. In alkaline hydrolysis, a hydroxide ion, which is a strong nucleophile, attacks the carbon in the peptide bond that carries a partial positive charge. In the enzyme-catalysed reaction, a deprotonated serine residue of the enzyme plays a role similar to that of the hydroxide ion. Now, you know that serine normally is a neutral amino acid – its OH-group does not spontaneously dissociate, no more than the-OH group of alcohol does. The question therefore is, how does the enzyme deprotonate its own serine side chain? This is brought about by placing the serine next to a histidine and an aspartate inside the active centre. Aspartate deprotonates histidine, which in turn deprotonates the serine residue. This motive – asp, his, ser – is very widespread among proteases and esterases, so much so that it is commonly called 'the catalytic triad'. *E.g.*, the protease trypsin and several lipases that occur in human metabolism have this motif and share the same mechanism of catalysis.

Structure and Mechanism of Chymotrypsin

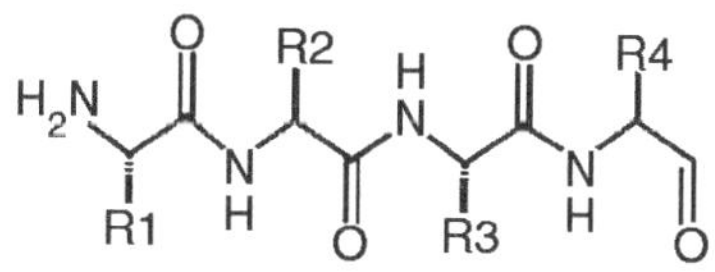

Fig. The Peptide Bond Cleaved by Chymotrypsin.

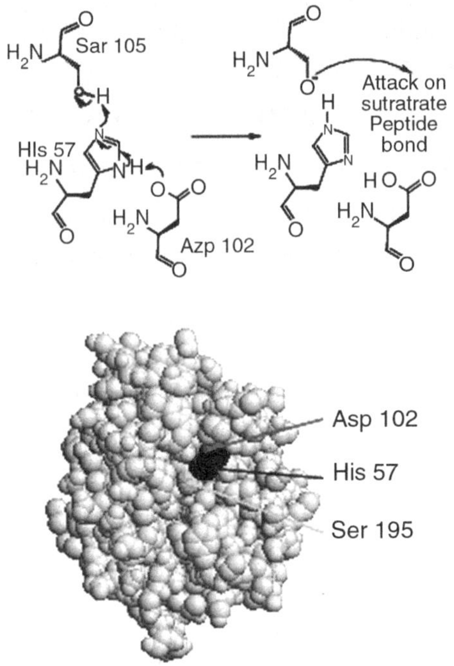

Fig. The Interplay of the Side Chains of Histidine 57, Aspartate 102, and Serine 195 that Results in the Deprotonation of Serine. All three side Chains are Close to each other in the Active Site of the Enzyme.

While with many enzymes the protein molecule and its amino acid side chains are sufficient for catalysis, many others require co-enzymes for their catalytic activity. Very often, both active-site amino acid side chains and coenzymes are required. For example, amino acid transaminases have a molecule of the coenzyme pyridoxalphosphate bound to the active site, which co-operates with an active site lysine in the enzyme's catalytic activity. Most enzymes have just one active site, or if they are multimeric one active site per subunit. However, there are exceptions: Fatty acid synthase has as many as eight different active sites on each subunit. Multi-enzyme complexes such as pyruvate dehydrogenase have one active site per subunit but combine different types of subunits and enzyme activities in one functional assembly.

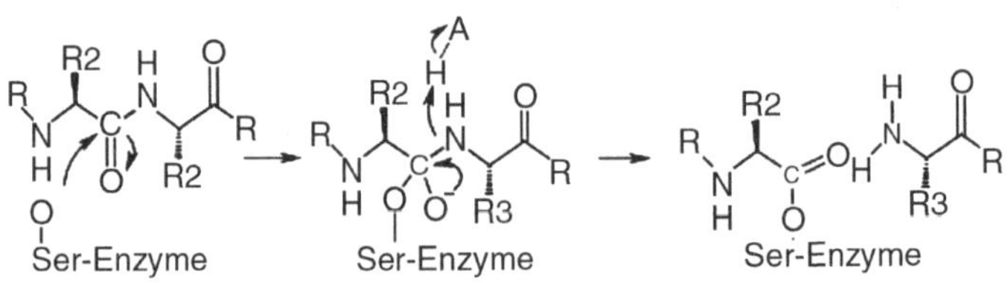

Fig. The Enzyme-catalised Reaction.

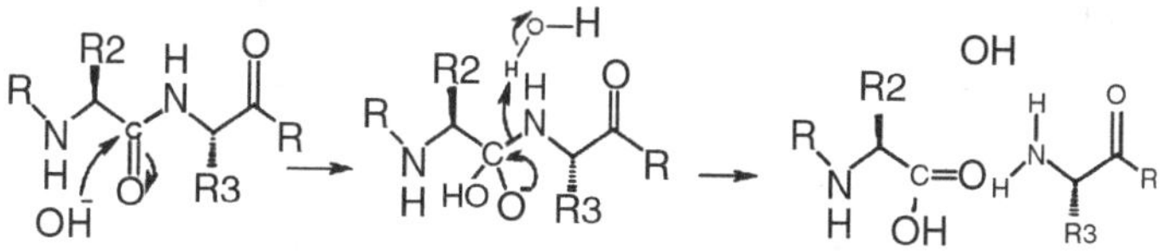

Fig. The Base-catalysed Reaction for Comparison.

TYPES OF ENZYMATIC REACTIONS

When looking at enzyme names such as 'transketolase' or 'phosphorylase', you will note that these names don't tell you exactly what reactions the enzymes may catalyse. A complete description should mention the coenzymes required, the substrates and the particular bonds in the substrates that are being severed or created. A nomenclature that meets these criteria has been developed by the Enzyme Commission of the IUBMB (International Union of Biochemistry and Molecular Biology). In the IUBMB nomenclature, the enzyme transketolase bears the formidable name:

Sedoheptulose-7-phosphate:

D-glyceraldehyde-3-phosphate glycolaldehydetransferase

Such names, of course, are rather lengthy, and their use is not very widespread. To make the tasks of tracking and bookkeeping more manageable, these names are supplemented with numeric codes. In the IUBMB scheme, enzymes are put into one out of six classes according to the reactions they catalyse. These classes are:

- *Oxidoreductases*: These catalyse redox reactions, frequently involving one of the coenzymes NAD^+, $NADP^+$, or FAD.
- *Transferases*: These bring about the transfer of functional groups – *e.g.*, phosphate groups from ATP to another metabolite, which activates the latter and sets it up for subsequent reaction steps.
- *Hydrolases*: These catalyse hydrolysis reactions – *e.g.*, such as those involved in digestion of foodstuffs.
- *Lyases:* These effect elimination reactions that result in the formation of double bonds.
- *Isomerases*: These facilitate the interconversion of isomers. We will meet two examples as soon as we get into glycolysis.
- Ligases, which form new covalent bonds at the expense of ATP hydrolysis.

Of course, within each of these main classes, there are subclasses and sub-sub classes that correspond to details of substrates and mechanisms of the enzyme reactions.

Each individual enzyme activity is assigned an individual number within a sub-sub class, so that we wind up with a four-figure designation, which is preceded by the letters 'EC' (Enzyme Commission).

One good thing about this classification is that it is rarely used–the 'recommended names', which most of the time happen to be the traditional ones, are used instead. The other good thing is that it has a sound appreciation of priorities. The single most important enzyme in student lifestyle – namely, alcohol dehydrogenase (or, as IUBMB puts it, alcohol:NAD oxidoreductase). This beneficial enzyme, residing in the liver, degrades ethanol, and without it you would be drunk all the time!

Enzyme Kinetics: Basic Enzyme Reactions

Enzymes are catalysts and increase the speed of a chemical reaction without themselves undergoing any permanent chemical change. They are neither used up in the reaction nor do they appear as reaction products. The basic enzymatic reaction can be represented as follows where E represents the enzyme catalyzing the reaction, S the substrate, the substance being changed, and P the product of the reaction.

Enzyme Kinetics: Energy Levels

Chemists have known for almost a century that for most chemical reactions to proceed, some form of energy is needed. They have termed this quantity of energy, "the energy of activation." It is the magnitude of the activation energy which determines just how fast the reaction will proceed. It is believed that enzymes lower the activation energy for the reaction they are catalyzing. Figure illustrates this concept.

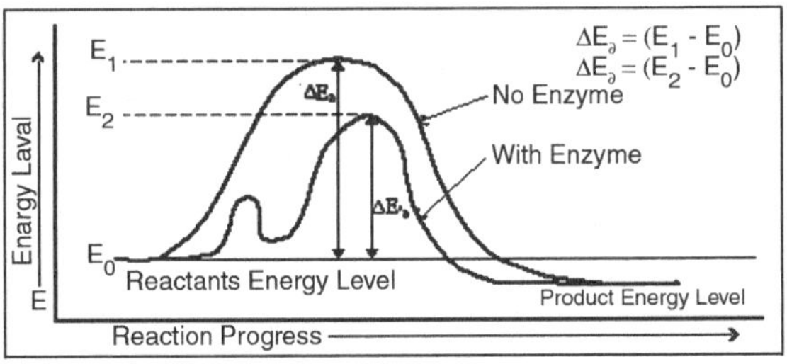

Fig. Energy of Activation D¶ E is less than D¶ E due to the effect of the Enzyme on the Substrate

The enzyme is thought to reduce the "path" of the reaction. This shortened path would require less energy for each molecule of substrate converted to product. Given a total amount of available energy, more molecules of substrate would be converted when the enzyme is present (the shortened "path") than when it is absent. Hence, the reaction is said to go faster in a given period of time.

Enzyme Kinetics: The Enzyme Substrate Complex

A theory to explain the catalytic action of enzymes was proposed by the Swedish chemist Savante Arrhenius in 1888. He proposed that the substrate and enzyme formed some intermediate substance which is known as the enzyme substrate complex. The reaction can be represented as:

$$\underset{\text{substrate}}{S} + \underset{\text{enzyme}}{E} \longrightarrow \underset{\substack{\text{enzyme}\\\text{substrate}\\\text{complex}}}{ES}$$

If this reaction is combined with the original reaction equation, the following results:

$$\underset{\text{substrate}}{S} + \underset{\text{enzyme}}{E} \longrightarrow \underset{\substack{\text{enzyme}\\\text{substrate}\\\text{complex}}}{ES} \longrightarrow \underset{\text{Product}}{P} + \underset{\text{enzyme}}{E}$$

The existence of an intermediate enzyme-substrate complex has been demonstrated in the laboratory, for example, using catalase and a hydrogen peroxide derivative. At Yale University, Kurt G. Stern observed spectral shifts in catalase as the reaction it catalysed proceeded. This experimental evidence indicates that the enzyme first unites in some way with the substrate and then returns to its original form after the reaction is concluded.

Chemical Equilibrium

The study of a large number of chemical reactions reveals that most do not go to true completion. This is likewise true of enzymatically-catalysed reactions. This is due to the reversibility of most reactions. In general:

$$A + B \xrightarrow{K+1} C + D \quad \text{forward reaction}$$

$$C + D \xrightarrow{K-1} A + B \quad \text{reverse reaction}$$

where K^{+1} is the forward reaction rate constant and K^{-1} is the rate constant for the reverse reaction.

Combining the two reactions gives:

$$A + B \underset{K-1}{\overset{K+1}{\rightleftharpoons}} C + D$$

Applying this general relationship to enzymatic reactions allows the equation:

$$E + S \underset{K-1}{\overset{K+1}{\rightleftharpoons}} ES \underset{K-2}{\overset{K+2}{\rightleftharpoons}} p + E$$

Equilbrium, a steady state condition, is reached when the forward reaction rates equal the backward rates. This is the basic equation upon which most enzyme activity studies are based.

Factors Affecting Enzyme Activity

Knowledge of basic enzyme kinetic theory is important in enzyme analysis in order both to understand the basic enzymatic mechanism and to select a method for enzyme analysis. The conditions selected to measure the activity of an enzyme would not be the same as those selected to measure the concentration of its substrate.

Several factors affect the rate at which enzymatic reactions proceed-temperature, pH, enzyme concentration, substrate concentration, and the presence of any inhibitors or activators.

Enzyme Concentration

In order to study the effect of increasing the enzyme concentration upon the reaction rate, the substrate must be present in an excess amount; *i.e.*, the reaction must be independent of the substrate concentration. Any change in the amount of product formed over a specified period of time will be dependent upon the level of enzyme present. Graphically this can be represented as:

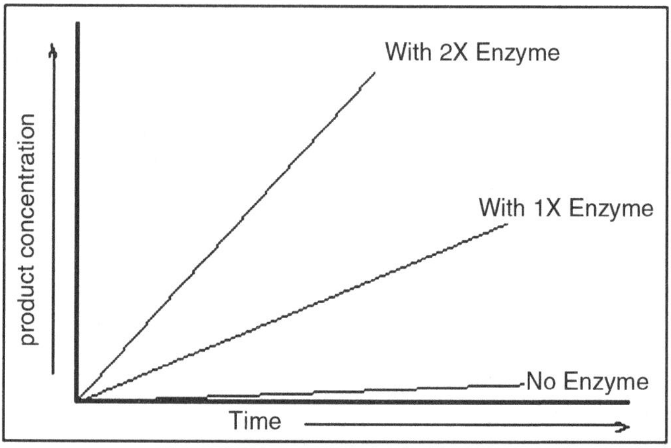

Fig. "Zer order" Reaction Rate is Independent of Substrate Concentration.

These reactions are said to be "zero order" because the rates are independent of substrate concentration, and are equal to some constant k. The formation of product proceeds at a rate which is linear with time. The addition of more substrate does not serve to increase the rate. In zero order kinetics, allowing the assay to run for double time results in double the amount of product.

The amount of enzyme present in a reaction is measured by the activity it catalyses. The relationship between activity and concentration is affected by many factors such as temperature, pH, etc. An enzyme assay must be designed so that the observed activity is proportional to the amount of enzyme present in order that the enzyme concentration is the only limiting factor.

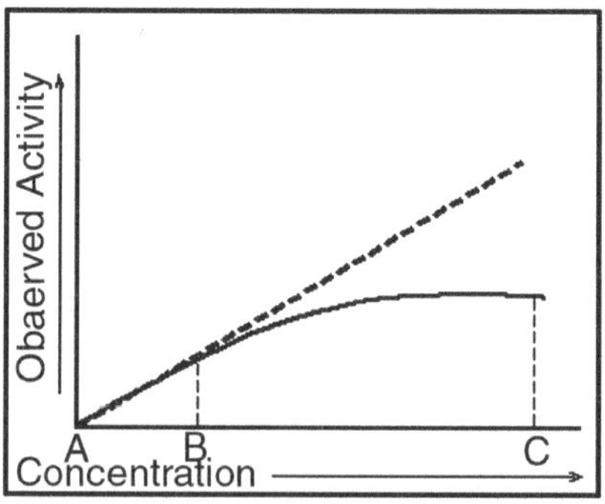

Fig. Activity Concentration.

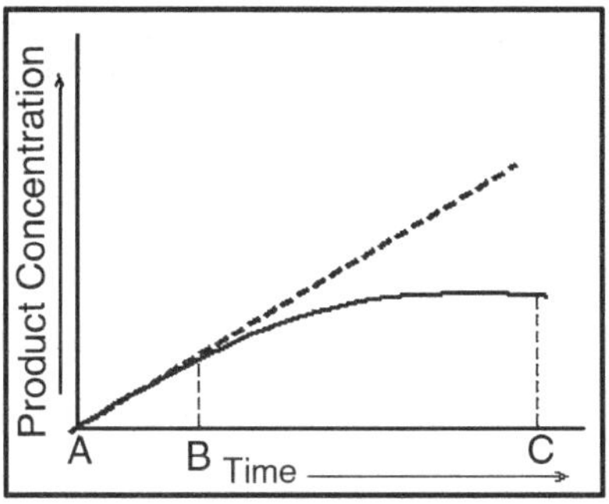

Fig. Reaction Rate Limited by Substrate Concentration

It is satisfied only when the reaction is zero order. In Figure, activity is directly proportional to concentration in the area AB, but not in BC. Enzyme activity is generally greatest when substrate concentration is unlimiting. When the concentration of the product of an enzymatic Reaction is plotted against time, a similar curve results.

Between A and B, the curve represents a zero order reaction; that is, one in which the rate is constant with time. As substrate is used up, the enzyme's active sites are no longer saturated, substrate concentration becomes rate limiting, and the reaction becomes first order between B and C. To measure enzyme activity ideally, the measurements must be made in that portion of the curve where the reaction is zero order.

A reaction is most likely to be zero order initially since substrate concentration is then highest. To be certain that a reaction is zero order, multiple measurements of product (or substrate) concentration must be made. Three types of reactions which might be encountered in enzyme assays and shows the problems which might be enountered if only single measurements are made.

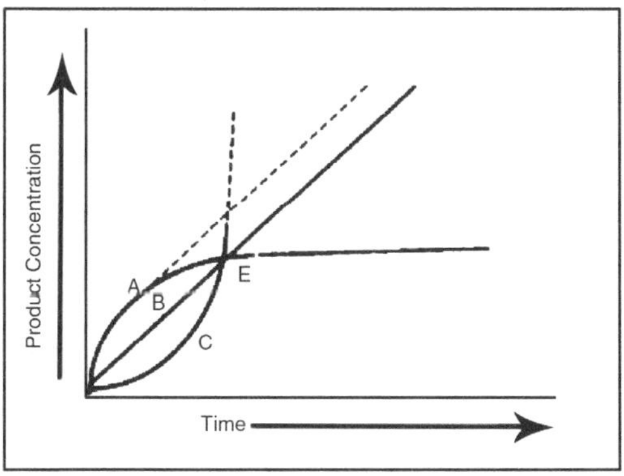

Fig. Leading, Lagging and Liner Reactions

B is a straight line representing a zero order reaction which permits accurate determination of enzyme activity for part or all of the reaction time.

This reaction is zero order initially and then slows, presumably due to substrate exhaustion or product inhibition. This type of reaction is sometimes referred to as a "leading" reaction.

True "potential" activity is represented by the dotted line. Curve C represents a reaction with an initial "lag" phase. Again the dotted line represents the potentially measurable activity. Multiple determinations of product concentration enable each curve to be plotted and true activity determined. A single end point determination at E would lead to the false conclusion that all three samples had identical enzyme concentration.

Substrate Concentration

It has been shown experimentally that if the amount of the enzyme is kept constant and the substrate concentration is then gradually increased, the reaction velocity will increase until it reaches a maximum. After this point, increases in substrate concentration will not increase the velocity (delta A/delta T).

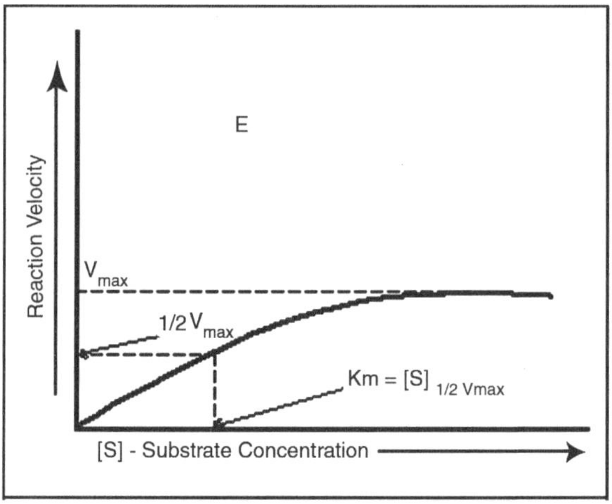

Fig. Effect of Substrate Concentration

It is theorised that when this maximum velocity had been reached, all of the available enzyme has been converted to ES, the enzyme substrate complex. This point on the graph is designated Vmax. Using this maximum velocity and equation, Michaelis developed a set of mathematical expressions to calculate enzyme activity in terms of reaction speed from measurable laboratory data.

$$E + S \underset{K-1}{\overset{K+1}{\rightleftarrows}} ES \underset{K-2}{\overset{K+2}{\rightleftarrows}} p + E$$

The Michaelis constant Km is defined as the substrate concentration at 1/ 2 the maximum velocity. Using this constant and the fact that Km can also be defined as:

$$K_m = K_{-1} + K_{2/}K_{+1}$$

K^{+1}, K^{-1} and K^{+2} being the rate constants from equation. Michaelis developed the following

$$V_1 \frac{V_{mx}[S]}{K_m + [S]}$$

V_1 = The velocity at any time
[S] = The substrate concentration at this time
V_{mx} = The highest under this set of experimental condition (pH, temperature, etc.)
K_M = The Michaselis constant for the particular enzyme being investigated Michaelis constants have been determined for many of the commonly used enzymes.

The size of Km tells us several things about a particular enzyme:

- A small Km indicates that the enzyme requires only a small amount of substrate to become saturated. Hence, the maximum velocity is reached at relatively low substrate concentrations.
- A large Km indicates the need for high substrate concentrations to achieve maximum reaction velocity.
- The substrate with the lowest Km upon which the enzyme acts as a catalyst is frequently assumed to be enzyme's natural substrate, though this is not true for all enzymes.

RNA AND DNA REVEALED: NEW ROLES, NEW RULES

For many years, when scientists thought about heredity, DNA was the first thing to come to mind. It's true that DNA is the basic ingredient of our genes and, as such, it often steals the limelight from RNA, the other form of genetic material inside our cells. But, while they are both types of genetic material, RNA and DNA are rather different. The chemical units of RNA are like those of DNA, except that RNA has the nucleotide uracil (U) instead of thymine (T). Unlike double-stranded DNA, RNA usually comes as only a single strand. And the nucleotides in RNA contain ribose sugar molecules in place of deoxyribose.

RNA is quite flexible—unlike DNA, which is a rigid, spiral-staircase molecule that is very stable. RNA can twist itself into a variety of complicated, three-dimensional shapes.

RNA is also unstable in that cells constantly break it down and must continually make it fresh, while DNA is not broken down often. RNA's instability lets cells change their patterns of protein synthesis very quickly in response to what's going on around them.

Many textbooks still portray RNA as a passive molecule, simply a "middle step" in the cell's gene-reading activities. But that view is no longer accurate. Each year, researchers unlock new secrets about RNA. These discoveries reveal that it is truly a remarkable molecule and a multi-talented actor in heredity.Today, many scientists believe that RNA evolved on the Earth long before DNA did. Researchers hypothesize—obviously, no one was around to write this down—that RNA was a major participant in the chemical reactions that ultimately spawned the first signs of life on the planet.

RNA World

At least two basic requirements exist for making a cell: the ability to hook molecules together and break them apart, and the ability to replicate, or copy itself, from existing information. Riboswitches are RNA sequences that control gene activity. The riboswitch shown here bends into a special shape when it grips tightly onto a molecule called a metabolite (coloured balls) that bacteria need to survive. RNA probably helped to form the first cell. The first organic molecules, meaning molecules containing carbon, most likely arose out of random collisions of gases in the Earth's primitive atmosphere, energy from the Sun, and heat from naturally occurring radioactivity. Some scientists think that in this primitive world, RNA was a critical molecule because of its ability to lead a double life: to store information and to conduct chemical reactions. In other words, in this world, RNA served the functions of both DNA and proteins. What does any of this have to do with human health? Plenty, it turns out. Today's researchers are harnessing some of RNA's flexibility and power. For example, through a strategy he calls directed evolution, molecular engineer Ronald R. Breaker of Yale University is developing ways to create entirely new forms of RNA and DNA that both work as enzymes.

Recently, Breaker and others have also uncovered a hidden world of RNAs that play a major role in controlling gene activity, a job once thought to be performed exclusively by proteins. These RNAs, which the scientists named riboswitches, are found in a wide variety of bacteria and other organisms. This discovery has led Breaker to speculate that new kinds of antibiotic medicines could be developed to target bacterial riboswitches.

Molecular Editor

Scientists are learning of another way to customise proteins: by RNA editing. Although DNA sequences spell out instructions for producing RNA and proteins, these instructions aren't always followed precisely. Editing a gene's mRNA, even by a single chemical letter, can radically change the resulting protein's function. Nature likely evolved the RNA editing function as a way to get more proteins out of the same number of genes. For example, researchers have found that the mRNAs for certain proteins important for the proper functioning of the nervous system are particularly prone to editing. It may be that RNA editing gives certain brain cells the capacity to react quickly to a changing environment.

Which molecules serve as the editor and how does this happen? Brenda Bass of the University of Utah School of Medicine in Salt Lake City studies one particular class of editors called adenosine deaminases. These enzymes "retype" RNA letters at various places within an mRNA transcript. They do their job by searching for characteristic RNA shapes. Telltale twists and bends in folded RNA molecules signal these enzymes to change the RNA sequence, which in turn changes the protein that gets made. Bass' experiments show that RNA editing occurs in a variety of organisms, including people. Another interesting aspect of editing is that certain disease-causing microorganisms, such as some forms of parasites, use RNA editing to gain a survival edge when living in a human host. Understanding the details of this process is an important area of medical research.

MICRO RNA

Recently, molecules called microRNAs have been found in organisms as diverse as plants, worms, and people. The molecules are truly "micro," consisting of only a few dozen nucleotides, compared to typical human mRNAs that are a few thousand nucleotides long. What's particularly interesting about microRNAs is that many of them arise from DNA that used to be considered merely filler material. How do these small but important RNA molecules do their work? They start out much bigger but get trimmed by cellular enzymes, including one aptly named Dicer. Like tiny pieces of Velcro®, microRNAs stick to certain mRNA molecules and stop them from passing on their protein-making instructions.

First discovered in a roundworm model system some microRNAs help determine the organism's body plan. In their absence, very bad things can happen. For example, worms engineered to lack a microRNA called let-7 develop so abnormally that they often rupture and practically break in half as the worm grows.

Perhaps it is not surprising that since microRNAs help specify the timing of an organism's developmental plan, the appearance of the microRNAs themselves is carefully timed inside a developing organism. Biologists, including Amy Pasquinelli of the University of California, San Diego, are currently figuring out how microRNAs are made and cut to size, as well as how they are produced at the proper time during development.

MicroRNA molecules also have been linked to cancer. For example, Gregory Hannon of the Cold Spring Harbour Laboratory on Long Island, New York, found that certain microRNAs are associated with the severity of the blood cancer B cell lymphoma in mice. Since the discovery of microRNAs in the first years of the 21st century, scientists have identified hundreds of them that likely exist as part of a large family with similar nucleotide sequences. New roles for these molecules are still being found.

Healthy Interference

RNA controls genes in a way that was only discovered recently: a process called RNA interference, or RNAi. Although scientists identified RNAi less than 10 years ago, they now know that organisms have been using this trick for millions of years. Researchers believe that RNAi arose as a way to reduce the production of a gene's encoded protein for purposes of fine-tuning growth or self-defence. When viruses infect cells, for example, they command their host to produce specialised RNAs that allow the virus to survive and make copies of itself. Researchers believe that RNAi eliminates unwanted viral RNA, and some speculate that it may even play a role in human immunity.

Oddly enough, scientists discovered RNAi from a failed experiment! Researchers investigating genes involved in plant growth noticed something strange: When they tried to turn petunia flowers purple by adding an extra "purple" gene, the flowers bloomed white instead. This result fascinated researchers, who could not understand how adding genetic material could somehow get rid of an inherited trait. The mystery remained unsolved until, a few years later, two geneticists studying development saw a similar thing happening in lab animals.

The researchers, Andrew Z. Fire, then of the Carnegie Institution of Washington in Baltimore and now at Stanford University, and Craig Mello of the University of Massachusetts Medical School in Worcester, were trying to block the expression of genes that affect cell growth and tissue formation in roundworms, using a molecular tool called antisense RNA.

To their surprise, Mello and Fire found that their antisense RNA tool wasn't doing much at all. Rather, they determined, a doublestranded contaminant produced during the synthesis of the single-stranded antisense RNA interfered with gene expression. Mello and Fire named the process RNAi, and in 2006 were awarded the Nobel Prize in physiology or medicine for their discovery.

Further experiments revealed that the doublestranded RNA gets chopped up inside the cell into much smaller pieces that stick to mRNA and block its action, much like the microRNA pieces of Velcro discussed above. Today, scientists are taking a cue from nature and using RNAi to explore biology. They have learned, for example, that the process is not limited to worms and plants, but operates in humans too. Medical researchers are currently testing new types of RNAi-based drugs for treating conditions such as macular degeneration, the leading cause of blindness, and various infections, including those caused by HIV and herpes virus.

Dynamic DNA

A good part of who we are is "written in our genes," inherited from Mom and Dad. Many traits, like red or brown hair, body shape, and even some personality quirks, are passed on from parent to offspring. But genes are not the whole story. Where we live, how much we exercise, what we eat: These and many other environmental factors can all affect how our genes get expressed. You know that changes in DNA and RNA can produce changes in proteins. But additional control happens at the level of DNA, even though these changes do not alter DNA directly. Inherited factors that do not change the DNA sequence of nucleotides are called epigenetic changes, and they too help make each of us unique.

Epigenetic means, literally, "upon" or "over" genetics. It describes a type of chemical reaction that can alter the physical properties of DNA without changing its sequence. These changes make genes either more or less likely to be expressed. Currently, scientists are following an intriguing course of discovery to identify epigenetic factors that, along with diet and other environmental influences, affect who we are and what type of illnesses we might get.

Secret Code

DNA is spooled up compactly inside cells in an arrangement called chromatin. This packaging is critical for DNA to do its work. Chromatin consists of long strings of DNA spooled around a compact assembly of proteins called histones. The "epigenetic code" controls gene activity with chemical tags that mark DNA (purple diamonds) and the "tails" of histone proteins (purple triangles). These markings help determine whether genes will be transcribed by RNA polymerase. Genes hidden from access to RNA polymerase are not expressed.

One of the key functions of chromatin is to control access to genes, since not all genes are turned on at the same time. Improper expression of growth-promoting genes, for example, can lead to cancer, birth defects, or other health concerns. Many years after the structure of DNA was determined, researchers used a powerful device known as an electron microscope to take pictures of chromatin fibres. Upon viewing chromatin up close, the researchers described it as "beads on a string," an image still used today. The beads were the histone balls, and the string was DNA wrapped around the histones and connecting one bead to the next.

Decades of study eventually revealed that histones have special chemical tags that act like switches to control access to the DNA. Flipping these switches, called epigenetic markings, unwinds the spooled DNA so the genes can be transcribed. The observation that a cell's gene-reading machinery tracks epigenetic markings led C. David Allis, who was then at the University of Virginia Health Sciences Centre in Charlottesville and now works at the Rockefeller University in New York City, to coin a new phrase, the "histone code." He and others believe that the histone code plays a major role in determining which proteins get made in a cell. Flaws in the histone code have been associated with several types of cancer, and researchers are actively pursuing the development of medicines to correct such errors.

IMPRINTING

A process called imprinting, which occurs naturally in our cells, provides another example of how epigenetics affects gene activity. Igf2 is an imprinted gene. A single copy of the abnormal, or mutant, form of the Igf2 gene causes growth defects, but only if the abnormal gene variant is inherited from the father.

With most genes, the two copies work exactly the same way. For some mammalian genes, however, only the mother's or the father's copy is switched on regardless of the child's gender. This is because the genes are chemically marked, or imprinted, during the process that generates eggs and sperm. As a result, the embryo that emerges from the joining of egg and sperm can tell whether a gene copy came from Mom or Dad, so it knows which copy of the gene to shut off.

One example of an imprinted gene is insulin-like growth factor 2 (Igf2), a gene that helps a mammalian fetus grow. In this case, only the father's copy of Igf2 is expressed, and the mother's copy remains silent (is not expressed) throughout the life of the offspring. Scientists have discovered that this selective silencing of Igf2 and many other imprinted genes occurs in all placental mammals (all except the platypus, echidna, and marsupials) examined so far, but not in birds.

Why would nature tolerate a process that puts an organism at risk because only one of two copies of a gene is working? The likely reason, many researchers believe, is that mothers and fathers have competing interests, and the battle-field is DNA!

The scenario goes like this: It is in a father's interest for his embryos to get bigger faster, because that will improve his offspring's chances of survival after birth. The better an individual's chance of surviving infancy, the better its chance of becoming an adult, mating, and passing its genes on to the next generation. Of course mothers want strong babies, but unlike fathers, mothers provide physical resources to embryos during pregnancy. Over her lifetime, a female is likely to be pregnant several times, so she needs to divide her resources among a number of embryos in different pregnancies.

Researchers have discovered over 200 imprinted genes in mammals since the first one was identified in 1991. We now know that imprinting controls some of the genes that have an important role in regulating embryonic and fetal growth and allocating maternal resources. Not surprisingly, mutations in these genes cause serious growth disorders. Marisa Bartolomei of the University of Pennsylvania School of Medicine in Philadelphia is trying to Fig out how Igf2 and other genes become imprinted and stay silent throughout the life of an individual. She has already identified sequences within genes that are essential for imprinting. Bartolomei and other researchers have shown that these sequences, called insulators, serve as "landing sites" for a protein that keeps the imprinted gene from being transcribed.

Starting at the End

When we think of DNA, we think of genes. However, some DNA sequences are different: They don't encode RNAs or proteins. Introns, described earlier are in this category. Another example is telomeres—the ends of chromosomes. There are no genes in telomeres, but they serve an essential function. Like shoelaces without their tips, chromosomes without telomeres unravel and fray. And without telomeres, chromosomes stick to each other and cause cells to undergo harmful changes like dividing abnormally. Researchers know a good deal about telomeres, dating back to experiments performed in the 1970s by Elizabeth Blackburn, a basic researcher who was curious about some of the fundamental events that take place within cells.

At the time, Blackburn, now at the University of California, San Francisco, was working with Joseph Gall at Yale University. For her experimental system, she chose a single-celled, pond-dwelling organism named Tetrahymena. These tiny, pear-shaped creatures are covered with hairlike cilia that they use to propel themselves through the water as they devour bacteria and fungi. Tetrahymena was a good organism for Blackburn's experiments because it has a large number of chromosomes—which means it has a lot of telomeres! Her research was also perfectly timed, because methods for sequencing DNA were just being developed. Blackburn found that Tetrahymena's telomeres had an unusual nucleotide sequence: TTGGGG, repeated about 50 times per telomere. Since then, scientists have discovered that the telomeres of almost all organisms have repeated sequences of DNA with lots of Ts and Gs. In human and mouse telomeres, for example, the repeated sequence is TTAGGG.

The number of telomere repeats varies enormously, not just from organism to organism but in different cells of the same organism and even within a single cell over time. Blackburn reasoned that the repeat number might vary if cells had an enzyme that added copies of the repeated sequence to the telomeres of some but not all chromosomes. With her then-graduate student Carol Greider, now at Johns Hopkins University, Blackburn hunted for the enzyme. The team found it and Greider named it telomerase.

The telomerase enzyme, it turns out, consisted of a protein and an RNA component, which the enzyme uses as a template for copying the repeated DNA sequence. What is the natural function of telomerase? As cells divide again and again, their telomeres get shorter.Most normal cells stop dividing when their telomeres wear down to a certain point, and eventually the cells die. Telomerase can counteract the shortening. By adding DNA to telomeres, telomerase rebuilds the telomere and resets the cell's molecular clock. The discovery of telomerase triggered new ideas and literally thousands of new studies. Many researchers thought that the enzyme might play important roles in cancer and aging. Researchers were hoping to find ways to turn telomerase on so that cells would continue to divide (to grow extra cells for burn patients, for example), or off so that cells would stop dividing (to stop cancer, for instance). So far, they have been unsuccessful. Although it is clear that telomerase and cellular aging are related, researchers do not know whether telomerase plays a role in the normal cellular aging process or in diseases like cancer. Recently, however, Blackburn and a team of other scientists discovered that chronic stress and the perception that life is stressful affect telomere length and telomerase activity in the cells of healthy women. Blackburn and her coworkers are currently conducting a long-term, follow-up study to confirm these intriguing results.

RECOMBINANT DNA TECHNOLOGY AND THE POLYMERASE CHAIN REACTION

Technical advances have played an important role in the advance of genetic understanding. In 1970, American microbiologists Daniel Nathans and Hamilton Othanel Smith discovered a specialised class of enzymes (called restriction enzymes) that cut DNA at specific nucleotide target sequences. That discovery allowed American biochemist Paul Berg in 1972 to make the first artificial recombinant DNA molecule by isolating DNA molecules from different sources, cutting them, and joining them together in a test tube.

These advances allowed individual genes to be cloned (amplified to a high copy number) by splicing them into self-replicating DNA molecules, such as plasmids (extragenomic circular DNA elements) or viruses, and inserting these into living bacterial cells.

From these methodologies arose the field of recombinant DNA technology that presently dominates molecular genetics. In 1977, two different methods were invented for determining the nucleotide sequence of DNA: one by American molecular biologists Allan Maxam and Walter Gilbert and the other by English biochemist Fred Sanger. Such technologies made it possible to examine the structure of genes directly by nucleotide sequencing, resulting in the confirmation of many of the inferences about genes originally made indirectly.

CLASSICAL GENETICS

Classical genetics, which remains the foundation for all other areas in genetics, is concerned primarily with the method by which genetic traits—classified as dominant (always expressed), recessive (subordinate to a dominant trait), intermediate (partially expressed), or polygenic (due to multiple genes)—are transmitted in plants and animals.

These traits may be sex-linked (resulting from the action of a gene on the sex, or X, chromosome) or autosomal (resulting from the action of a gene on a chromosome other than a sex chromosome). Classical genetics began with Mendel's study of inheritance in garden peas and continues with studies of inheritance in many different plants and animals. Today, a prime reason for performing classical genetics is for gene discovery—the finding and assembling of a set of genes that affects a biological property of interest.

CYTOGENETICS

Cytogenetics, the microscopic study of chromosomes, blends the skills of cytologists, who study the structure and activities of cells, with those of geneticists, who study genes. Cytologists discovered chromosomes and the way in which they duplicate and separate during cell division at about the same time when geneticists began to understand the behaviour of genes at the cellular level. The close correlation between the two disciplines led to their combination. Plant cytogenetics early became an important subdivision of cytogenetics because, as a general rule, plant chromosomes are larger than those of animals. Animal cytogenetics became important after the development of the so-called squash technique, in which entire cells are pressed flat on a piece of glass and observed through a microscope; the human chromosomes were numbered using this technique. Today, there are multiple ways to attach molecular labels to specific genes and chromosomes, as well as to specific RNAs and proteins, that make these molecules easily discernible from other components of cells, thereby greatly facilitating cytogenetics research.

Microbial Genetics

Microorganisms were generally ignored by the early geneticists because they are small in size and were thought to lack variable traits and the sexual reproduction necessary for a mixing of genes from different organisms. After it was discovered that microorganisms have many different physical and physiological characteristics that are amenable to study, they became objects of great interest to geneticists because of their small size and the fact that they reproduce much more rapidly than larger organisms. Bacteria became important model organisms in genetic analysis, and many discoveries of general interest in genetics arose from their study. Bacterial genetics is the centre of cloning technology. Viral genetics is another key part of microbial genetics. The genetics of viruses that attack bacteria were the first to be elucidated. Since then, studies and findings of viral genetics have been applied to viruses pathogenic on plants and animals, including humans. Viruses are also used as vectors (agents that carry and introduce modified genetic material into an organism) in DNA technology.

Molecular Genetics

Molecular genetics is the study of the molecular structure of DNA, its cellular activities (including its replication), and its influence in determining the overall makeup of an organism. Molecular genetics relies heavily on genetic engineering (recombinant DNA technology), which can be used to modify organisms by adding foreign DNA, thereby forming transgenic organisms. Since the early 1980s, these techniques have been used extensively in basic biological research and are also fundamental to the biotechnology industry, which is devoted to the manufacture of agricultural and medical products. Transgenesis forms the basis of gene therapy, the attempt to cure genetic disease by addition of normally functioning genes from exogenous sources.

Genomics

The development of the technology to sequence the DNA of whole genomes on a routine basis has given rise to the discipline of genomics, which dominates genetics research today. Genomics is the study of the structure, function, and evolutionary comparison of whole genomes.

Genomics has made it possible to study gene function at a broader level, revealing sets of genes that interact to impinge on some biological property of interest to the researcher. Bioinformatics is the computer-based discipline that deals with the analysis of such large sets of biological information, especially as it applies to genomic information.

Population Genetics

The study of genes in populations of animals, plants, and microbes provides information on past migrations, evolutionary relationships, and extents of mixing among different varieties and species, and methods of adaptation to the environment. Statistical methods are used to analyse gene distributions and chromosomal variations in populations.

Population genetics is based on the mathematics of the frequencies of alleles and of genetic types in populations. For example, the Hardy-Weinberg formula, $p^2 + 2pq + q^2 = 1$, predicts the frequency of individuals with the respective homozygous dominant (*AA*), heterozygous (*Aa*), and homozygous recessive (*aa*) genotypes in a randomly mating population. Selection, mutation, and random changes can be incorporated into such mathematical models to explain and predict the course of evolutionary change at the population level. These methods can be used on alleles of known phenotypic effect, such as the recessive allele for albinism, or on DNA segments of any type of known or unknown function.

Human population geneticists have traced the origins and migration and invasion routes of modern humans, *Homo sapiens*. DNA comparisons between the present peoples on the planet have pointed to an African origin of *Homo sapiens*.

Tracing specific forms of genes has allowed geneticists to deduce probable migration routes out of Africa to the areas colonised today. Similar studies show to what degree present populations have been mixed by recent patterns of travel.

DNA REPLICATION AND PROTEIN SYNTHESIS

MUTATIONS

Occasionally mistakes occur during DNA replication and protein synthesis. Any alteration in the structure of a gene results in a mutation. Mutations occur during DNA replication when the chemical structure of genes undergoes random modifications. Once a change has occurred, the altered genes continue to replicate in their changed form unless another mutation occurs. Sometimes mutations occur during transcription or translation, causing protein synthesis to go awry. Although mutations may occur in any living cell, they are most important when they occur in gametes because then the change affects the traits of following generations.

Most mutations harm an organism. If a mutation occurs in a gene sequence that codes for a particular protein, the mutation may result in a change in the amino acid sequence directed by the gene. This change, in turn, may affect the function of the protein. The implications can be significant: The amino acid sequence distinguishing normal hemoglobin from the altered form of hemoglobin responsible for sickle-cell anemia differs by only a single amino acid.

Some mutations may be neutral or silent and do not affect the function of a protein. Occasionally a mutation benefits an organism. Over the course of evolutionary time, however, mutations serve the crucial role of providing organisms with previously non-existent proteins. In this way, mutations are a driving force behind genetic diversity and the rise of new or more competitive species better able to adapt to changes, such as climate variations, depletion of food sources, or the emergence of new types of disease.

Mutations can produce a change in any region of a DNA molecule. In a *point mutation*, for example, a single nucleotide replaces another nucleotide. Although a point mutation produces a small change to the DNA sequence, it may cause a change in the amino acid sequence, and thus the function, of a protein.

Far more serious are mutations that involve the addition or deletion of one or more bases from a DNA molecule. Adding or subtracting even a single base from a normal sequence during transcription can disrupt translation by shifting the "reading frame" of every subsequent codon. For example, an mRNA strand may include two codons in the following sequence: AUG UGA. The addition of a cytosine base at the beginning of this sequence shifts the "spelling" of these codons so that they read: CAU GUG. This may result in an incorrect amino acid sequence during translation, or the protein may be truncated. Known as *frameshift mutation*, this type of alteration could result in the production of a protein with no real function or one with a harmful effect.

Sometimes mutations are caused by *transposition*, in which long stretches of DNA (containing one or more genes) move from one chromosome to another.

These jumping genes, called transposons, can disrupt transcription and change the type of amino acids inserted into a protein. Transposons rearrange and interrupt genes in a way that generally improves the genetic variation of a species. While mutations can occur spontaneously, some can be caused by exposure to physical or chemical agents in the environment called mutagens. Common environmental mutagens include ultraviolet rays from the sun and various chemicals, such as asbestos, cigarette smoke, and nitrous acid. High-energy radiation, such as medical X rays, can cause DNA strands to break, leading to the deletion of potentially important genetic information.

Radiation damage can also affect an entire chromosome, disrupting the function of many genes. In chromosomal translocation, a piece of one chromosome breaks off and merges with another chromosome. In some cases, large sections of chromosomes may break off and be lost. The cell has highly effective self-repair mechanisms that can correct the harmful changes made by mutations and prevent some mutations from being passed on. Some 50 specialized enzymes locate different types of faulty sequences in the DNA and clip out those flaws. Another repair mechanism scans DNA after replication and marks mismatched base pairs for repair.

PROTEIN SYNTHESIS

DNA replication ensures that the genetic instructions encoded in DNA can be used continuously through generations to produce the proteins that build and operate the cells of an organism. The process of tapping the genetic code to create proteins, known as protein synthesis, has two crucial steps: transcription and translation.

Transcription

Transcription transfers the genetic code from a molecule of DNA to an intermediary molecule called ribonucleic acid (RNA). The basic nucleotide structure of RNA resembles that of DNA, but the two compounds have three critical differences. First, the structure of RNA incorporates the sugar ribose rather than deoxyribose, the sugar in DNA. Second, RNA uses the base uracil (U) instead of thymine (T). In RNA uracil binds with adenine just as thymine does in DNA. Third, RNA usually exists as a single strand, unlike the double-helix structure that normally characterizes DNA. Transcription involves the production of a special kind of RNA known as messenger RNA (mRNA). The process begins when the two strands of a DNA molecule separate, a task directed by the enzyme RNA polymerase. After the double helix splits apart, one of the strands serves as a template, or pattern, for the formation of a complementary mRNA molecule. Free-floating individual bases within the cell bind to the bases on the DNA template using complementary base pairing. The individual bases then link together to form a strand of mRNA.

In eukaryotes (organisms whose cells have a nucleus), the mRNA strand undergoes an additional step before the next stage of protein synthesis can occur. The mRNA strand consists of coding regions called exons separated by regions called introns. The introns do not contribute to protein synthesis. Special enzymes in the nucleus remove the introns from the mRNA strand. The remaining exons then link together to form an mRNA strand that contains the entire code for making a protein.

Translation

Once transcription is complete and the genetic code has been copied onto mRNA, the genetic code must be converted into the language of proteins. That is, the information coded in the four bases found in mRNA must be translated into the instructions encoded by the 20 amino acids used in the formation of proteins. This process, called translation, takes place in cellular organelles called ribosomes. In eukaryotes, mRNA travels out of the nucleus into the cell body to attach to a ribosome. In prokaryotes (organisms without a nucleus), the ribosome clasps mRNA and starts translation before these strands have finished transcription and separated from the DNA. In both eukaryotes and prokaryotes, the ribosome acts like a workbench and clamp that holds the mRNA strand and coordinates the activity of enzymes and other molecules essential to translation. Another form of RNA called transfer RNA (tRNA) is found in the cytoplasm of the cell.

There are many different types of tRNA, and each type binds with one of the 20 amino acids used in protein formation. One end of a tRNA binds with a specific amino acid. The other end carries three bases, known as an anticodon. The tRNA with an amino acid attached travels to the ribosome where the mRNA is stationed. The anticodon of the tRNA undergoes complementary base pairing with a series of three bases on the mRNA, known as the codon. The mRNA codon codes for the type of amino acid carried by the tRNA. A second tRNA bonds with the next codon on the mRNA. The resident tRNA transfers its amino acid to the amino acid of the incoming tRNA and then leaves the ribosome. This process continues repeatedly, with new tRNA receiving the growing chain of amino acids, known as a polypeptide chain, from a resident tRNA. The ribosome moves the mRNA strand one codon at a time, making new codons available to bind with tRNAs. The process ends when the entire sequence of mRNA has been translated. The polypeptide chain falls away from the ribosome as a newly formed protein, ready to go to work in the cell.

2

Genetics and DNA Technology

INTRODUCTION

People have known for many years that living things inherit traits from their parents. That common-sense observation led to agriculture, the purposeful breeding and cultivation of animals and plants for desirable characteristics. Firming up the details took quite some time, though. Researchers did not understand exactly how traits were passed to the next generation until the middle of the 20th century.

Now it is clear that genes are what carry our traits through generations and that genes are made of deoxyribonucleic acid (DNA). But genes themselves do not do the actual work. Rather, they serve as instruction books for making functional molecules such as ribonucleic acid (RNA) and proteins, which perform the chemical reactions in our bodies.

Proteins do many other things, too. They provide the body's main building materials, forming the cell's architecture and structural components. But one thing proteins cannot do is make copies of themselves.

When a cell needs more proteins, it uses the manufacturing instructions coded in DNA. The DNA code of a gene—the sequence of its individual DNA building blocks, labelled as A (adenine), T (thymine), C (cytosine), and G (guanine) and collectively called nucleotides—spells out the exact order of a protein's building blocks, amino acids. Occasionally, there is a kind of typographical error in a gene's DNA sequence. This mistake— which can be a change, gap or duplication—is called a mutation. A mutation can cause a gene to encode a protein that works incorrectly or that does not work at all. Sometimes, the error means that no protein is made.

DEVELOPMENT OF GENETICS

The significance of Mendel's laws was not recognised by his contemporaries but became evident 35 years later, just after the turn of the twentieth century. At this time other important events marked the progress of genetics: Sir Archibald Garrod, a physician, made important contributions to the chemistry of diseases especially porphyria, cystinuria and alkaptonuria, noting their recurrence among sibs in the same family and the effect of consanguineous marriage. He introduced the term "inborn errors of metabolism" and the concept of "one gene, one enzyme" and thus opened the way to biochemical genetics, which became one of the most important aspects in the development of genetics.

Theodor Boveri and Walter's. Sutton in 1903 independently proposed the chromosome theory of inheritance, stating that the chromosomes carried the "hereditary factors" proposed by Mendel and that their behaviour during cell division provided an explanation for their segregation during meiosis.

In the twentieth century the fruit fly, Drosophila was the organism most extensively used in the study of genetics. It was extremely useful because it was easily bred in the labouratory, reproduced very rapidly (about 20 times in one year) and produced thousands of offspring that could be analysed. Besides, it had only four chromosome pairs that were exceptionally large in the salivary glands and showed a pattern of transverse bands on which individual genes could be localised. THOMAS HUNT MORGAN is one of the most famous of the hundreds of scientists who worked on Drosophila. He described sex-linked characteristics that were manifest in the male flies. He is best known for proposing methods for estimating distances between genes located on the same chromosome through the phenomenon of recombination. The unit of gene distance, the centimorgan, is named after him. Morgans work opened up the important field of gene mapping.

Advances in human genetics did not parallel the great advances in the genetics of Drosophila since the beginning of the twentieth century.

This was mainly due to:

- The inability to perform breeding experiments in man;
- The very long generation time
- The small size and relatively large number of human chromosomes, which made their microscopic examination very difficult.

However, several hundreds of human genes were known to exist through the visible characteristics they produced.

In the 1950s two great advances completed the phase of growth of classical genetics and paved the way for a new era of genetics.

The molecular structure of DNA, which explained how the molecule could replicate itself and how the same molecule could represent an almost infinite variety of different genes. From then the gene could be visualised as a structural entity and was no longer an invisible hereditary particle that could be identified only by its effects on the characteristics of an organism. The discovery of the molecular structure of DNA thus paved the way for the study of the gene itself and the direct analysis of DNA in the new era of molecular genetics that was to follow.

TIJO and LEVAN devised a technique for the study of human chromosomes and established that the human chromosome number was 46. Earlier techniques were suitable for organisms that had few chromosomes but not for human chromosomes, which were small and numerous.

This was immediately followed by the discovery of chromosomal abnormalities and marked the beginning of human cytogenetics. It was one of the most important advances that brought human and medical genetics into prominence.

THE NEW GENETICS

A new era of genetics began in the 1970s with the development of three new techniques, which together enabled the direct analysis and manipulation of genes. At this stage studies on DNA were performed mainly on bacteria as they had a very short generation time and very simple DNA. Most of the new discoveries arose from the study of bacterial genetics.

Hamilton Smith discovered that bacteria produced "restriction endonucleases", enzymes that had the ability to cut open the DNA molecule at specific sites.

ED Southern at Edinburgh University developed a technique by which specific genes could be isolated from the whole complement or genome of DNA. The process was called Southern blotting.

Stanley Cohen at Stanford University developed the use of plasmids, which are naturally occurring circular loops of DNA capable of entering bacterial cells, as vectors for carrying foreign DNA into bacterial cells.

These three techniques of cutting, isolating and inserting genes form the basis of recombinant DNA technology, popularly called genetic engineering. They also formed the basis of modern genetic technology.

We are now on the threshold of a new era that promises the treatment of genetic diseases.

SOME BASIC DEFINITIONSS

The genetic material is contained within the nucleus of all cells. It consists of extremely fine thread-like structures composed of DNA and associated proteins. Within it is stored the genetic information, a whole library of genes which constitute the genome.

These thread-like structures become visible as chromosomes during cell division as they gradually coil up and become thick and prominent.

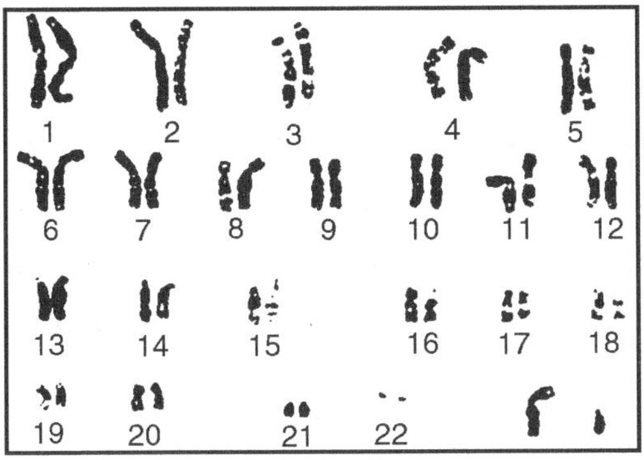

For the purpose of analysis the chromosomes can be arranged as a karyotype. They are classified in order of decreasing size and according to their pattern of bands. Chromosomes occur in pairs. There are 23 pairs of chromosomes: 22 pairs of autosomes and one pair of sex chromosomes.

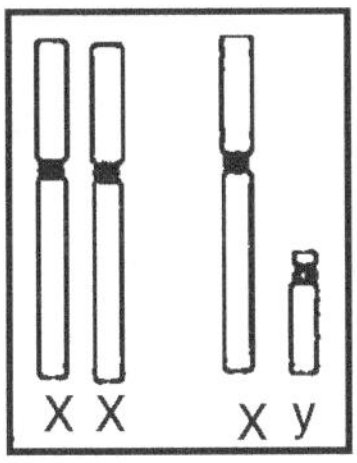

Fig. The sex Chromosomes are XX in Females and XY in Males.

The Y chromosome is much smaller than the X chromosome. It carries the genes responsible for the male sexual characteristics. The X chromosome carries genes that are not related to sex and are equally important in both males and females. The genetic material is made of DNA. The DNA molecule consists of two spirally arranged chains in which each unit consists of a sugar (deoxyribose), a phosphate group and a nitrogenous base. The base can be Adenine, Guanine, Thymine or Cytosine (A,G,T or C).

A gene is a segment of DNA that carries a genetic message. One gene carries instructions for the synthesis of one protein or enzyme.Enzymes are the functioning units of the cell.

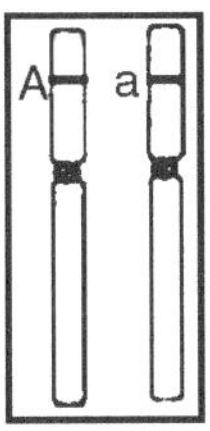

Fig. The Genes are Situated on the Chromosomes.

A locus is the point on a chromosome at which a gene is situated.

Genes occur in pairs that occupy corresponding loci on a pair of chromosomes.

The genes at a locus are called alleles.

An individual is homozygous if the two alleles are identical *e.g.* AA or aa.

An individual is heterozygous if the two alleles are different *e.g.* Aa.

A A	A a	a a
Homozygous	Heterozygous	Homozygous

The genotype of an individual indicates the type of alleles present (heterozygous or homozygous).

The phenotype of an individual indicates the identifiable characteristics present in an individual. These may be visible or biochemical characteristics. Examples of phenotypic characteristics are eye colour, height, blood group.

A dominant characteristic is one that manifests itself in heterozygotes (*i.e.* only one gene is required for its manifestation in the phenotype). A recessive characteristic is one that appears only in homozygotes (*i.e.* two identical alleles are required for its manifestation in the phenotype).

An X-linked gene is one that is situated on the X chromosome. An X-linked recessive characteristic is usually manifest in males while females are carriers of the gene. An X-linked dominant characteristic is manifest in both sexes. The following is an example of an inherited trait for brown or blue eyes.

Alleles	B - gene for brown eyes b - gene for blue eyes		
Genotype	B B Homozygote	B b Heterozygote	b b Homozygote
Phenotype	Brown eyes	Brown eyes	Blue eyes

"Brown eyes" is the dominant characteristic because it appears in the heterozygote (Bb). "Blue eyes" is the recessive characteristic because it appears only in the homozygote (bb).

Genotypes of Parents

↓ →	B	b
B	BB	Bb
b	Bb	bb

It is possible to work out how genes are inherited by constructing a Punnett square. In this example both parents were heterozygous. The phenotypes of the offspring were 3 brown: 1 blue eyes. The genotypes of the offspring were 1 dominant homozygote (BB): 2 heterozygote: 1 recessive homozygote (bb)

The inheritance of the phenotype can be illustrated as a pedigree. The following pedigree shows inheritance of eye colour.

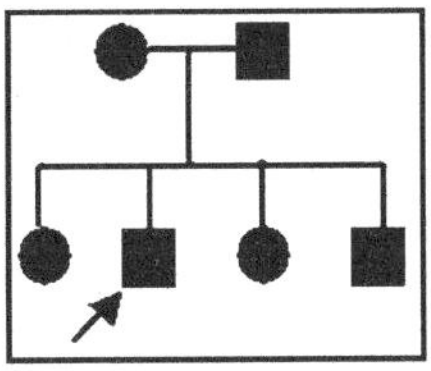

Fig. Males are Represented by Squares, Females by Circles.

The arrow indicates the propositus or proband *i.e.* the individual with a particular characteristic through whom the family came to be investigated. In this case the propositus was a boy with blue eyes.

GENETIC ALGORITHMS

Genetic programming is a branch of genetic algorithms. The main difference between genetic programming and genetic algorithms is the representation of the solution. Genetic programming creates computer programmes in the lisp or scheme computer languages as the solution. Genetic algorithms create a string of numbers that represent the solution.

Genetic programming, uses four steps to solve problems:

1. Generate an initial population of random compositions of the functions and terminals of the problem (computer programmes).
2. Execute each programme in the population and assign it a fitness value according to how well it solves the problem.

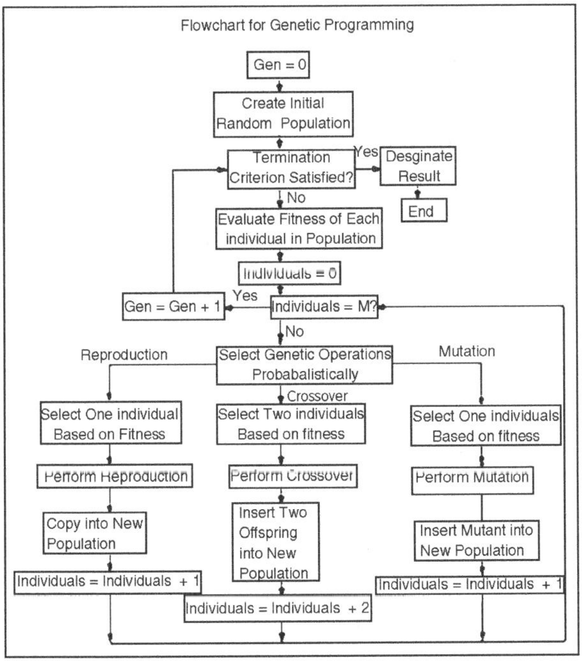

Fig. Genetic Programming Flowchart.

3. Create a new population of computer programmes.
   - Copy the best existing programmes
   - Create new computer programmes by mutation.
   - Create new computer programmes by crossover(sexual reproduction).
4. The best computer programme that appeared in any generation, the best-so-far solution, is designated as the result of genetic programming.

FITNESS FUNCTION

The most difficult and most important concept of genetic programming is the fitness function. The fitness function determines how well a programme is able to solve the problem. It varies greatly from one type of programme to the next.

For example, if one were to create a genetic programme to set the time of a clock, the fitness function would simply be the amount of time that the clock is wrong. Unfortunately, few problems have such an easy fitness function; most cases require a slight modification of the problem in order to find the fitness.

GUN FIRING PROGRAMME

A more complicated example consists of training a genetic programme to fire a gun to hit a moving target. The fitness function is the distance that the bullet is off from the target. The programme has to learn to take into account a number of variables, such as wind velocity, type of gun used, distance to the target, height of the target, velocity and acceleration of the target. This problem represents the type of problem for which genetic programmes are best. It is a simple fitness function with a large number of variables.

WATER SPRINKLER SYSTEM

Consider a programme to control the flow of water through a system of water sprinklers. The fitness function is the correct amount of water evenly distributed over the surface. Unfortunately, there is no one variable encompassing this measurement. Thus, the problem must be modified to find a numcrical fitness. One possible solution is placing water-collecting measuring devices at certain intervals on the surface.

The fitness could then be the standard deviation in water level from all the measuring devices. Another possible fitness measure could be the difference between the lowest measured water level and the ideal amount of water; however, this number would not account in any way the water marks at other measuring devices, which may not be at the ideal mark.

MAZE SOLVING PROGRAMME

If one were to create a programme to find the solution to a maze, first, the programme would have to be trained with several known mazes. The ideal solution from the start to finish of the maze would be described by a path of dots. The fitness in this case would be the number of dots the programme is able to find. In order to prevent the programme from wandering around the maze too long, a time limit is implemented along with the fitness function.

FUNCTIONS AND TERMINALS

The terminal and function sets are also important components of genetic programming. The terminal and function sets are the alphabet of the programmes to be made. The terminal set consists of the variables and constants of the programmes. In the maze example, the terminal set would contain three commands: forward, right and left.

The function set consists of the functions of the programme. In the maze example the function set would contain: If "dot" then do x else do y. In the gun firing programme is the terminal set would be composed of the different variables of the problem.

Some of these variables could be the velocities and accelerations of the gun, the bullet and target. The functions are several mathematical functions, such as addition, subtraction, division, multiplication and other more complex functions.

CROSSOVER OPERATION

Two primary operations exist for modifying structures in genetic programming. The most important one is the crossover operation. In the crossover operation, two solutions are sexually combined to form two new solutions or offspring. The parents are chosen from the population by a function of the fitness of the solutions. Three methods exist for selecting the solutions for the crossover operation. The first method uses probability based on the fitness of the solution. If f $(s_i(t))$ is the fitness of the solution Si and,

$$\sum_{j=1}^{M} f\left(s_j(t)\right)$$

is the total sum of all the members of the population, then the probability that the solution Si will be copied to the next generation:

$$\frac{f\left(s_i(t)\right)}{\sum_{j=1}^{M} f\left(s_j(t)\right)}$$

Another method for selecting the solution to be copied is tournament selection. Typically the genetic programme chooses two solutions random. The solution with the higher fitness will win. This method simulates biological mating patterns in which, two members of the same sex compete to mate with a third one of a different sex. Finally, the third method is done by rank.

In rank selection, selection is based on the rank, (not the numerical value) of the fitness values of the solutions of the population. The creation of the offsprings from the crossover operation is accomplish by deleting the crossover fragment of the first parent and then inserting the crossover fragment of the second parent.

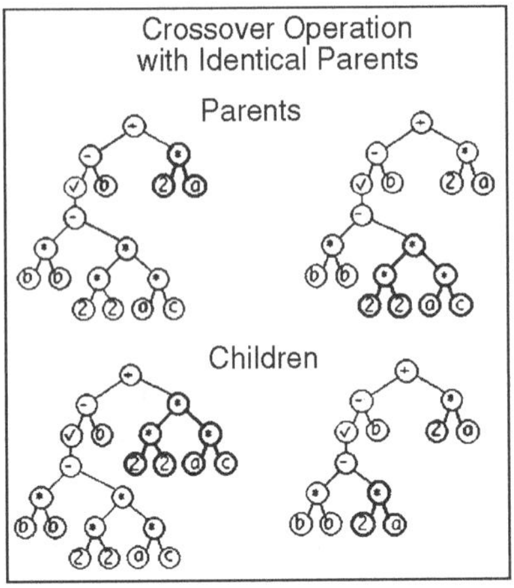

Fig. One of the Main Advantages of Genetic Programming Over Genetic Algorithms.

In genetic programming identical parents can yield different offspring, while in genetic algorithms identical parents would yield identicaloffspring. The bold selections indicate the subtrees to be swapped. An important improvement that genetic programming displays over genetic algorithms is its ability to create two new solutions from the same solution. In the Figure the same parent is used twice to create two new children.

MUTATION

Mutation is another important feature of genetic programming. Two types of mutations are possible. In the first kind a function can only replace a function or a terminal can only replace a terminal. In the second kind an entire subtree can replace another subtree. Figure explains the concept of mutation:

The top parse tree is the original agent. The mutation of a single terminal for another single terminal (a). It also illustrates a mutation of a single function (–) for another single function (+). The parse tree on the bottom right illustrates a the replacement of a subtree by another subtree.

Genetic programming is much more powerful than genetic algorithms. The output of the s is a quantity, while the output of the genetic programming is a another computer programme. In essence, this is the beginning of computer programmes that programme themselves. Genetic programming works best for several types of problems. The first type is where there is no ideal solution, (for example, a programme that drives a car). There is no one solution to driving a car. Some solutions drive safely at the expense of time, while others drive fast at a high safety risk.

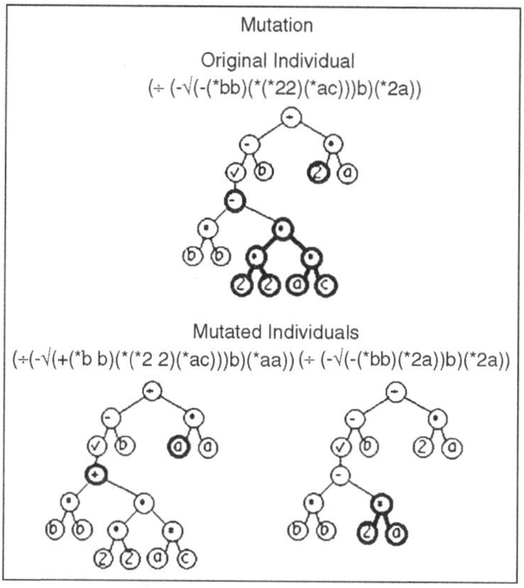

Fig. Two Different Types of Mutations.

Therefore, driving a car consists of making compromises of speed versus safety, as well as many other variables. In this case genetic programming will find a solution that attempts to compromise and be the most efficient solution from a large list of variables. Furthermore, genetic programming is useful in finding solutions where the variables are constantly changing. In the previous car example, the programme will find one solution for a smooth concrete highway, while it will find a totally different solution for a rough unpaved road.

MENDEL'S GENETICS

Fig. Hybridised domesticated horses

For thousands of years, farmers and herders have been selectively breeding their plants and animals to produce more useful hybrids. It was somewhat of a hit or miss process since the actual mechanisms governing inheritance were unknown. Knowledge of these genetic mechanisms finally came as a result of careful laboratory breeding experiments carried out over the last century and a half.

By the 1890s, the invention of better microscopes allowed biologists to discover the basic facts of cell division and sexual reproduction. The focus of genetics research then shifted to understanding what really happens in the transmission of hereditary traits from parents to children. A number of hypotheses were suggested to explain heredity, but Gregor Mendel, a little-known Central European monk, was the only one who got it more or less right. His ideas had been published in 1866 but largely went unrecognised until 1900, which was long after his death. His early adult life was spent in relative obscurity doing basic genetics research and teaching high-school mathematics, physics, and Greek in Brno (now in the Czech Republic). In his later years, he became the abbot of his monastery and put aside his scientific work.

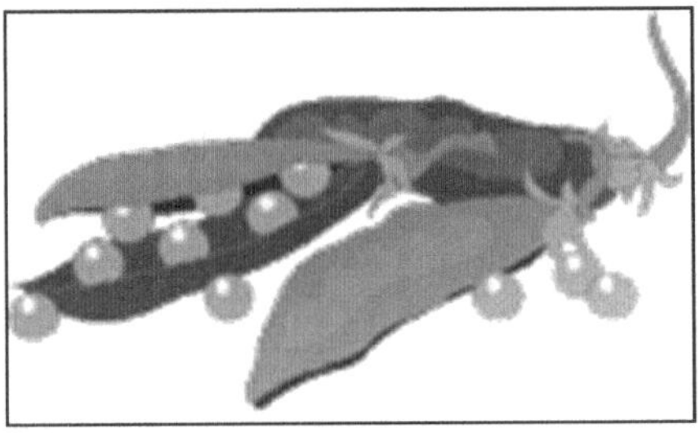

Fig. Common edible peas

While Mendel's research was with plants, the basic underlying principles of heredity that he discovered also apply to people and other animals, because the mechanisms of heredity are essentially the same for all complex life forms.

Through the selective cross-breeding of common pea plants (*Pisum sativum*) over many generations, Mendel discovered that certain traits show up in offspring without any blending of parent characteristics. For instance, the pea flowers are either purple or white—intermediate colours do not appear in the offspring of cross-pollinated pea plants. Mendel observed seven traits that are easily recognised and apparently only occur in one of two forms:

1. flower colour is purple or white,
2. flower position is axil or terminal,
3. stem length is long or short,
4. seed shape is round or wrinkled,
5. seed colour is yellow or green,
6. pod shape is inflated or constricted, and
7. pod colour is yellow or green.

This observation that these traits do not show up in offspring plants with intermediate forms was critically important because the leading theory in biology at the time was that inherited traits blend from generation to generation. Most of the leading scientists in the 19th century accepted this "*blending theory*." Charles Darwin proposed another equally wrong theory known as "*pangenesis*." This held that hereditary "*particles*" in our bodies are affected by the things we do during our lifetime.

These modified particles were thought to migrate via blood to the reproductive cells and subsequently could be inherited by the next generation. This was essentially a variation of Lamarck's incorrect idea of the "*inheritance of acquired characteristics*." Mendel picked common garden pea plants for the focus of his research because they can be grown easily in large numbers and their reproduction can be manipulated. Pea plants have both male and female reproductive organs. As a result, they can either self-pollinate themselves or cross-pollinate with another plant. In his experiments, Mendel was able to selectively cross-pollinate pure-bred plants with particular traits and observed the outcome over many generations. This was the basis for his conclusions about the nature of genetic inheritance.

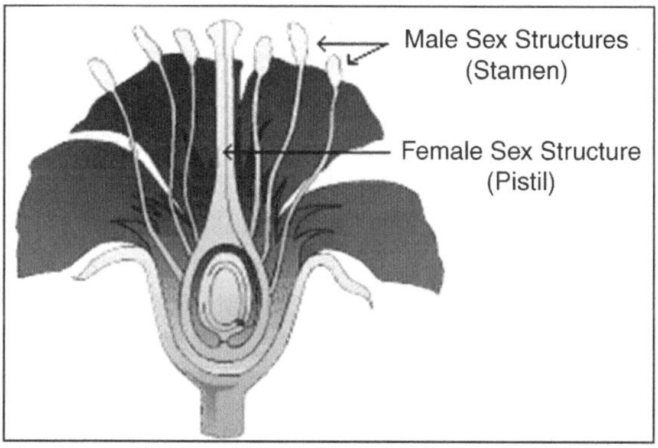

Fig. Reproductive structures of flowers

In cross-pollinating plants that either produce yellow or green pea seeds exclusively, Mendel found that the first offspring generation (f1) always has yellow seeds. However, the following generation (f2) consistently has a 3:1 ratio of yellow to green.

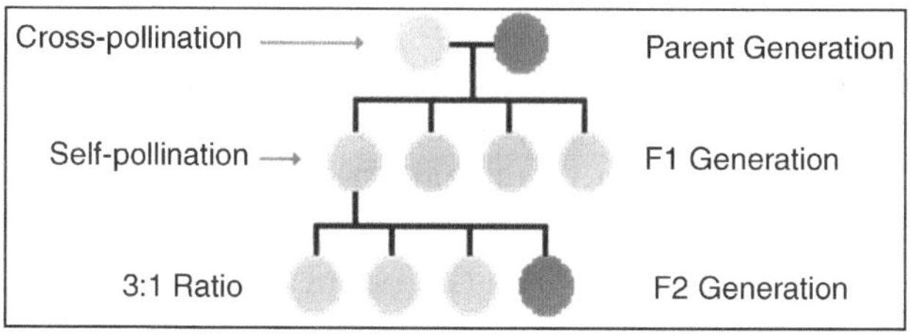

This 3:1 ratio occurs in later generations as well. Mendel realised that this underlying regularity was the key to understanding the basic mechanisms of inheritance.

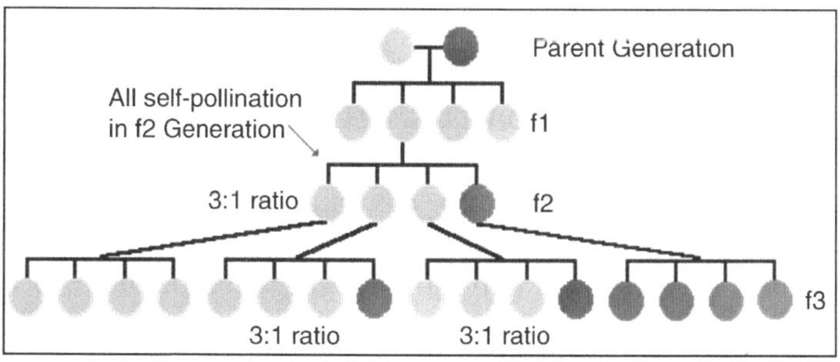

He came to three important conclusions from these experimental results:

1. The inheritance of each trait is determined by *"units"* or *"factors"* that are passed on to descendents unchanged (these units are now called genes),
2. An individual inherits one such unit from each parent for each trait, and
3. A trait may not show up in an individual but can still be passed on to the next generation.

It is important to realise that, in this experiment, the starting parent plants were homozygous for pea seed colour. That is to say, they each had two identical forms of the gene for this trait—2 yellows or 2 greens. The plants in the f1 generation were all heterozygous. In other words, they each had inherited two different alleles—one from each parent plant. It becomes clearer when we look at the actual genetic makeup, or genotype, of the pea plants instead of only the phenotype, or observable physical characteristics.

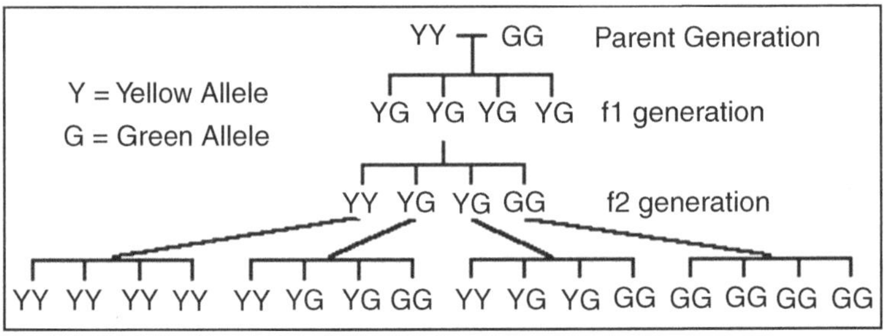

Note that each of the f1 generation plants inherited a Y allele from one parent and a G allele from the other. When the f1 plants breed, each has an equal chance of passing on either Y or G alleles to each offspring.

With all of the seven pea plant traits that Mendel examined, one form appeared dominant over the other, which is to say it masked the presence of the other allele. For example, when the genotype for pea seed colour is YG (heterozygous), the phenotype is yellow. However, the dominant yellow allele does not alter the recessive green one in any way. Both alleles can be passed on to the next generation unchanged.

Mendel's observations from these experiments can be summarised in two principles:

1. The principle of segregation and
2. The principle of independent assortment.

According to the principle of segregation, for any particular trait, the pair of alleles of each parent separate and only one allele passes from each parent on to an offspring. Which allele in a parent's pair of alleles is inherited is a matter of chance. We now know that this segregation of alleles occurs during the process of sex-cell formation.

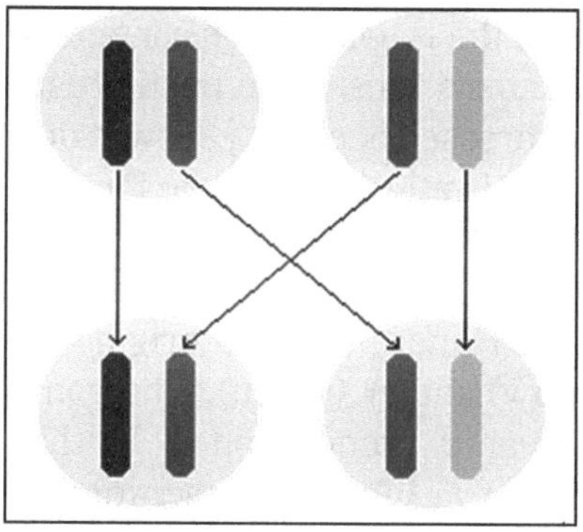

Fig. Segregation of alleles in production of sex cells

According to the principle of independent assortment, different pairs of alleles are passed to offspring independently of each other. The result is that new combinations of genes present in neither parent are possible. For example, a pea plant's inheritance of the ability to produce purple flowers instead of white ones does not make it more likely that it will also inherit the ability to produce yellow pea seeds in contrast to green ones. Likewise, the principle of independent assortment explains why the human inheritance of a particular eye colour does not increase or decrease the likelihood of having 6 fingers on each hand. Today, we know this is due to the fact that the genes for independently assorted traits are located on different chromosomes.

These two principles of inheritance, along with the understanding of unit inheritance and dominance, were the beginnings of our modern science of genetics. However, Mendel did not realise that there are exceptions to these rules. Some of these exceptions will be explored in the third section of this tutorial and in the Synthetic Theory of Evolution tutorial.

By focusing on Mendel as the father of genetics, modern biology often forgets that his experimental results also disproved Lamarck's theory of the inheritance of acquired characteristics described in the Early Theories of Evolution tutorial. Mendel rarely gets credit for this because his work remained essentially unknown until long after Lamarck's ideas were widely rejected as being improbable.

THE WORK OF MENDEL

Before Gregor Mendel, theories for a hereditary mechanism were based largely on logic and speculation, not on experimentation. In his monastery garden, Mendel carried out a large number of cross-pollination experiments between variants of the garden pea, which he obtained as pure-breeding lines. He crossed peas with yellow seeds to those with green seeds and observed that the progeny seeds (the first generation, F_1) were all yellow. When the F_1 individuals were self-pollinated or crossed among themselves, their progeny (F_2) showed a ratio of 3:1 (3/4 yellow and 1/4 green).

He deduced that, since the F_2 generation contained some green individuals, the determinants of greenness must have been present in the F_1 generation, although they were not expressed because yellow is dominant over green. From the precise mathematical 3:1 ratio (of which he found several other examples), he deduced not only the existence of discrete hereditary units (genes) but also that the units were present in pairs in the pea plant and that the pairs separated during gamete formation. Hence, the two original lines of pea plants were proposed to be *YY* (yellow) and *yy* (green). The gametes from these were *Y* and *y*, thereby producing an F_1 generation of *Yy* that were yellow in colour because of the dominance of *Y*. In the F_1 generation, half the gametes were *Y* and the other half were *y*, making the F_2 generation produced from random mating 1/4 *Yy*, 1/2 *YY*, and 1/4 *yy*, thus explaining the 3:1 ratio. The forms of the pea colour genes, *Y* and *y*, are called alleles.

Mendel also analysed pure lines that differed in pairs of characters, such as seed colour (yellow *versus* green) and seed shape (round *versus* wrinkled). The cross of yellow round seeds with green wrinkled seeds resulted in an F_1 generation that were all yellow and round, revealing the dominance of the yellow and round traits. However, the F_2 generation produced by self-pollination of F_1 plants showed a ratio of 9:3:3:1 (9/16 yellow round, 3/16 yellow wrinkled, 3/16 green round, and 1/16 green wrinkled; note that a 9:3:3:1 ratio is simply two 3:1 ratios combined). From this result and others like it, he deduced the independent assortment of separate gene pairs at gamete formation.

Mendel's success can be attributed in part to his classic experimental approach. He chose his experimental organism well and performed many controlled experiments to collect data. From his results, he developed brilliant explanatory hypotheses and went on to test these hypotheses experimentally. Mendel's methodology established a prototype for genetics that is still used today for gene discovery and understanding the genetic properties of inheritance.

How the Gene Idea Became Reality

Mendel's genes were only hypothetical entities, factors that could be inferred to exist in order to explain his results. The 20th century saw tremendous strides in the development of the understanding of the nature of genes and how they function. Mendel's publications lay unmentioned in the research literature until 1900, when the same conclusions were reached by several other investigators. Then, there followed hundreds of papers showing Mendelian inheritance in a wide array of plants and animals, including humans. It seemed that Mendel's ideas were of general validity. Many biologists noted that the inheritance of genes closely paralleled the inheritance of chromosomes during nuclear divisions, called meiosis, that occur in the cell divisions just prior to gamete formation.

THE DISCOVERY OF LINKED GENES

It seemed that genes were parts of chromosomes. In 1910, this idea was strengthened through the demonstration of parallel inheritance of certain *Drosophila* (a type of fruit fly) genes on sex-determining chromosomes by American zoologist and geneticist Thomas Hunt Morgan. Morgan and one of his students, Alfred Henry Sturtevant, showed not only that certain genes seemed to be linked on the same chromosome, but that the distance between genes on the same chromosome could be calculated by measuring the frequency at which new chromosomal combinations arose (these were proposed to be caused by chromosomal breakage and reunion, also known as crossing-over).

In 1916, another student of Morgan's, Calvin Bridges, used fruit flies with an extra chromosome to prove beyond reasonable doubt that the only way to explain the abnormal inheritance of certain genes was if they were part of the extra chromosome. American geneticist Hermann Joseph Müller showed that new alleles (called mutations) could be produced at high frequencies by treating cells with X-rays, the first demonstration of an environmental mutagenic agent (mutations can also arise spontaneously). In 1931, American botanist Harriet Creighton and American scientist Barbara McClintock demonstrated that new allelic combinations of linked genes were correlated with physically exchanged chromosome parts.

EARLY MOLECULAR GENETICS

In 1908, British physician Archibald Garrod proposed the important idea that the human disease alkaptonuria and certain other hereditary diseases were caused by inborn errors of metabolism, suggesting for the first time that linked genes had molecular action at the cell level. Molecular genetics did not begin in earnest until 1941 when American geneticist George Beadle and American biochemist Edward Tatum showed that the genes they were studying in the fungus *Neurospora crassa* acted by coding for catalytic proteins called enzymes. Subsequent studies in other organisms extended this idea to show that genes generally code for proteins. Soon afterwards, American bacteriologist Oswald Avery, Canadian American geneticist Colin M. MacLeod, and American biologist Maclyn McCarty showed that bacterial genes are made of DNA, a finding that was later extended to all organisms.

DNA AND THE GENETIC CODE

A major landmark was attained in 1953 when American geneticist and biophysicist James D. Watson and British biophysicists Francis Crick and Maurice Wilkins devised a double-helix model for DNA structure. This model showed that DNA was capable of self-replication by separating its complementary strands and using them as templates for the synthesis of new DNA molecules. Each of the intertwined strands of DNA was proposed to be a chain of chemical groups called nucleotides, of which there were known to be four types. Because proteins are strings of amino acids, it was proposed that a specific nucleotide sequence of DNA could contain a code for an amino acid sequence and hence protein structure. In 1955, American molecular biologist Seymour Benzer, extending earlier studies in *Drosophila*, showed that the mutant sites within a gene could be mapped in relation to each other. His linear map indicated that the gene itself is a linear structure.

In 1958, the strand-separation method for DNA replication (called the semi-conservative method) was demonstrated experimentally for the first time by American molecular biologist Matthew Meselson and American geneticist Franklin W. Stahl. In 1961, Crick and South African biologist Sydney Brenner showed that the genetic code must be read in triplets of nucleotides, called codons.

American geneticist Charles Yanofsky showed that the positions of mutant sites within a gene matched perfectly the positions of altered amino acids in the amino acid sequence of the corresponding protein. In 1966, the complete genetic code of all 64 possible triplet coding units (codons), and the specific amino acids they code for, was deduced by American biochemists Marshall Nirenberg and Har Gobind Khorana.

Subsequent studies in many organisms showed that the double-helical structure of DNA, the mode of its replication, and the genetic code are the same in virtually all organisms, including plants, animals, fungi, bacteria, and viruses. In 1961, French biologist François Jacob and French biochemist Jacques Monod established the prototypical model for gene regulation by showing that bacterial genes can be turned on (initiating transcription into RNA and protein synthesis) and off through the binding action of regulatory proteins to a region just upstream of the coding region of the gene.

BEAUTIFUL DNA

Up until the 1950s, scientists knew a good deal about heredity, but they did not have a clue what DNA looked like. In order to learn more about DNA and its structure, some scientists experimented using X rays as a form of molecular photography.

Rosalind Franklin, a physical chemist working with Maurice Wilkins at King's College in London, was among the first to use this method to analyse genetic material. Her experiments produced what were referred to at the time as *"the most beautiful X-ray photographs of any substance ever taken."*

Other scientists, including zoologist James Watson and physicist Francis Crick, both working at Cambridge University in the United Kingdom, were trying to determine the shape of DNA too. Ultimately, this line of research revealed one of the most profound scientific discoveries of the 20th century: that DNA exists as a double helix.

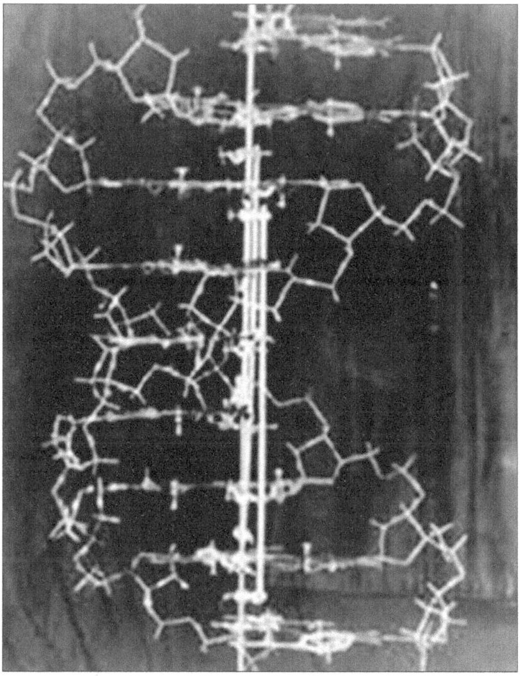

Fig. In 1953, Watson and Crick created their historic model of the shape of DNA: the double helix.

The 1962 Nobel Prize in physiology or medicine was awarded to Watson, Crick, and Wilkins for this work. Although Franklin did not earn a share of the prize due to her untimely death at age 38, she is widely recognised as having played a significant role in the discovery.

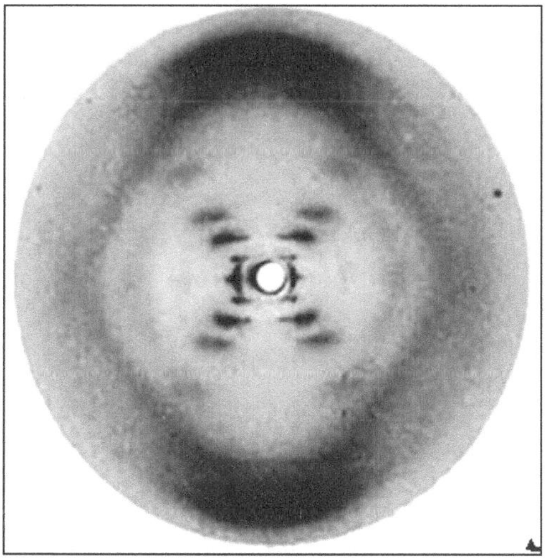

Fig. Rosalind Franklin's original X-ray diffraction photo revealed the physical structure of DNA.

The spiral staircase-shaped double helix has attained global status as the symbol for DNA. But what is so beautiful about the discovery of the twisting ladder structure is not just its good looks.

Rather, the structure of DNA taught researchers a fundamental lesson about genetics. It taught them that the two connected strands—winding together like parallel handrails—were complementary to each other, and this unlocked the secret of how genetic information is stored, transferred, and copied.

In genetics, complementary means that if you know the sequence of nucleotide building blocks on one strand, you know the sequence of nucleotide building blocks on the other strand: A always matches up with T and C always links to G.

Long strings of nucleotides form genes, and groups of genes are packaged tightly into structures called chromosomes. Every cell in your body except for eggs, sperm, and red blood cells contains a full set of chromosomes in its nucleus.

If the chromosomes in one of your cells were uncoiled and placed end to end, the DNA would be about 6-feet long. If whole of the DNA in your body was connected in this way, it would stretch approximately 67 billion miles! That is nearly 150,000 round trips to the Moon.

DNA STRUCTURE

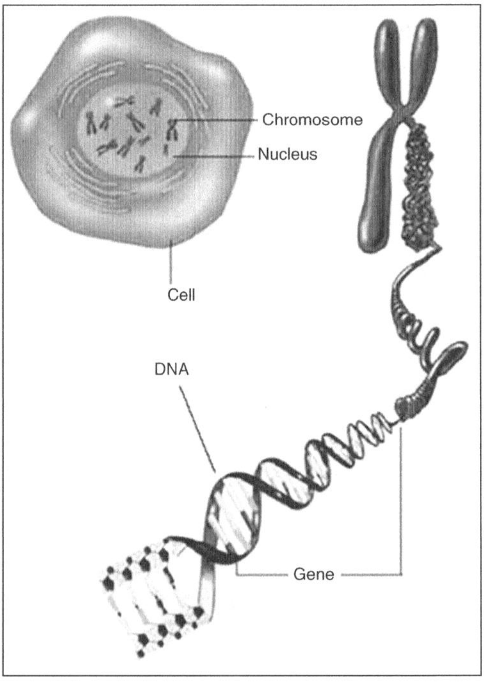

Fig. The long, stringy DNA that makes up genes is spooled within chromosomes inside the nucleus of a cell.

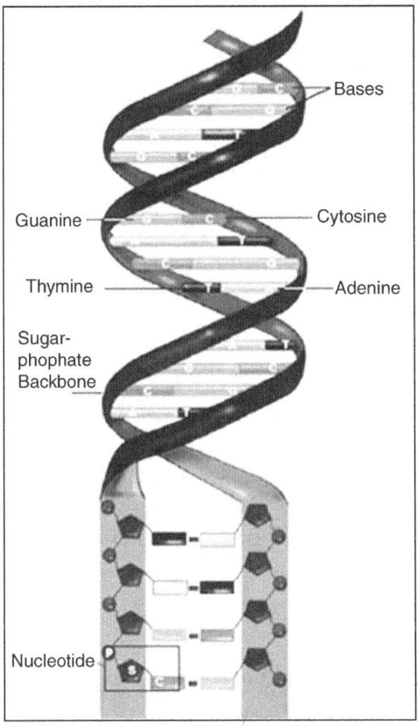

Fig. DNA consists of two long, twisted chains made up of nucleotides. Each nucleotide contains one base, one phosphate molecule, and the sugar molecule deoxyribose. The bases in DNA nucleotides are adenine, thymine, cytosine, and guanine.

Copycat

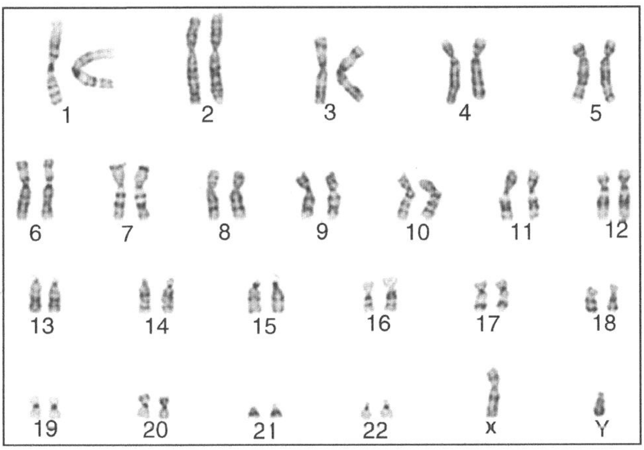

Fig. Humans have 23 pairs of chromosomes. Male DNA contains an X and a Y chromosome, whereas female DNA contains two X chromosomes.

It is astounding to think that your body consists of trillions of cells. But what is most amazing is that it all starts with one cell. How does this massive expansion take place? As an embryo progresses through development, its cells must reproduce. But before a cell divides into two new, nearly identical cells, it must copy its DNA so there will be a complete set of genes to pass on to each of the new cells. To make a copy of itself, the twisted, compacted double helix of DNA has to unwind and separate its two strands. Each strand becomes a pattern, or template, for making a new strand, so the two new DNA molecules have one new strand and one old strand.

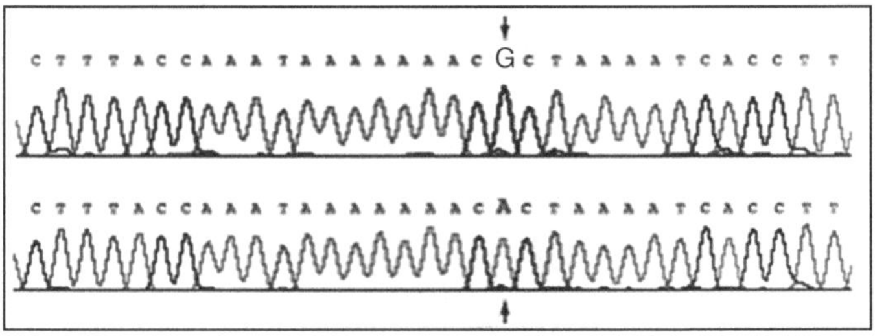

Fig. When DNA polymerase makes an error while copying a gene's DNA sequence, the mistake is called a mutation. In this example, the nucleotide G has been changed to an A.

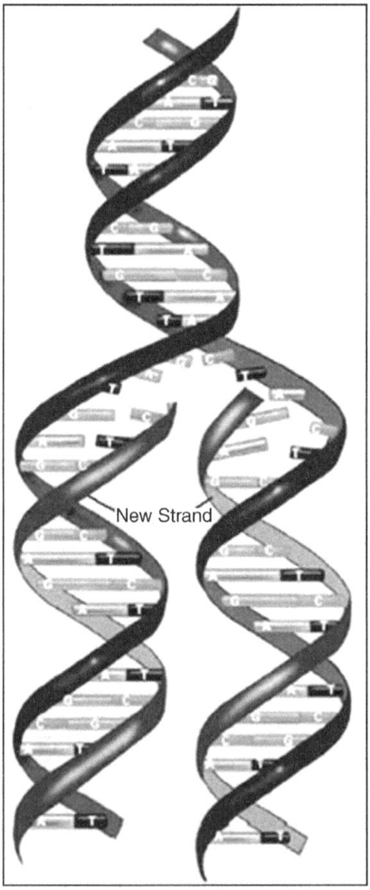

Fig. During DNA replication, each strand of the original molecule acts as a template for the synthesis of a new, complementary DNA strand.

The copy is courtesy of a cellular protein machine called as DNA polymerase, which reads the template DNA strand and stitches together the complementary new strand. The process, called replication, is astonishingly fast and accurate, although occasional mistakes, such as deletions or duplications, occur. Fortunately, a cellular spell-checker catches and corrects nearly all of these errors.

Mistakes that are not corrected can lead to diseases such as cancer and certain genetic disorders. Some of these include Fanconi anaemia, early aging diseases, and other conditions in which people are extremely sensitive to sunlight and some chemicals.

DNA copying is not the only time when DNA damage can happen. Prolonged, unprotected sun exposure can cause DNA changes that lead to skin cancer, and toxins in cigarette smoke can cause lung cancer.

It may seem ironic, then, that many drugs used to treat cancer work by attacking DNA. That is because these chemotherapy drugs disrupt the DNA-copying process, which goes on much faster in rapidly dividing cancer cells than in other cells of the body. The trouble is that most of these drugs do affect normal cells that grow and divide frequently, such as cells of the immune system and hair cells. Understanding DNA replication better could be a key to limiting a drug's action to cancer cells only.

Let us Call it Even

After copying its DNA, a cell's next challenge is getting just the right amount of genetic material into each of its two offsprings.

Most of your cells are called diploid ("*di*" means two, and "*ploid*" refers to sets of chromosomes) because they have two sets of chromosomes (23 pairs). Eggs and sperm are different; these are known as haploid cells. Each haploid cell has only one set of 23 chromosomes so that at fertilization the math will work out: a haploid egg cell will combine with a haploid sperm cell to form a diploid cell with the right number of chromosomes: 46.

Chromosomes are numbered 1 to 22, according to size, with 1 being the largest chromosome. The 23rd pair, known as the sex chromosomes, are called X and Y. In humans, abnormalities of chromosome number usually occur during meiosis, the time when a cell reduces its chromosomes from diploid to haploid in creating eggs or sperm. What happens if an egg or a sperm cell gets the wrong number of chromosomes, and how often does this happen?

Molecular biologist Angelika Amon of the Massachusetts Institute of Technology in Cambridge says that mistakes in dividing DNA between daughter cells during meiosis are the leading cause of human birth defects and miscarriages. Current estimates are that 10 per cent of all embryos have an incorrect chromosome number. Most of these do not go to full term and are miscarried. In women, the likelihood that chromosomes would not be apportioned properly increases with age. One of every 18 babies born to women over the age of 45 has three copies of chromosome 13, 18, or 21 instead of the normal two, and this improper balancing can cause trouble. For example, three copies of chromosome 21 leads to Down syndrome.

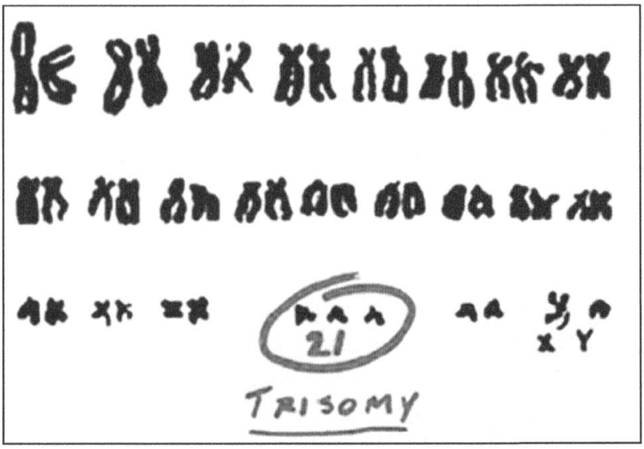

Fig. Trisomy, the hallmark of Down syndrome, results when a baby is born with three copies of chromosome 21 instead of the usual two.

To make her work easier, Amon—like many other basic scientists—studies yeast cells, which separate their chromosomes almost exactly the same way human cells do, except that yeast does it much faster. A yeast cell copies its DNA and produces daughter cells in about 1.5 hours, compared to a whole day for human cells.

The yeast cells she uses are the same kind bakeries use to make bread and breweries use to make beer!

Amon has made major progress in understanding the details of meiosis. Her research shows how, in healthy cells, glue-like protein complexes called cohesins release pairs of chromosomes at exactly the right time. This allows the chromosomes to separate properly. These findings have important implications for understanding and treating infertility, birth defects, and cancer.

MEIOSIS

Getting the Message

So, we have described DNA—its basic properties and how our bodies make more of it. But how does DNA serve as the language of life? How do you get a protein from a gene?

There are two major steps in making a protein. The first is transcription, where the information coded in DNA is copied into RNA. The RNA nucleotides are complementary to those on the DNA: a C on the RNA strand matches a G on the DNA strand.

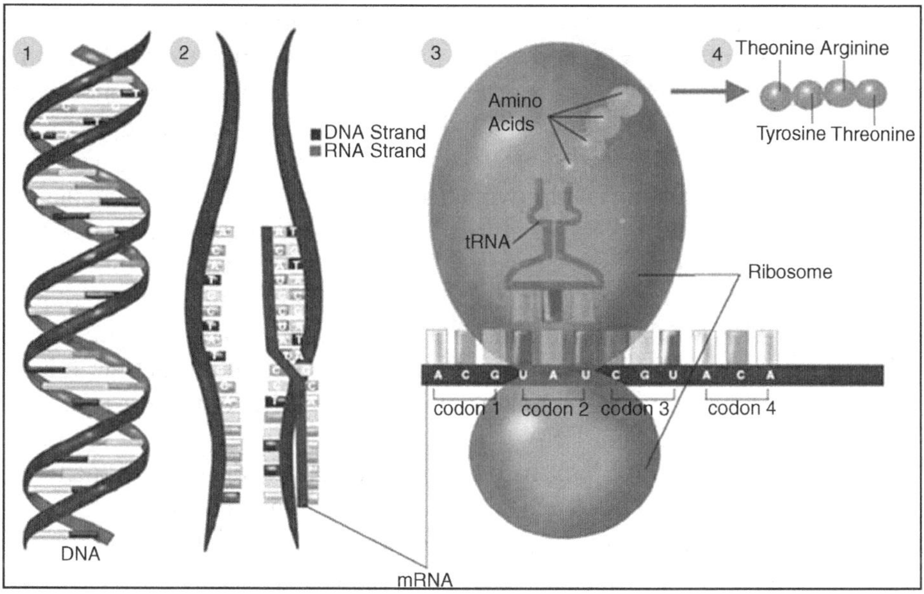

Fig. RNA polymerase transcribes DNA to make messenger RNA (mRNA). 2. The mRNA sequence (dark red strand) is complementary to the DNA sequence (blue strand). 3. On ribosomes, transfer RNA (tRNA) helps convert mRNA into protein. 4. Amino acids link up to make a protein.

The only difference is that RNA pairs a nucleotide called uracil (U), instead of a T, with an A on the DNA.

A protein machine called RNA polymerase reads the DNA and makes the RNA copy. This copy is called messenger RNA, or mRNA, because it delivers the gene's message to the protein-producing machinery.

At this point, you may be wondering why all of the cells in the human body are not exactly alike, as they all contain the same DNA. What makes a liver cell different from a brain cell? How do the cells in the heart make the organ contract, but those in skin allow us to sweat?

Cells can look and act differently, and do entirely different jobs, because each cell *"turns on,"* or expresses, only the genes appropriate for what it needs to do.

STEVEN BLOCK

That is because RNA polymerase does not work alone, but rather function with the aid of many helper proteins. Although the core part of RNA polymerase is the same in all cells, the helpers vary in different cell types throughout the body.

You would think that for a process so essential to life, researchers would know a lot about how transcription works. Although it is true that the basics are clear—biologists have been studying gene transcribing by RNA polymerases since these proteins were first discovered in 1960; some of the details are actually still murky. The biggest obstacle to learning more has been lack of tools. Until recently, researchers were unable to get a picture at the atomic level of the giant RNA polymerase protein assemblies inside cells to understand how the many pieces of this amazing, living machine do what they do, and do it so well.

But our understanding is improving fast, thanks to spectacular technological advances. We have new X-ray pictures that are far more sophisticated than those that revealed the structure of DNA. Roger Kornberg of Stanford University in California used such methods to determine the structure of RNA polymerase. This work earned him the 2006 Nobel Prize in chemistry. In addition, very powerful microscopes and other tools that allow us to watch one molecule at a time provide a new look at RNA polymerase while it is at work reading DNA and producing RNA.

For example, Steven Block, also of Stanford, has used a physics technique called optical trapping to track RNA polymerase as it inches along DNA. Block and his team performed this work by designing a specialised microscope sensitive enough to watch the real-time motion of a single polymerase travelling down a gene on one chromosome.

The researchers discovered that molecules of RNA polymerase behave like battery-powered spiders as they crawl along the DNA ladder, adding nucleotides one at a time to the growing RNA strand. The enzyme works much like a motor, Block believes, powered by energy released during the chemical synthesis of RNA.

Nature's Cut-and-Paste Job

Several types of RNA play key roles in making a protein. The gene transcript (the mRNA) transfers information from DNA in the nucleus to the ribosomes that make the protein. Ribosomal RNA forms about 60 per cent of the ribosomes. Lastly, transfer RNA carries amino acids to the ribosomes. All three types of cellular RNAs come together to produce new proteins.

But the journey from gene to protein is not quite as simple as we have just made it out to be. After transcription, several things need to happen to mRNA before a protein can be made. For example, the genetic material of humans and other eukaryotes (organisms that have a nucleus) includes a lot of DNA that does not encode proteins. Some of this DNA is stuck right in the middle of genes.

To distinguish the two types of DNA, scientists call the coding sequences of genes exons and the pieces in between them introns (for intervening sequences).

If RNA polymerase was to transcribe DNA from the start of an intron-containing gene to the end, the RNA would be complementary to the introns as well as the exons.

To get an mRNA molecule that yields a working protein, the cell needs to trim out the intron sections and then stitch only the exon pieces together. This process is called RNA splicing.

Splicing has to be extremely accurate. An error in the splicing process, even one that results in the deletion of just one nucleotide in an exon or the addition of just one nucleotide in an intron, will throw the whole sequence out of alignment. The result is usually an abnormal protein, or no protein at all. One form of Alzheimer's disease, for example, is caused by this kind of splicing error.

Molecular biologist Christine Guthrie of the University of California, San Francisco, wants to understand more fully the mechanism for removing intron RNA and find out how it stays so accurate.

She uses yeast cells for these experiments. Just like human DNA, yeast DNA has introns, but they are fewer and simpler in structure and are therefore easier to study. Guthrie can identify which genes are required for splicing by finding abnormal yeast cells that mangle splicing.

So why do introns exist, if they are just going to be chopped out? Without introns, cells would not need to go through the splicing process and keep monitoring it to be sure it's working right.

As it turns out, splicing also makes it possible for cells to create more proteins.

Think about all the exons in a gene. If a cell stitches together exons 1, 2, and 4, leaving out exon 3, the mRNA will specify the production of a particular protein. But instead, if the cell stitches together exons 1, 2, and 3, this time leaving out exon 4, then the mRNA will be translated into a different protein.

By cutting and pasting the exons in different patterns, which scientists call alternative splicing, a cell can create different proteins from a single gene. Alternative splicing is one of the reasons why human cells, which have about 20,000 genes, can make hundreds of thousands of different proteins.

All Together Now

Until recently, researchers looked at genes, and the proteins they encode, one at a time. Now, they can look at how large numbers of genes and proteins act, as well as how they interact. This gives them a much better picture of what goes on in a living organism.

Already, scientists can identify all of the genes that are transcribed in a cell—or in an organ, such as the heart. And although researchers cannot tell you, right now, what is going on in every cell of your body while you read a book or walk down the street, they can do this sort of "*whole-body*" scan for simpler, single-celled organisms like yeast.

Using a technique called genome-wide location analysis, Richard Young of the Massachusetts Institute of Technology unravelled a "*regulatory code*" of living yeast cells, which have more than 6,000 genes in their genome. Young's technique enabled him to determine the exact places where RNA polymerase's helper proteins sit on DNA and tell RNA polymerase to begin transcribing a gene.

Since he did the experiment with the yeast exposed to a variety of different conditions, Young was able to figure out how transcription patterns differ when the yeast cell is under stress (say, in a dry environment) or thriving in a sugary-rich nutrient solution. Done one gene at a time, using methods considered state-of-the-art just a few years ago, this kind of analysis would have taken hundreds of years.

After demonstrating that his technique worked in yeast, Young then took his research a step forward. He used a variation of the yeast method to scan the entire human genome in small samples of cells taken from the pancreases and livers of people with type 2 diabetes. He used the results to identify genes that are not transcribed correctly in people with the disease.

This information provides researchers with an important tool for understanding how diabetes and other diseases are influenced by defective genes. By building models to predict how genes respond in diverse situations, researchers may be able to learn how to stop or jump-start genes on demand, change the course of a disease, or prevent it from ever happening.

RNA SURPRISES

But which ribosomal RNAs are doing the work? Most scientists assumed that RNA nucleotides buried deep within the ribosome complex—the ones that have the same sequence in every species from bacteria to people—were the important ones for piecing the growing protein together.

However, recent research by Rachel Green, who worked with Noller before moving to Johns Hopkins University in Baltimore, Maryland, showed that this is not the case. Green discovered that those RNA nucleotides are not needed for assembling a protein. Instead, she found, the nucleotides do something else entirely: they help the growing protein slip off the ribosome once it is finished.

Noller, Green, and hundreds of other scientists work with the ribosomes of bacteria. Why should you care about how bacteria create proteins from their genes? One reason is that this knowledge is important for learning how to disrupt the actions of disease-causing microorganisms. For example, antibiotics like erythromycin and neo-mycin work by attacking the ribosomes of bacteria, which are different enough from human ribosomes that our cells are not affected by these drugs.

As researchers gain new information about bacterial translation, the knowledge may lead to more antibiotics for people. New antibiotics are urgently needed because many bacteria have developed resistance to the current arsenal.

This resistance is sometimes the result of changes in the bacteria's ribosomal RNA. It can be difficult to find those small, but critical, changes that may lead to resistance, so it is important to find completely new ways to block bacterial translation.

Green is working on that problem too. Her strategy is to make random mutations to the genes in a bacterium that affect its ribosomes. But what if the mutation disables the ribosome so much that it cannot make proteins? Then the bacterium would not grow, and Green would not find it.

Using clever molecular tricks, Green figured out a way to rescue some of the bacteria with defective ribosomes so they could grow. While some of the rescued bacteria have changes in their ribosomal RNA that make them resistant to certain antibiotics (and thus would not make good antibiotic targets), other RNA changes that do not affect resistance may point to promising ideas for new antibiotics.

AN INTERESTING DEVELOPMENT

In the human body, one of the most important jobs for proteins is to control how embryos develop. Scientists discovered a hugely important set of proteins involved in development by studying mutations that cause bizarre malformations in fruit flies.

The most famous such abnormality is a fruit fly with a leg, rather than the usual antenna, growing out of its head. According to Thomas C. Kaufman of Indiana University in Bloomington, the leg is perfectly normal—it is just growing in the wrong place.

In this type of mutation and many others, something goes wrong with the genetic programme that directs some of the cells in an embryo to follow developmental pathways, which are a series of chemical reactions that occur in a specific order. In the antenna-into-leg problem, it is as if the cells growing from the fly's head, which normally would become an antenna, mistakenly believe that they are in the fly's thorax, and therefore ought to grow into a leg. And so they do.

Thinking about this odd situation taught scientists an important lesson—that the proteins made by some genes can act as switches. Switch genes are master controllers that provide each body part with a kind of identification card. If a protein that normally instructs cells to become an antenna is disrupted, cells can receive new instructions to become a leg instead.

Scientists determined that several different genes, each with a common sequence, provide these anatomical identification card instructions. Kaufman isolated and described one of these genes, which became known as *Antennapedia*, a word that means "*antenna feet.*"

Kaufman then began looking a lot more closely at the molecular structure of the *Antennapedia* gene. In the early 1980s, he and other researchers made a discovery that has been fundamental to understanding evolution as well as developmental biology.

The scientists found a short sequence of DNA, now called the homeobox, that is present not only in *Antennapedia* but in the several genes next to it and in genes in many other organisms. When geneticists find very similar DNA sequences in the genes of different organisms, it is a good clue that these genes do something so important and useful that evolution uses the same sequence over and over and permits very few changes in its structure as new species evolve.

Researchers quickly discovered nearly identical versions of homeobox DNA in almost every non-bacterial cell they examined—from yeast to plants, frogs, worms, beetles, chickens, mice, and people. Hundreds of homeobox-containing genes have been identified, and the proteins they make turn out to be involved in the early stages of development of many species. For example, researchers have found that abnormalities in the homeobox genes can lead to extra fingers or toes in humans.

3

DNA Fingerprinting and Genetic Map Construction

INTRODUCTION

The rapid evolution of repeated sequences results in high levels of polymorphism that can be detected by methods based on Southern hybridization or polymerase chain reaction (PCR). Repeated sequences are, therefore, potentially useful as fingerprinting markers. Polymorphism in repeated sequences can also be used for genetic map construction. For example, it has recently been possible to map very rapidly almost 500 interspersed repetitive sequences in mouse.

Larger sequences, such as retrotransposons, can also be used for fingerprinting or genetic mapping. Specific PCR primers enable some repeats to be used as single locus markers.

The most useful of these are the simple sequence repeats (SSRs or microsatellites), which are usually di- or tri-nucleotide repeats, since these are often flanked by unique sequence DNA. This allows the use of flanking primers for PCR detection of length polymorphism within the SSR itself. SSRs are highly polymorphic and are therefore useful for studies of genetic diversity as well as for mapping.

Although repeated sequences can be used for genetic mapping, the genetic maps of most crop species and model organisms are based on polymorphisms of low- or single-copy sequences. Most reasonably detailed maps are based on analysis of restriction fragment length polymorphism (RFLP), which is detected by the hybridization of cloned DNA segments to genomic DNA digested with restriction enzymes.

Most RFLP probes are either small fragments of genomic DNA (selected for low copy number) or cDNA clones which represent transcribed genes of the tissue from which they were isolated. In future, however, most maps will probably be based largely on amplified fragment length polymorphisms (AFLP), simple sequence repeats (SSRs or microsatellites) and, to a lesser extent, random amplified polymorphic DNAs (RAPD). These PCR methods will be favoured because of their speed of assay, degree of polymorphism and their requirement for only small amounts of DNA from the plants of interest. This is illustrated by the analysis of the haploid megasporophytes of conifers. These yield sufficient DNA for the construction of genetic maps using RAPD markers.

Genetic maps are usually constructed to locate genes of biological or economic interest in relation to molecular markers. For single genes of large effect this is straightforward since a chromosome location can easily be established by segregational analysis. However, many traits of interest to plant breeders and researchers are controlled by several to many genes acting in combination. The availability of high-resolution genetic maps and increasingly sophisticated statistical methods has greatly increased the ability to identify the number and chromosome locations of such quantitative trait loci (QTL).

There are two main applications for genetic maps. The first is the use of DNA markers linked to the gene or genes of interest to select individuals in breeding programmes without the need to assess the phenotype at each stage of development. Such markerassisted selection (MAS) is already used in breeding programmes. It is particularly useful for the manipulation of quantitative traits, which are the most difficult to evaluate phenotypically in conventional breeding programmes. The second application of genetic maps is the isolation of genes by a map-based cloning approach such as chromosome walking or chromosome landing. This approach has been used, for example, in the isolation of genes conferring disease resistance from tomato.

COMPARATIVE MAPS AND CONSERVED SYNTENY

The cloned DNA fragments used to detect RFLP often hybridize well to DNA of related species, giving a series of common markers that enable the respective genetic maps to be aligned. This is termed comparative mapping. cDNA clones are particularly useful for inter-species work, probably because the accumulation of mutations in expressed gene sequences is limited by selection.

In contrast to the plasticity of genome structure revealed by studies of repeated sequences, including nucleolus organizer regions, the order of single-copy sequences on genetic maps of different species often shows a high degree of conservation. Conservation of marker order is termed conserved synteny, since two genes or markers are syntenic if they lie on the same piece of DNA, usually a chromosome or chromosome fragment, irrespective of whether they are genetically linked.

COMPARATIVE MAPPING IN GRASSES

Comparative mapping shows that the three genomes of hexaploid wheat have very similar marker orders with the exception of a few major chromosome rearrangements. Additional translocations are found when comparison is made with the rye or barley genome, but marker order is still highly conserved. Comparative maps of maize and sorghum are well established and comparative maps of *Oryza* species are being developed.

Remarkably, comparison of the genetic maps of wheat and rice, rice and maize and wheat and maize also reveals large linkage segments that can be aligned. Furthermore, equivalent linkage blocks can be recognized in rice, wheat, maize, sorghum, sugar cane and millet. Thus a model of cereal genomes can be constructed using segments of the rice genetic map as the building blocks. This aligns the maps of the various species, revealing the genetic relationships between their chromosomes. This analysis also confirms the tetraploid nature of the maize genome since the duplicated chromosome segments can be organized into two separate genomes. An additional important result from comparative mapping is that RFLP detected by cDNA clones from single rice YAC (yeast artificial chromosome) clones either cosegregate in barley or are tightly linked in the same map order as in rice.

Thus, fine-scale organization of gene order is also conserved. Mapping studies also suggest that there are duplicated regions in diploid species, suggesting that duplication and subsequent divergence of chromosome segments has been an important feature of plant evolution even in the absence of polyploidy.

Examples include regions containing dwarfing genes in rye, segments of rice chromosomes 1 and 5, duplications in sorghum and soybean (*Glycine max*) that may reflect ancient polyploidy, and duplications in the diploids *Brassica oleracea* and *B. nigra*.

COMPARATIVE MAPPING IN OTHER PLANT GROUPS

Comparative mapping in the *Solanaceae* showed that when RFLP maps of potato (*Solanum tuberosum*) and tomato (*Lycopersicon esculentum*) were compared, the two genomes showed a high degree of conserved synteny. Five whole-arm inversions were identified that distinguished the two genomes.

Comparative mapping has also been reported extensively in legumes, where several studies show conserved linkage groups between species such as mungbean (*Vigna radiata*) and cowpea (*V. unguiculata*) and pea (*Pisum sativum*) and lentil (*Lens culinaris*). However, the soybean genome appears to show considerable internal duplication and rearrangement relative to other legume genomes.

Comparison of *Brassica* species shows that there are a greater number of chromosome rearrangements than commonly found in cereal species, but there are still large segments with conserved marker orders. Importantly, *Brassica* species and the model plant species *Arabidopsis thaliana* are both in the *Cruciferae* family and are sufficiently closely related for *Arabidopsis* clones to be used directly for mapping and gene isolation in *Brassica*. Comparison of the genomes of *B. oleracea* and *Arabidopsis* reveals extensive rearrangement, including at least 17 translocations and nine inversions, but homoeologous segments can still be recognized. Thus, *Brassica* researchers can readily exploit the vast mapping, sequencing and genetic resources of *Arabidopsis*.

MANIPULATING THE GENETIC INFORMATION STORED IN DNA

Breakthroughs in decoding and manipulating the genetic information stored in DNA promise a world of benefits. These scientific advances already help to diagnose and treat disease, develop new medicines, bring criminals to justice, improve our food supply, and clean up the environment. At the same time, however, genetic technologies also present society with the potential for new and serious social or environmental problems. Many of the developments that worry critics of genetic technologies remain on the horizon, but the debate over their inevitable arrival is already in full swing.

GENES AND THE ENVIRONMENT

Humans have tampered with the genetic composition of other organisms for thousands of years. Most of this manipulation has been decidedly low-tech: domestication of animals and selective breeding of desirable food crops. The development and use of genetic engineering techniques has accelerated the pace at which humans can alter nature, creating some products that have unquestionable benefits and others that raise serious concerns.

Some scientists fear that genetic engineering techniques will damage genetic diversity. Domestication and selective breeding, which aim to produce many similar organisms with particular desirable traits, reduce the natural variation of genes within a species. Genetic engineering techniques take this effect a giant step further. Many crop species—including wheat, corn, tomatoes, and strawberries—have been manipulated to maximize yield, appearance, resistance to pests and chemicals, hardiness, and other commercially valuable traits.

Once attractive varieties have been developed, agricultural techniques favour wide-scale planting of such genetically similar or identical stocks. The impact of this practice on biodiversity can be ominous because older plant varieties that carry diverse and useful alleles may be lost forever. The forfeited genetic material could leave a species vulnerable to annihilation from a single factor in the environment, such as an insect pest or an infectious virus.

Fortunately, efforts to combat decreasing biodiversity are under way. Farmers, gardeners, government agencies, and other interested parties have collaborated to create seed banks to maintain genetic diversity. These banks catalogue, store, and distribute the seeds of rare or endangered plants, enabling gardeners and farmers to continue cultivating rare plant varieties so that the unique genetic makeup of these plants does not disappear. In the same vein, zoos and other institutions breed endangered species of animals that may no longer be able to survive in their native habitats.

Other animal programmes seek to preserve or enhance the genetic variation within certain endangered animal populations. For example, all cheetahs are almost identical genetically, most likely due to their near extinction about 12,000 years ago. Inbreeding among the few remaining individuals has resulted in a loss of genetic diversity in modern cheetahs that may have affected the cheetah's immune system, leaving the animal vulnerable to disease. Scientists hope to use genetic engineering techniques to introduce new genes into the cheetah population to increase the genetic diversity of the species.

The broad use of genetic engineering techniques in agriculture has raised other concerns beyond issues of biodiversity. From a consumer point of view, for instance, new technologies may have compromised a food's taste or nutritional value in exchange for a plant or animal that can be grown faster or at less cost. In addition, some critics question the safety of genetically engineered foods. They fear that plants or animals that have received new genes will produce proteins that would not be present in non-manipulated organisms.

Such changes could have a serious effect: causing allergies or toxicity in humans who eat these foods, for example, or disrupting a plant's production of key nutrients. Though there has been much speculation about the potential health risks of genetically engineered foods, rigorous scientific investigation into the effects of these foods on humans is just beginning.

The potential environmental impact of genetically engineered agriculture is equally controversial. Critics fear that transgenic organisms, which contain DNA from other species, could give rise to populations of genetically altered life forms that could cause disease, displace native species, or otherwise harm delicate ecosystems. Consider the example of a genetically engineered form of oilseed rape, the plant that yields canola oil.

The altered form, which has been grown commercially in the United States since 1993, contains inserted genes that increase its resistance to herbicides (weed-killing chemicals). The altered plants can grow unabated even when a farmer sprays a field with enough herbicide to kill troublesome weeds. Yet a curious problem has emerged: The transgenic rape can interbreed with weedy relatives that grow nearby. Some scientists fear that this interbreeding could create weed varieties that also have a genetic resistance to known herbicides. The biological and economic impact of such a development could be enormous.

Genes and Society

Advances in genetic technologies allow scientists to take an unprecedented glimpse into the genetic makeup of every person. The information derived from this testing can serve many valuable purposes: It can save lives, assist couples trying to decide whether or not to have children, and help law-enforcement officials solve a crime. Yet breakthroughs in genetic testing also raise some troubling social concerns about privacy and discrimination.

For example, if an individual's genetic information becomes widely available, it could give health insurers cause to deny coverage to people with certain risk factors or encourage employers to reject certain high-risk job applicants. Furthermore, many genetically linked problems are more common among certain racial and ethnic groups—for example, the BRCA1 breast cancer allele is more common in Ashkenazi Jews, and the blood disorder sickle-cell anemia is more prevalent among blacks of African ancestry. Many minority groups fear that the expansion of genetic testing could create whole new avenues of discrimination.

Of particular concern are genetic tests that shed light on traits such as personality, intelligence, and mental health or potential abilities. Genetic tests that indicate a person is unlikely to get along with other people could be used to limit a person's professional advancement. In other cases, tests that identify a genetic risk of heart failure could discourage a person from competing in sports.

New technologies that allow the manipulation of genes have raised even more disturbing possibilities. Gene therapy advances, which allow scientists to replace defective genes with normal alleles, give people with typically fatal diseases new hope for healthy lives. To date, gene therapy has focused on manipulating the genetic material in body cells other than gametes, so the changes will not be passed on to future generations. However, the application of gene therapy techniques to gametes—the cells involved in reproduction—seems inevitable.

Such manipulation might help prevent the transmission of disease from one generation to another, but it could also produce unforeseen problems with long-lasting consequences. For instance, many people worry that new genetic techniques could be used to alter or encourage traits now viewed as part of normal human variability, such as shortness or baldness. At various times in the past century, people have advocated efforts to improve the human condition by promoting the perpetuation of certain genes.

This concept, known as eugenics, typically involves encouraging people with "positive" genes to reproduce and discouraging those with "inferior" genes from having offspring. Many people fear that new genetic technologies used to manipulate the human genome could give people previously unattainable methods to resort to extreme forms of eugenics.

Advances in genetic technologies have turned some genes into valuable commercial commodities, spawning a host of controversial questions. Who owns a genetically altered organism or the genes it contains? Is it right to patent the use of a naturally occurring gene? Some people feel that genetic material should not be owned or used for profit. Costa Rica has enacted laws to prevent foreign companies from patenting and then profiting from genes of native Costa Rican plant and animal species. Balancing the need to limit patents on genes are concerns that the profit motive of companies must be protected to maintain incentives to make new discoveries for medical products. The citizens of Iceland, for example, are cooperating with a biotechnology company in a study of the genetic makeup of the Icelandic people. The information compiled will be used to learn about genetic diseases.

HISTORY

Humans have had some understanding of heredity since prehistoric times, observing how similar traits pass from parent to offspring and noting that differences arise with each generation. Most of the mechanisms of heredity, however, were shrouded in mystery until early in the 20th century. Since that time, the rate of discovery has reached a feverish pace, enabling the advancement of modern molecular biology and the current Human Genome Project.

Early Views of Heredity

In ancient times, people understood some basic rules of heredity and used this knowledge to breed domestic animals and crops. By about 5000 BC, for example, people in different parts of the world had begun applying selective breeding techniques to grow new plant varieties, including types of wheat, maize, rice, and date palms, that had never existed in the wild.

Ancient people understood that the rules of inheritance also applied to humans. The ancient Greeks were particularly interested in human heredity and evolution. Greek scientists and philosophers hotly debated whether a male or female parent contributed more to an offspring. In the 4th century BC, Aristotle speculated that acquired characteristics, such as a scar that was incurred during life, could be passed on to offspring.

He also believed in a widely held theory known as pangenesis. This theory proposed that particles in the body, called gemmules, reside in the limbs and organs. The gemmules become imprinted with any changes acquired by the body, such as muscle development from exercise. The gemmules then move to the reproductive cells and transfer information about the body's alterations to these cells. The reproductive cells transmit the acquired traits to offspring through particles called pangenes.

The theories about the inheritance of acquired characteristics and pangenesis persisted until the middle of the 19th century. French zoologist Jean-Baptiste Lamarck formalized the theory of acquired characteristics in his treatise *Philosophie Zoologique* (1809). Lamarck proposed that organisms evolve by responding to changes in their environment. When organisms undergo a change in order to adjust to their environment, that change acts as a trait that can be passed on to offspring.

INFLUENCES OF DARWIN AND MENDEL

A surprising supporter of pangenesis was the British naturalist Charles Robert Darwin, who believed that the theory accounted for the process of heredity and the wide variety of traits seen among offspring.

Despite his mistaken belief in pangenesis, Darwin nonetheless had an enormous impact on human understanding of heredity. During his years of extensive worldwide travel, Darwin collected many observations of how related species adapt to their local environments. Darwin and British naturalist Alfred Wallace independently formulated the theory of natural selection, which holds that members of a given species born with more favorable characteristics to deal with their environment would be most likely to survive to pass on these traits to the next generation. This important theory was popularized by Darwin's publication *On the Origin of Species* (1859). The book was an immediate sensation, but it raised many questions. Foremost among these was the mystery of how organisms could appear with modified or entirely new traits.

At roughly the same time that Darwin published his natural selection theories, the answer to many questions about the mechanisms of heredity were being unraveled by Gregor Mendel, a reclusive Austrian monk. Mendel conducted a long series of experiments on pea plants during the 1850s and 1860s. Mendel crossbred plants that expressed differing traits, such as height and flower colour.

His conclusions from these experiments helped him formulate a comprehensive theory of how such traits pass from one generation to another. In his studies, Mendel recognized that characteristics were inherited as discrete units, and that each of these was inherited independently of the others. He speculated that each parent has pairs of these units but passes only one to an offspring. He also noted that certain forms of one trait were always dominant over others. Today the units that Mendel described are known as genes.

EMERGENCE OF THE SCIENCE OF GENETICS

Mendel published his findings in 1866, but they went largely unnoticed for more than three decades. In the year 1900, however, Dutch botanist Hugo Marie de Vries, German botanist Karl Correns, and Austrian botanist Erich Tschermak independently rediscovered the monk's works and verified his conclusions.

Advances in cytology, the science of the structure and function of cells, enabled scientists to more deeply appreciate Mendel's work. In 1902 American biologist Walter S. Sutton and German cell biologist Theodor Boveri separately noted the parallels between Mendel's units and chromosomes. The demonstration of the chromosomal basis of inheritance gave rise to the modern science of genetics. The term *genetics* itself was coined in 1905 by British biologist William Bateson. The terms *gene* and *genotype* were contributed in 1909 by German scientist Wilhelm Johannsen.

In 1905 American biologists Edmund B. Wilson and Nettie Stevens independently discovered and identified the sex chromosomes. Wilson discovered the X chromosome in a butterfly, and Stevens discovered the Y chromosome in a beetle. The discoveries of the X and Y chromosomes helped scientists begin to unravel new patterns of inheritance. Foremost among this research was the work of American biologist Thomas Hunt Morgan on fruit flies. In 1910 Morgan identified the first proof of a sex-linked trait, an eye-colour characteristic that resides on the X chromosome of fruit flies. With this finding, Morgan became the first scientist to pin down the location of a gene to a specific chromosome. Morgan was also the first to explain the implications of linkage, unusual patterns of inheritance that occur when multiple genes found on the same chromosome are inherited together.

A student of Morgan's, American biologist Alfred Sturtevant, found early evidence of the mechanisms of crossing over, the phenomenon in which chromosomes interchange genes. More definitive proof emerged in the 1930s with work by American geneticists Harriet Creighton and Barbara McClintock. The pair demonstrated gene recombination with experiments on seed colour in corn. McClintock later gained notice for her work on transposable elements, large genetic segments that move within a chromosome or even between chromosomes. Her research into these elements, commonly known as jumping genes, earned McClintock the 1983 Nobel Prize in physiology or medicine.

BREAKTHROUGHS IN DNA STUDIES

While cytologists and geneticists were studying the properties and location of genes on chromosomes, other scientists focused their studies on the composition of genes. In 1928 British microbiologist Frederick Griffith ran a series of experiments on two strains of bacteria, one that kills mice and another that is harmless to them. When Griffith injected mice with killed cells of the virulent bacteria, all of the mice survived. But in a second trial, when Griffith injected a combined cocktail of dead virulent bacteria and live "harmless" bacteria, the mice all died. He concluded that something in the dead virulent cells "transformed" the hereditary material of normally harmless bacteria so that they became killers. Most scientists at the time theorized that the transforming factor was composed of a protein.

The real identity of the transforming factor in this experiment was not identified until 1944, when American geneticists Oswald Avery, Colin MacLeod, and Maclyn McCarty revisited Griffith's research. After isolating different molecular components from dead bacterial cells, Avery and his colleagues determined that DNA was the agent that transformed the live harmless bacteria into killers. Despite a growing body of evidence about the function of DNA, many scientists were not ready to reject proteins as the hereditary material. The debate was largely quieted in 1952 by American geneticists Alfred Hershey and Martha Chase. Hershey and Chase showed that when a type of virus called a bacteriophage infects a bacterium, it is the virus's DNA—not protein—that enters the bacterium to cause infection. Their studies confirmed that DNA contained the virus's genetic information, which triggered viral replication within the bacteria.

The experiments of Hershey and Chase convinced most scientists that DNA was the molecule of heredity, but many questions about the structure and mechanisms of DNA remained. In the early 1950s researchers began to apply techniques of X-ray diffraction to learn about the basic structure of DNA. X-ray diffraction can determine molecular structures by measuring patterns of scattered X rays after they pass through a crystalline substance. British physical chemist Rosalind Franklin and British biophysicist Maurice Wilkins used X-ray diffraction to obtain DNA images of unprecedented clarity.

Yet the exact three-dimensional structure of DNA remained unclear. The groundbreaking work of American biochemist James Watson and British biophysicist Francis Crick solved that mystery. In 1953 the two proposed a model of DNA that is still accepted today: A double helix molecule formed by two chains, each composed of alternating sugar and phosphate groups, connected by nitrogenous bases. Watson and Crick (along with Wilkins) were awarded the 1962 Nobel Prize in physiology or medicine for their discoveries.

Watson and Crick speculated that the structure of DNA provided some obvious clues about how the molecule could replicate itself. They proposed a replication model in which each strand of DNA serves as a template for making exact copies. This model of replication, called semi-conservative replication, was demonstrated in 1958 by American molecular biologists Matthew Meselson and Franklin Stahl. Their experiments demonstrated the mechanisms of replication by tracking DNA containing a heavy nitrogen isotope through a series of replications.

With DNA's structure and replication mechanisms largely solved, scientists turned their attention to identifying the genetic code—learning how a gene's nucleotide sequence determines what type of protein is made. In the late 1950s, South African geneticist Sydney Brenner and other scientists confirmed that RNA acted as an intermediary between DNA and protein production. Researchers still were uncertain how the sequence of nucleotides in DNA corresponded to the production of specific amino acids. In 1961 Crick and Brenner determined that groups of three nucleotides, now known as codons, code for the 20 amino acids that form the foundation of proteins.

The exact relationship between codons and amino acids was clarified after several important discoveries. American biochemists Marshall Nirenberg and

J. Heinrich Matthaei synthesized repeated nucleotide sequences that led to the production of repeated single amino acids. They identified how certain codon combinations code for a specific amino acid. A process developed by American geneticist Har Gobind Khorana helped scientists create a "dictionary" of codons that defined specific amino acids, thus resolving the remaining ambiguities in the genetic code. Only 12 years after the structure of DNA was deduced, the genetic code was solved.

Learning to Manipulate DNA

After scientists had unraveled the structure and replication mechanisms of DNA, many felt that the major discoveries of genetic research were resolved. They predicted that the only task left in genetics was to sort out the molecular details of how genes work. But in the process of studying gene function, researchers developed powerful new molecular techniques, enabling them to analyse and manipulate genes with a speed and precision never before possible.

A number of discoveries made during the 1960s and 1970s shed light on how distinct fragments of DNA could be isolated. The work of Swiss molecular biologist Werner Arber focused on specialized enzymes that digest, or "restrict," the DNA of viruses infecting bacteria. These enzymes were subsequently dubbed restriction enzymes. In the following decade, scientists learned that restriction enzymes could also act like molecular scissors to cut DNA. In 1970 American molecular biologist Hamilton Smith and colleagues determined that restriction enzymes could cleave DNA molecules at precise and predictable locations. Hamilton concluded that the enzymes were able to recognize specific nucleotide sequences.

Scientists quickly realized that restriction enzymes could be used in the laboratory to manipulate DNA. In 1973 American biochemist Herb Boyer used restriction enzymes to produce a DNA molecule with genetic material from two different sources. This splicing technique is now known as recombinant DNA. Boyer inserted foreign genes into plasmids and observed that the plasmids could replicate to make many copies of the inserted genes.

In subsequent experiments, Boyer, American biochemist Stanley Cohen, and other researchers demonstrated that inserting a recombinant DNA molecule into a host bacteria cell would lead to extremely rapid replication and the production of many identical copies of the recombinant DNA. This process, known as cloning, gave scientists the power to make many copies of desired DNA for molecular study.

The speed and efficiency of DNA cloning were vastly improved in the 1980s with the invention of polymerase chain reaction (PCR). Developed by American biochemist Kary Mullis, PCR enables scientists to produce large amounts of DNA sequences in a test tube. In a matter of hours, the process can produce millions of cloned DNA molecules. Yet all of the advances in isolating and replicating DNA would not be possible or be of much use if researchers could not determine the nucleotide sequence of genetic material. In the late 1970s and early 1980s, British biochemist Frederick Sanger and his associates developed DNA sequencing techniques. Sanger's methods, which used special compounds called dideoxy nucleotides, rapidly yielded the exact nucleotide sequence of a desired sample. With the use of automated equipment, the new techniques transformed genetic sequencing into a speedy, routine laboratory procedure.

Many of the new techniques for isolating, sequencing, and replicating DNA have been put to practical use through the field of genetic engineering. The Human Genome Project and the new field of proteomics have both benefited from continuing technical advances and have accelerated the development of new genetic technologies. Modern genetics is poised to radically change the practice of medicine and the biotechnology industry.

COMPARATIVE TRAIT MAPPING

Analysis using molecular markers can be used to study the genetical control of any aspect of plant phenotype. This approach is currently being pursued in many species, including all major crops. The major implication of comparative mapping is that genes controlling specific aspects of plant phenotype will have corresponding genetic map locations. Comparative trait mapping in related species should therefore identify homoeologous genes which can be defined as genes that share a common evolutionary ancestor, sequence similarity and equivalent function.

This is illustrated by analyses of wheat, rye and barley which show that all three species have a major vernalization response gene in a similar map location on the long arm of the homoeologous group 5 chromosomes. This implies that vernalization response in these three species is regulated by a common gene which forms a homoeoallelic series across the *Triticeae*. Similarly, major photoperiod response genes have been mapped in equivalent regions of the group 2 chromosomes of wheat and barley.

Figure also illustrates an example of equivalent phenotypes spanning a much greater taxonomic distance. Comparative mapping shows that single recessive mutations conferring a liguleless phenotype are found in corresponding chromosome locations in barley, rice and maize. The group 2 chromosomes of wheat and rye have *liguleless* mutants, which have not been mapped with RFLP markers but whose genetic map location can be predicted with some certainty. The recently identified phenol reaction gene in barley may be equivalent to a gene conferring a similar phenotype in rice.

These and other results suggest that much of the biology of cereal species is conserved. Similar results could be quoted from other plant families. Thus, traits mapped in one species should provide plant breeders and researchers with valuable information on the map locations of equivalent genes in related species. A comparative approach will therefore simplify and accelerate the genetic analysis of crop species.

Genes of major agronomic importance are likely to be cloned from crop species in the near future, particularly in rice, maize and *Brassica* (in the latter via *Arabidopsis*). Map-based cloning will not be easy in large genome species such as wheat and barley due to high amounts of repetitive DNA, but exploitation of conserved synteny with rice and other grass species offers an attractive alternative route. Once genes are identified, novel alleles could be identified in germplasm collections by PCR screening. An implication of comparative mapping is that this approach will be usable in a range of related species, even those in which the primary trait may not have been mapped. This may therefore offer a rapid means of identifying useful accessions in species where little mapping information is available.

Several authors have proposed that allelic variation may be manifest either as a qualitative major gene effect, or as a more subtle effect which can only be detected statistically, depending on the particular allele combination under study. In the latter case the effect would be termed a quantitative trait locus or QTL. Comparative mapping suggests that this analysis can be extended across species and that flanking markers for major genes will provide markers for QTL manipulation in other crosses and in other species, thus increasing selection efficiencies and the ability to construct new QTL allele combinations. Identification of qualitative differences between alleles would also allow map-based cloning of at least some genes that usually affect quantitative traits.

The added value offered by comparative mapping means that it will be highly beneficial to crop genetics if standard sets of cross-hybridizing 'anchor' probes are developed and made widely available, as these will provide the necessary basis for genetic map alignment. cDNA clones for detecting RFLP would be suitable for such anchor sets, although the analysis of RFLP is slow and cumbersome. Alternatively, it may be possible to develop PCR anchor markers as there is a high level of transferability.

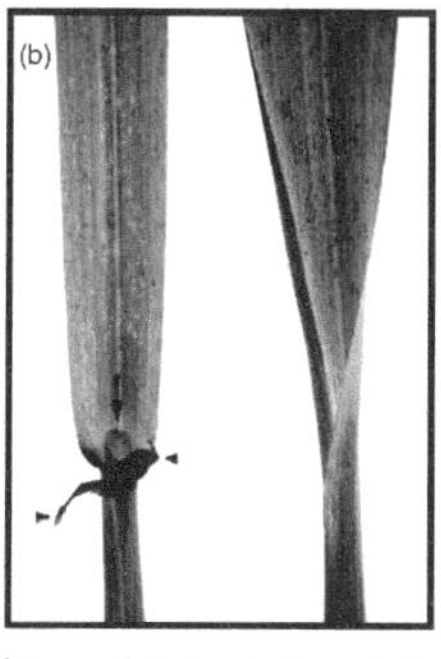

Fig. (b) Wild-type (left) and Liguleless Mutant (right) Leaves of Barley Showing Absence of the Ligule (Arrow) and Auricles (Arrowheads).

EXPERIMENTAL ADVANTAGES OF INDIVIDUAL CEREAL SPECIES

A comparative approach to crop genetics will benefit from careful utilization of the particular experimental advantages of individual species. In cereals, for example, transposon tagging is available in maize and has been used extensively to isolate mutations. Moreover, with the availability of cDNA sequence information it is now possible to design strategies to select for mutants affecting a specific cloned sequence, even if the sequence is of unknown function. For example, PCR screening of lines using a gene sequence primer and a primer for the *Mutator* transposon has been used to select mutant alleles of the *Anther ear1* gene containing *Mu* insertions. This is an extremely powerful strategy for determining the biological role of cloned sequences.

The small genome size of rice and its detailed genetic map make it attractive for chromosome walking and the assembly of complete physical genome maps using YAC and bacterial artificial chromosome (BAC) libraries. Chromosome manipulations, including single chromosome assays, are easily undertaken in hexaploid wheat, taking advantage of its tolerance of aneuploidy. Recessive mutants are easily produced in diploids like barley and rye. Wheat and barley also have simple and efficient systems for the production of doubled haploids, which makes them particularly suited to the genetical analysis of quantitative traits.

In order to identify genes controlling characters of interest, it would be valuable to have detailed knowledge of all the genes expressed in a plant. cDNA sequencing and the development of EST (expressed sequence tagged) databases can be undertaken with similar efficiency in any diploid, but an advantage of using small genome species such as *Arabidopsis* or rice is that they have comprehensive physical maps (contigs) built up of overlapping YAC and BAC clones. Hybridizing probes to such contigs will, potentially, enable all cDNAs to be mapped to specific chromosome locations without the need to identify polymorphism or to analyse segregating populations. This will greatly accelerate cDNA mapping which will, in turn, help in the identification of candidate genes for mapped traits.

COMPARATIVE GENETICS IN PLANTS AS A WHOLE

The above sections concern gene conservation between related species. It is also important to consider what proportion of genes might be common to plants as a whole. It is not straightforward to determine what this proportion might be, especially from comparisons of nucleotide sequence data, which may be affected by factors such as codon usage. For example, monocotyledons tend to have a more GC-rich codon usage than dicotyledons. A more effective strategy is to compare the predicted amino-acid sequences of known function or random cDNA clones. This routinely detects significant homologies, and it is expected that a proportion of genes will be highly conserved. Examples include essential 'housekeeping' genes, cell-cycle genes and genes controlling recombination or photosynthesis.

In many cases, however, homology is restricted to part of a cloned sequence, indicating a conserved protein domain, and the biological function of the gene may remain obscure. This is typically the case for members of gene families. In such cases, homologies may be detected that span widely diverged plant groups or even the plant and animal kingdoms. For example, transcription factors of the MADS-box and homeobox gene families have been identified in species as diverged as yeast, *Drosophila*, plants and humans. Such gene families probably evolved early in eukaryote evolution and subsequently were recruited for many purposes by duplication and divergence.

Despite these complications, it should be possible to clone many genes from crop plants using information from other species. Furthermore, this is likely to become progressively easier as strategies for exploiting comparative genetics become refined. One of the major problems will be the need to access and collate large amounts of information from different species. This will only be possible by developing comparative databasing facilities on computer networks with sophisticated search-and-compare programmes.

COMPARATIVE MAPPING AND GERMPLASM UTILIZATION

Comparative trait mapping suggests that sets of key genes can be recognized that are predicted to have important effects on plant phenotype in many species. Genes for photoperiod response, general pathogen defence or abiotic stress response are examples. Once mapped, markers for these genes could be used to screen germplasm for intraspecific variation and for variation in the equivalent chromosome regions of other species.

Where cloned genes are available, variation in the gene itself could be assayed by PCR. However, germplasm assessment requires far more than the analysis of 'key gene sets' and it is necessary to develop efficient methods for identifying, quantifying and making available the widest possible range of useful variation.

It is therefore important to consider what 'useful variation' might mean. Genome studies, comparative mapping, and sequence database comparisons suggest that a large proportion of genes are common to many if not all plant species and that related species must therefore share a very high proportion of their genetic make-up. This suggests that biodiversity is essentially due to allelic variation rather than to the creation of novel genes. For example, the distinct morphological differences between maize and teosinte are probably due to allelic differences at a relatively small number of genes. Novel interactions between alleles might also be major determinants of species differences. If allelic variation is the major source of variation in germplasm collections, how can it be assayed effectively? Relatively few genes will show easily identifiable allelic variation in any given cross. Thus, the number of accessions that can be analysed genetically is limited. Most researchers therefore feel that molecular methods that fingerprint accessions are needed to complement phenotypic analysis. Such methods would also be valuable for the development of core collections that represent the range of genetic variation and exclude duplicate, or closely related, samples.

A property of any molecular technique used for the evaluation of genetic resources is that the markers provide reliable measures of genetic variability. There must be a high correlation between variability at the molecular level and phenotypic variability since there is an implicit assumption that molecular markers can be used as indicators of genetic variability between accessions without the need for extensive crossing and mapping. To date, there is little evidence that this is so. Furthermore, until very recently, diversity studies using genetic fingerprinting methods such as RAPD have adopted a black box approach with the implicit assumption that the markers used are randomly distributed throughout the genome and do not, for example, target heterochromatin. Few attempts have been made to map RAPD markers from diversity studies, so there is little knowledge of the degree of genome coverage.

Ideally, a map-based approach should be used so that the molecular markers systematically scan the entire genome, or at least a large part of it, in order to provide an overall measure of genetic diversity. A map-based approach would also enable the genome to be partitioned into domains, giving a polymorphism profile along the chromosomes which should reflect the degree of genetic and phenotypic diversity. By selecting variation at spaced markers it may be possible to preserve polymorphism at intervening loci by linkage, especially in inbreeding species where linkage disequilibrium may result in fixed associations between markers and QTL through founder effects.

It is theoretically possible, using PCR marker technology such as AFLP, to fingerprint a collection of material and simultaneously assess the level of polymorphism across the genome by relation to markers mapped in one or more segregating populations. In practice it remains to be proven to what extent markers with dominant modes of inheritance, such as RAPD and AFLP, can be used in this way. RFLPs, though cumbersome to use, have the advantage of 'transportability', due to their usually codominant inheritance and their strong tendency to transcend species barriers. PCR primers derived from the cloned DNA fragments used as RFLP probes (sequence-tagged-site or STS markers) provide a valuable compromise between PCR and RFLP methods.

PCR methods will probably become the methods of choice because of their high throughput and requirement for small amounts of starting material. PCR methods also have the distinct advantage of allowing the retrieval of the amplified sequence for cloning and sequencing. PCR-based approaches may therefore be especially valuable for comparative analysis of specific genic regions. Conventional PCR can only amplify target sequences of limited size, but recent developments in 'long PCR' allow amplification of larger sequences. This may allow the isolation of coding sequences of genes from different species together with their promoters and other flanking regions.

Whatever the methods used, a map-based screening method would result in more rational decision-making concerning the maintenance of genetic polymorphism within a collection. In grasses, the use of markers to detect specific genetic 'building blocks' is a logical approach to germplasm screening, since these appear to form discrete segments of genomes. If so, markers for specific segments may be useful even in species for which detailed maps have not been developed since there would be a high likelihood of good coverage of the genome.

GENETIC VARIATION

Variation means the differences in characteristics (phenotype) within a species. There are many causes of variation as this chart shows:

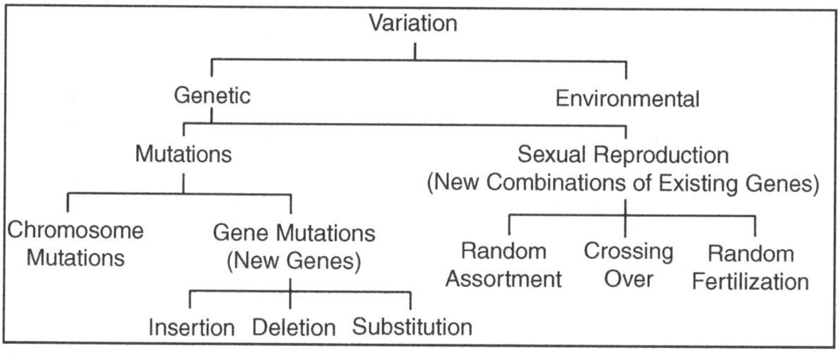

Variation consists of differences between species as well as differences within the same species. Each individual is influenced by the environment, so this is another source of variation.

Genotype + environment = phenotype

The alleles that are expressed in the phenotype can only perform their function efficiently if they have a supply of suitable substances and have appropriate conditions. Any study of variation inevitably involves the collection of large quantities of data.

When collecting the data there is a need for *random sampling*. This involves choosing a sample "*at random*" from a population. This means that every member of the population has the same chance of being chosen, and that the choices are independent of one another. In choosing a random sample some things are very important:

- The selection criteria must not correspond to what you are studying, *i.e.*, not introduce any systematic bias.
- You must try to maintain independence of the sample points, and

- You must be careful in making inference. The population might not be what you wish it were.

Variation in a population can be studied by measuring the characteristic (height, eye colour, seed shape, or whatever) in a large number of different individuals and by then plotting a *frequency histogram*. This graph has the values of the characteristic on the X-axis (grouped into bins if necessary) and the number of individuals showing that characteristic on the Y-axis. These histograms show that there are two major types of variation: discontinuous and continuous.

DISCONTINUOUS VARIATION

Sometimes the characteristic has just a few discrete categories (like blood group). The frequency histogram has separate bars (or sometimes peaks).

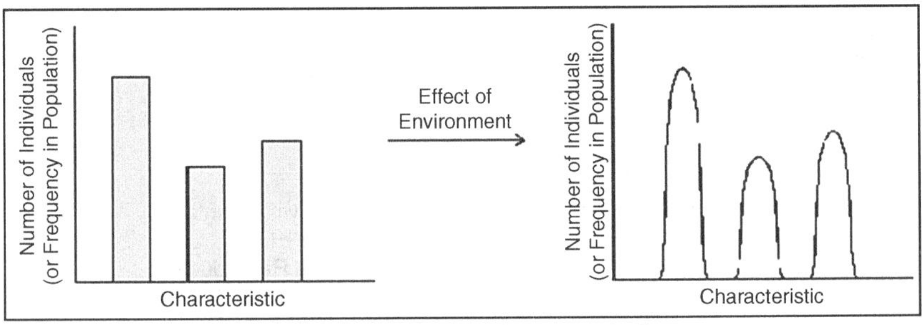

This is discontinuous variation. The characteristics:

- Have distinct categories into which individuals can be placed,
- Tend to be qualitative, with no overlap between categories,
- Are controlled by one gene, or a small number of genes, and
- Are largely unaffected by the environment.

Discontinuous characteristics are rare in humans and other animals, but are more common in plants. Some examples are human blood group, detached ear lobes, flower colour, seed colour, etc.

These characteristics are very useful for geneticists because they give clear-cut results.

CONTINUOUS VARIATION

Sometimes, the character has a continuous range of values (like height). The frequency histogram is a smooth curve (usually the bell-shaped normal distribution curve).

This is continuous variation. The characteristics:

- Have no distinct categories into which individuals can be placed,
- Tend to be quantitative, with overlaps between categories,
- Are controlled by a large number of genes (polygenic), and
- Are significantly affected by the environment.

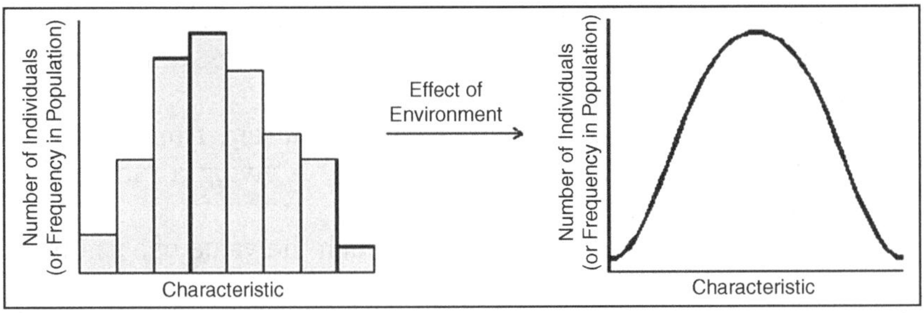

Continuous characteristics are very common in humans and other animals. Some examples are height, hair colour, heart rate, muscle efficiency, intelligence, growth rate, rate of photosynthesis, etc. Characteristics that show continuous variation are controlled not by one, but by the combined effect of a number of genes, called *polygenes*.

Therefore, any character that results from the interaction of many genes is called a *polygenic character*. The random assortment of genes during prophase I of meiosis ensures that individuals possess a range of genes from any polygenic complex.

Sometimes, you can see the effect of both variations. For example, the histogram of height of humans can be bimodal (*i.e.* it has got two peaks). This is because the two sexes (a discontinuous characteristic) each have their own normal distribution of height (a continuous characteristic).

Standard Deviation and Standard Error

Standard deviation is a measure of the spread of results at either side of the mean (average).

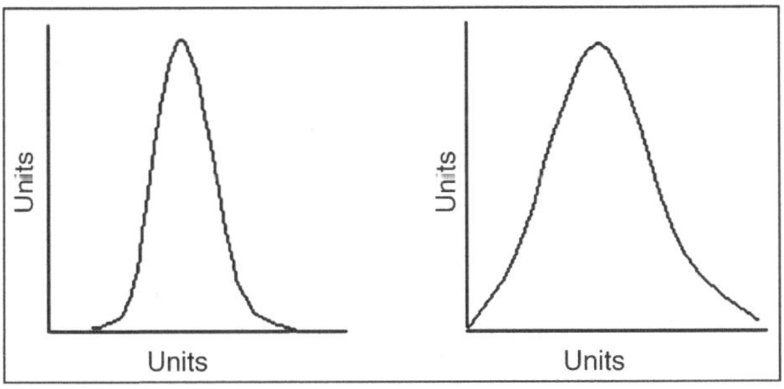

- These sets of data have the same mean (average).
- The data shows a *normal distribution* about the mean value – there is a bell-shaped and even distribution of values above and below the mean.
- The diagram on the left shows a smaller standard deviation, indicating there is less variation between individuals for the character mentioned.
- The diagram on the right shows a greater standard deviation, indicating there is greater variation.

The standard error is the standard deviation of the mean:

- If individuals in a number of samples are measured, then each sample will have its own mean.
- These means will usually be slightly different from each other – reflecting chance differences in samples – giving a range of values for sample means.
- Standard error is a measure of how much the value of a sample mean is likely to vary.
- The greater the standard error, the greater the variation of the mean.

Variation, Gene Pool, and Populations

It is at the population level that evolution occurs. A population is a group of individuals of the same species in a given area whose members can interbreed. Because the individuals of a population can interbreed, they share a common group of genes known as the *gene pool*. Each gene pool contains all the alleles for all the traits of all the population.

For evolution to occur in real populations, some of the gene frequencies must change with time. The *gene frequency* of an allele is the number of times an allele for a particular trait occurs compared to the total number of alleles for that trait.

$$\text{Gene Frequency} = \frac{\text{the number of specifice types of allele}}{\text{the total number of alles in the gene pool}}$$

Hardy–Weinberg Equation

G.H. Hardy, an English mathematician, and W.R. Weinberg, a German physician, independently worked out the effects of random mating in successive generations on the frequencies of alleles in a population. This is important for biologists because it is the basis of hypothetical stability from which real change can be measured.

An important way of discovering why real populations change with time is to construct a model of a population that does not change. This is just what Hardy and Weinberg did. Their principle describes a hypothetical situation in which there is no change in the gene pool (frequencies of alleles), hence no evolution.

Consider a population whose gene pool contains the alleles A and a. Hardy and Weinberg assigned the letter p to the frequency of the dominant allele A and the letter q to the frequency of the recessive allele a. Since the sum of all the alleles must equal 100 per cent, then $p + q = 1$. They then reasoned that all the random possible combinations of the members of a population would equal $(p + q)^2$ *or* $p^2 + 2pq + q^2$.

The Hardy-Weinberg equation:

$$p^2 + 2pq + q^2 = 1$$

The frequencies of A and a will remain unchanged generation after generation if the following conditions are met:

1. *Large population:* The population must be large to minimize random sampling errors.
2. *Random mating:* There is no mating preference. For example, an *AA* male does not prefer an *aa* female.
3. *No mutation:* The alleles must not change.
4. *No migration:* Exchange of genes between the population and another population must not occur.
5. *No natural selection:* Natural selection must not favour any particular individual.

Let us look at an analogy to help demonstrate Hardy-Weinberg Equilibrium. Imagine a "*swimming*" pool of genes.

Find: Frequencies of *A* and *a*. and the genotypic frequencies of *AA*, *Aa*, and *aa*.

Solution:

$f(A) = 12/30 = 0.4 = 40\%$

$f(a) = 18/30 = 0.6 = 60\%$

then, $p + q = 0.4 + 0.6 = 1$

and $p^2 + 2pq + q^2 = AA + Aa + aa$

$= .24 + .48 + .30 = 1$

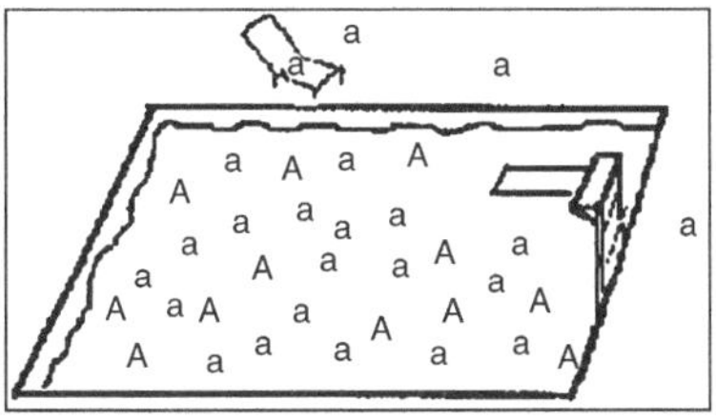

As long as the conditions of Hardy-Weinberg are met, the population can increase in size and the gene frequencies of *A* and *a* will remain the same. Thus, the gene pool does not change. Now, suppose more "*swimmers*" dive. What will the gene and genotypic frequencies be?

Solution:

$f(A) = 12/34 = .35 = 35\ \%$

$f(a) = 22/34 = .65 = 65\%$

$f(AA) = .12$, $f(Aa) = .23$ and $f\ (aa)$

$= .42$

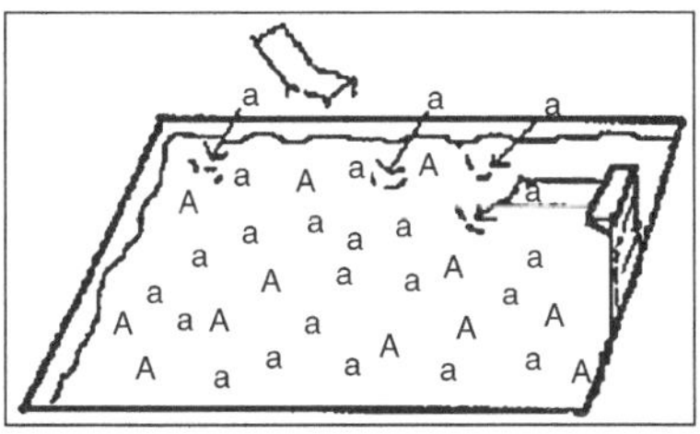

The results show that Hardy-Weinberg equilibrium was not maintained. The migration of swimmers (genes) into the pool (population) resulted in a change in the population's gene frequencies. If the migrations were to stop and the other agents of evolution (*i.e.*, mutation, natural selection, and non-random mating) did not occur, then the population would maintain the new gene frequencies generation after generation. It is important to note that a fifth factor affecting gene frequencies is ***population size***. The larger a population is, the number of changes that occur by chance alone becomes insignificant. In the analogy above, a small population was deliberately used to simplify the explanation.

GENOME RELATIONSHIPS AND WIDE-HYBRIDIZATION

Domestication of crops is likely to have captured only part of the allele diversity present in the wild progenitor gene pools, leaving many desirable traits absent from the crop. In some cases, such as barley 3 *Hordeum spontaneum* or maize 3 teosinte, crops can be crossed easily with wild relatives and there are few barriers to the incorporation of novel alleles since the genomes are essentially homologous.

In contrast, recently evolved polyploids such as hexaploid wheat will have had few opportunities to enlarge the gene pool. This is reflected by RFLP studies of wheat, barley and maize which show that wheat is less polymorphic.

The frequency with which allopolyploid species occur in nature shows that the processes of pollen germination, correct pollen tube growth, fertilization of the egg cell, and embryo development are often conserved in related species. This can be exploited to synthesize new combinations of genomes. Such interspecific or intergeneric hybrids are often referred to as wide-hybrids or wide-crosses. Wide-hybrids can be used to enlarge crop gene pools for any character but are most valuable where suitable genetic variation does not exist within a species, or where screening for a desired character would be difficult, time consuming or expensive. Disease resistance or abiotic stress tolerances are the characters most frequently targeted for transfer.

BARRIERS TO WIDE-HYBRIDIZATION

The ease of producing wide-hybrids, and the ease with which genetic material may be transferred by backcrossing to a crop parent, varies greatly from cross to cross. Successful gene transfer may be restricted by prefertilization, postfertilization or chromosome instability barriers. These are not necessarily directly related to the degree of genetic divergence between two species and the genetical control of hybridization barriers may be simple.

Prefertilization barriers usually involve failure of pollen germination or inhibition of pollen tube growth, which may be at the surface of the stigma, in the stigma, or in the transmitting tissue of the ovary wall. The best studied incompatibility reactions affecting cereal wide-hybrids are due to the *Kr* genes of wheat where dominant alleles inhibit the growth of pollen tubes of other species at the base of the style. These genes are best studied in crosses with rye and *Hordeum bulbosum*. No method is known for overcoming the effect of these genes, so wide-hybrids involving bread or durum wheat usually use genotypes with recessive, non-inhibiting alleles. In grasses there are, of course, well-characterized genes for self-incompatibility (the *S* and *Z* loci) but it is not known whether there is homology between these and the *Kr* loci or whether a similar mechanism of inhibition operates.

Postfertilization barriers leading to seed abortion are common in widehybrids. In cereals, this is frequently due to defective development and subsequent collapse of the endosperm. Embryos, however, are often viable, and hybrid plants can be recovered if the embryos are excised from developing seeds and cultured on suitable nutrient agar. In the *Triticeae*, using these techniques, a huge catalogue of wide-hybrids have been produced.

For example, hexaploid, tetraploid and diploid wheats and their wild relatives can be intercrossed but embryo rescue is usually unavoidable, especially if different ploidy levels are crossed. Similarly, ovule or embryo culture has been used to recover hybrid plants from *Brassica* crosses. In some crosses, for example between bread and durum wheats, hybrid necrosis kills the developing seedling. This is due to the epistatic interaction of independent genes in the respective species. Crosses within the separate gene pools are unaffected, probably because selection has eliminated necrotic combinations of alleles. Incompatibilities between nuclear genomes and the cytoplasm donated by the female parent may also contribute to lethality or sterility in certain crosses.

In vitro fertilization using isolated gametes, and the subsequent recovery of plants, is a means by which prefertilization and some postfertilization barriers might be overcome. Currently, *in vitro* fertilization has only been used successfully in maize, but such systems have potential for wide-crosses.

A further barrier to the production of wide-hybrids is the phenomenon of chromosome elimination. In such cases fertilization results in the formation of a hybrid zygote, but the complete chromosome complement of one of the parents is lost during subsequent mitotic divisions, leading to the production of a haploid embryo. The best characterized system is the cross between barley and its wild relative *Hordeum bulbosum*. However, the phenomenon occurs in a wide range of other cross combinations in the *Triticeae* including wheat 3 *H. bulbosum*, wheat 3 barley and wheat 3 maize, sorghum or pearl millet.

The mechanism of chromosome elimination is not understood but involves the inability of the chromosomes of one genome to successfully attach to the mitotic spindle. In some crosses, the occurrence of chromosome elimination is affected by the parental genomes since some barley 3 *H. bulbosum* crosses produce a high frequency of barley haploids while others give a high frequency of relatively stable hybrids. The genetical control of this regulation is probably fairly simple. Although chromosome elimination is detrimental to gene transfer, efficient haploid production methods benefit breeding and genetic analysis programmes and are extensively used by workers in these areas.

CUTTING AND MEASURING DNA

The first step in gene cloning is cutting DNA into appropriately sized pieces. The size of DNA molecules is measured by subjecting the DNA to a technique called "electrophoresis". An electrophoretic apparatus gives DNA an electrical shove (DNA is negatively charged because of its phosphate groups. The electrophoresis apparatus puts the DNA in an electric field, and it moves towards the positive pole), forcing it to move through a porous gel. Two types of gels are commonly used.

AGAROSE GELS

The most popular variant of electrophoresis is carried out with agarose gels. In the laboratory, suspensions of agarose in buffer are heated to boiling (often in microwave ovens), and the hot solution is poured on plastic or glass plates to a height of a few millimeters or so. Agarose is a derivative of agar — an edible polysaccharide extracted from seaweed.

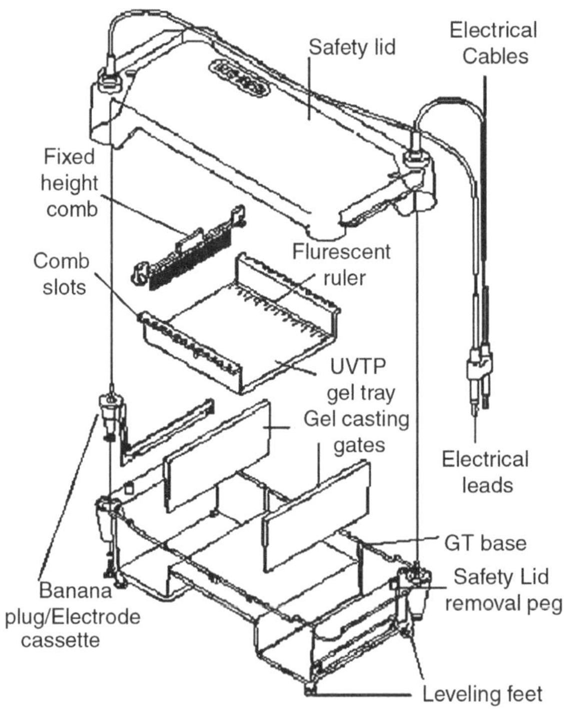

When cooled from heated solutions, solutions of agarose form a nearly transparent gel that looks like a slab of gelatin dessert. After the gel sets, DNA solutions are placed in little depressions formed by leaving a comb in the gel. The gel is then flooded with a weak salt solution, and a voltage is applied. Since DNA is negatively charged (due to its phosphate groups), it moves towards the positive pole.

Large DNA molecules advance slowly in the gel because they are impeded by the gel matrix. Smaller ones encounter fewer barriers and move more quickly. By plotting the distance migrated against the reciprocal of size (in base pairs) of a group of DNA standards, a fairly straight line can be obtained. From these data, the length of unknowns can be easily calculated. For ease of comparison, the standards are usually run in the same gel beside the unknowns.

Gels of differing porosities can be made by adjusting the concentration of agarose. With 2 and 3 per cent solutions, double-stranded DNA's as small as 50 or 100 base pairs can be resolved. More dilute gels can resolve fragments as large as about 30,000 base pairs (30 kilobase pairs or 30kb).

Polyacrylamide Gels

When very small DNA molecules need to be analyzed or when high-resolution separations are required, electrophoresis is carried out in polyacrylamide gels. They, like the gels made from agarose, are water white. Unlike agarose, polyacrylamide gels are polymers of acrylamide, a small synthetic organic compound. Polyacrylamide gel electrophoresis (PAGE) is most often used to separate fragments of DNA that are between 6 and a few thousand base pairs in length. The technique is used mostly in DNA sequencing, a topic that we'll come back to later in the course.

Detecting DNA

DNA doesn't have any colour; it only absorbs light in the ultraviolet range. How is it visualized in a clear gel after electrophoresis? DNA can be detected in both agarose and polyacrylamide gels after staining with various dyes, incuding ethidium bromide, a dye that forms a fluorescent complex upon binding to DNA.

Usually, the ethidium bromide is added to the gel before electrophoresis. After the run, the gel is examined and often photographed under an ultraviolet light.

Yellow-orange zones (bands) of fluorescence appear (similar to those in the picture at the right), indicating migration of discrete pieces of DNA. Other agents — like methylene blue or silver stains — can also be used to visualize DNA, but they are either less sensitive or less convient, and aren't used as often. By the way, many dyes that bind to DNA are mutagenic; *i.e*, they cause mutations.

Separating Large Fragments of DNA

Contrary to what you might expect, very large fragments of DNA readily move in agarose gels when an electrical field is applied. But they don't separate according to size. In fact, very big pieces of DNA all migrate at about the same velocity regardless of whether they are 20kb in length or 2000kb.

PFE

Since many techniques, particulary recently derived ones, depend on working with very large DNA fragments, even pieces the size of some chromosomes, several newly developed electrophoretic techniques have been developed. In one of these, called pulsed field electrophoresis (PFE), DNA molecules are analyzed on agarose gels but with a modified apparatus. Instead of having only two sets of electrodes situated at opposite ends of the electrophoresis device as in the picture shown above, the pulsed field apparatus bears four sets of electrodes.

Two are designated A- and A+ and the other two are called B- and B+. The two A's and the two B's are set at an angle of 120° with respect to one another. During electrophoresis, the DNA molecules are subjected to alternating bursts of current from the two A and B pairs, thereby pulling the DNA alternately to the right and to the left as it advances (from south to north in this case) through the gel. The DNA seems to reorient itself each time the current is switched, and the time that it takes to turn itself around is dependent on its length. The result is that very large pieces of DNA can be separated from another on the basis of their size. The technique also allows one to estimate the length of an unknown when run beside standards of known size. One of the most impressive accomplishments of this technique is the separation of the 16 chromosomes of yeast in a single electrophoretic run.

FIE

Another procedure, field inversion electrophoresis, is also effective at separating large molecules of DNA. In it, the electrodes are set up as in ordinary electrophoresis, but the current is switched so that the DNA first moves forward and then backward. Of course, if the switching were to be done in equal time intervals — say, 1 second forward and 1 second back — the DNA wouldn't move at all. To get it to move at all, electrophoresis is conducted forward for a longer time than backward (or at a greater voltage forward versus backward). One might expect that moving ahead three steps and back two would be equivalent to moving forward one step at a time, but that's not what happens. Apparently the DNA reorients itself during the switching cycles, just as it does in pulsed field electrophoresis. Again, the ability to change directions seems to be dependent on size. This allows different lengths of DNA to be separated from one another.

Restriction Endonucleases

One critically important advance that has greatly stimulated the rapid progress in molecular biology and genetic engineering was the discovery of a set of enzymes that are capable of cutting DNA at defined sequences. These enzymes are found in a variety of microorganisms and are called restriction endonucleases, or more simply, restriction enzymes. The first specific restriction enzyme was discovered by Hamilton Smith in 1970, an accomplishment for which he (and two others) were awarded the Nobel Prize. Since then, some 3,059 similar enzymes have been reported (as of April 6, 1999), many of which have different specificities.

Nomenclature

Smith and another Nobel laureate, Daniel Nathans, devised a nomenclature for these enzymes. In brief, the name of each restriction enzyme derives from the organism from which it is isolated. The first letter of the genus name plus the first two letters of the species name form the first three letters of the restriction enzyme's name.

If necessary, a letter indicating strain designation is added, and finally a number is appended that stands for the order in which the enzyme was discovered in each organism. For example, BamHI is the name of an enzyme that is isolated from the bacterium *Bacillus amyloliquifaciens,* strain H, and it was, presumably, the first restriction endonuclease identified from that source. The first three letters of the name should be italicized, but because italicized letters are often hard to read on a computer, I have left them in plain text throughout these notes.

What they Do

The restriction enzymes owe their usefulness to the fact that they bind to DNA at specific DNA sequences, four to eight nucleotides in size, called recognition sites. Once bound, restriction enzymes cut the DNA at or near this site. With a little thought, it should be clear that an enzyme that has a six base pair recognition site will, on the average, produce larger pieces of DNA than one that recognizes a four base site. Expressed quantitatively, the approximate size of the fragments produced by a particular enzyme, given that it is cutting a DNA containing an equal proportion of all four nucleotides, can be calculated from the formula:

Average size of fragment = 4^N

where N is the number of bases that the enzyme recognizes. Hence, a four-cutter (the shorthand name for an enzyme that recognizes a site containing four base pairs) is expected to cleave random DNA into fragments of about 4^4 (256) base pairs while an enzyme with a recognition site of six bases will produce pieces (on the average) of about 4^6 (4096) base pairs.

PROPERTIES OF RESTRICTION ENZYMES

Ends

Another interesting property of restriction enzymes is that while they often recognize a symmetrical site, they do not always cut at the axis of symmetry. For instance, the enzyme EcoRI (from *Escherichia coli* strain RY13 (I guess they didn't want to put all those letters in the name of the enzyme) and pronounced "echo are one"), recognizes the site GAATTC and cuts the DNA between the G's and the A's in a manner depicted below.

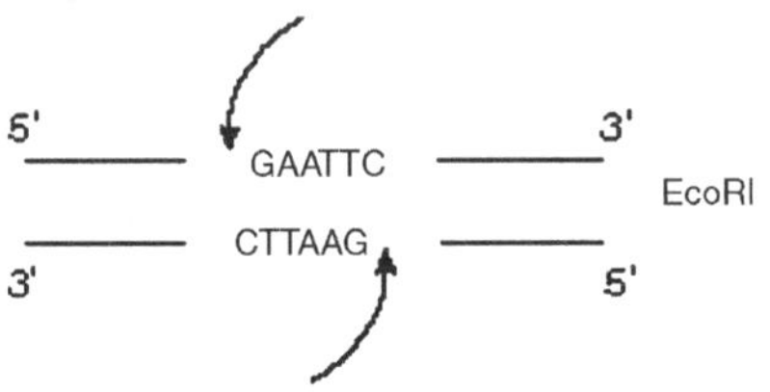

The arrows indicate where the enzyme cuts the DNA

Similarly, the enzyme BglII (from the microorganisms, *Bacillus globiggi* and universally and irreverently pronounced BAGEL TWO) recognizes the sequence AGATCT and cuts between the first A and G residues.

In fact, the four nucleotide single-stranded ends are the same for both BglII and BamHI. Moreover, there are at least two other six cutting enzymes that have been discovered that leave the same four nucleotide overhang: BclI and XhoII.

These overhanging ends are very useful because — under the proper conditions — they may base pair with each other. In fact, because of their affinity for one another, they are often called cohesive or sticky ends. Moreover, if molecules with these ends are treated with the appropriate enzyme — DNA ligase — their phosphodiester bonds may be rejoined (ligated).

When two ends that originate from digestion by a single enzyme are ligated, the resulting molecule can be cut by the same enzyme again. But if the ends of a DNA molecule that originated with a BamHI cut and a BglII cut are joined together, the new sequence will not be cut with either enzyme (I'll ask you to explain why this is so in class). All restriction endonucleases do not generate 5' single-strand overhangs.

In fact, some don't even produce an overhang at all. Several enzymes — like SacI — produce 3' single-stranded sticky ends. And some enzymes — like PvuII — cut at the axis of symmetry, leaving perfectly aligned ends. DNA molecules without overhangs are said to have blunt ends.In addition there are restriction enzymes that cleave DNA some distance away from the sequence that they recognize. For example the enzyme HgaI makes staggered cuts that lie 5 and 10 nucleotides away from a 5 base pair sequence, GACGC. This leaves

5' overhanging ends, but, in contrast to the enzymes described above, these will be different almost every time the enzyme cuts.

The Utility of the Restriction Enzymes

The discovery of these many restriction endonucleases have allowed genetic engineers to cut pieces of DNA at specific sites and into defined sizes. The result has been that a scientist can work with a collection of molecules all of the same size and with ends of known sequence. Restriction enzymes have proved to be valuable analytical and diagnostic tools as well.

RESTRICTION ENZYMES

Restriction enzymes are DNA-cutting enzymes found in bacteria (and harvested from them for use). Because they cut within the molecule, they are often called restriction endonucleases. A restriction enzyme recognizes and cuts DNA only at a particular sequence of nucleotides. For example, the bacterium Hemophilus aegypticus produces an enzyme named HaeIII that cuts DNA wherever it encounters the sequence

5'GGCC3'
3'CCGG5'

The cut is made between the adjacent G and C. This particular sequence occurs at 11 places in the circular DNA molecule of the virus phiX174. Thus treatment of this DNA with the enzyme produces 11 fragments, each with a precise length and nucleotide sequence. These fragments can be separated from one another and the sequence of each determined.

HaeIII and AluI cut straight across the double helix producing "blunt" ends. However, many restriction enzymes cut in an offset fashion. The ends of the cut have an overhanging piece of single-stranded DNA. These are called "sticky ends" because they are able to form base pairs with any DNA molecule that contains the complementary sticky end. Any other source of DNA treated with the same enzyme will produce such molecules. Mixed together, these molecules can join with each other by the base pairing between their sticky ends. The union can be made permanent by another enzyme, DNA ligase, that forms covalent bonds along the backbone of each strand. The result is a molecule of recombinant DNA (rDNA). The ability to produce recombinant DNA molecules has not only revolutionized the study of genetics, but has laid the foundation for much of the biotechnology industry. The availability of human insulin (for diabetics), human factor VIII (for males with hemophilia A), and other proteins used in human therapy all were made possible by recombinant DNA.

ADVERSE EFFECTS AS A CONSEQUENCE OF GENETIC MODIFICATION AND CLONING

Where GA animals are created as models of specific diseases, they can experience a range of adverse effects associated with the condition in question, which can have profound effects on their health and welfare. However, being genetically altered does not necessarily compromise the welfare of individual animals, and indeed some GA animals are indistinguishable from their non-GA siblings.

It also needs to be acknowledged that adverse effects may only become apparent when animals are subsequently maintained in a less well defined or controlled, environment than that of the laboratory or experimental farm. Nevertheless, in many cases, genetic alteration can have a deleterious effect on animal welfare and the harms caused depend on a number of factors. Studies have shown that some *in vitro* culture procedures carried out during genetic manipulation or cloning protocols can lead to unpredictable complications in the animals subsequently produced (ECVAM, 1998).

An example of this is Large Offspring Syndrome (LOS), which affects sheep, mice and cows following nuclear transfer. As well as causing complications for the mother during birth, LOS encompasses a range of debilitating pathologies for the offspring including malformations in the liver, brain and urogenital tract, immune dysfunction, placental abnormalities, stillbirth, fetal overgrowth, respiratory failure and circulatory problems (Vajta and Gjerris, 2006). A further concern is that there is currently no mechanism to ensure consistency in the training of personnel in specific GA technologies and procedures, or to ensure that experimental and other refinements are disseminated throughout the biotechnology community. This can have a profound influence on the levels of pain, suffering and distress experienced by animals undergoing these procedures. It can also have a negative impact on the success rates achieved and levels of animal wastage.

CONCERNS REGARDING MUTAGENESIS PROGRAMMES

Mutagenesis (using chemicals such as ENU, or physical mutagens such as radiation, to increase the natural mutation rate of DNA) is the quickest method for producing large numbers of GA mice. This technique is being widely applied both in Europe and beyond, with mutagenesis programmes aiming to explore the function of every gene in the mouse genome.

This project presents a number of animal welfare concerns:

- The number of animals involved in mutagenesis programmes is vast; *e.g.* around 35-36,000 for a 3 year project, or around 50 animals per mutant line established (Mammalian Genetics Unit, Harwell, 2006);
- The process of mutagenising animals has significant animal welfare implications for the animals involved, for example males require 12-14 weeks to recover their fertility following treatment with a mutagen. Furthermore, only around 50 per cent of mutagenised males are able to go on to sire offspring with the remainder being culled (Goldwitz et al, 2004);

- The mutations induced are by their very nature unpredictable and the scientific usefulness of mutagenised animals cannot therefore be predicted, nor can the effect of any mutation on the health and welfare of the animals. This makes the justification for producing GA mice in this way highly questionable.

Concerns Regarding Knock-in and Knock-out Technology

A more targeted approach to the production of GA animals uses DNA constructs inserted into the animals genome to either completely remove a gene of interest (knock-out), or to replace a given gene with an altered version (knock-in). This is done either in the whole animal, or within specific tissues of an animal (conditional).

The animal welfare concerns include:

- At the current level of efficiency, the numbers of animals used is high - at least 200 animals will be used in the production of a single GA animal (Mammalian Genetics Unit, Harwell, 2006);
- The site of insertion into the host genome and the number of copies of the DNA construct inserted cannot always be controlled unless ES cell manipulation is used. This can be a problem because the incorporation of the construct DNA at an incorrect location can result in the random inactivation of other genes or alterations in the expression of surrounding genes, both of which can impact on the health and welfare of the animals produced;
- When a knockout animal is generated the level, or pattern of expression observed for other genes can be altered to compensate for the lost gene (Okkenhaug, 2003). Thus any effects observed in the GA animal may not only reflect the loss of the gene of interest, which means that investigation of the effects is not always straightforward. The use of knock-out as an alternative to knock-in or conditional technology, therefore needs careful consideration and justification.

Special Concerns for Non-human Primates

This submission relates to all animals used in biotechnologies, but Eurogroup has particular concerns about the application of such technologies to non-human primates. Macaques have already been produced by nuclear transfer cloning in the USA (Chan et al, 2001) and recent technical advances show that the production of GA primates is currently increasing (*e.g.*: Genetically altered primates with Huntingdon's disease produced in 2008; Yang et al, 2008). Eurogroup firmly believes that the genetic modification of non-human primates should not be allowed for any purpose, given the ethical issues raised by such developments, the large number of animals required to produce each GA animal, and the associated potential for harms.

Animal models of disease

A common justification for the creation of GA animals is that they will provide, or contribute to, 'improved', more predictive models of disease. This is an oversimplification because a) this is not necessarily always true, and b) even where a new GA model is more appropriate, it may be used alongside other, older models by different researchers. There is no mechanism for ensuring only the most relevant are used and that newer models are available to all researchers. In such cases the GA animal is not the new *definitive* model but just an *additional* one.

The motivation for the research may merely be an interest in the model for its own sake, but a medical application may be used to justify the work since this is likely to be more acceptable publicly and politically. Basic, fundamental research carried out within academic research establishments is not regulated under Directive 86/609, so it may not undergo an ethical review with appropriate assessment and weighing of harms and benefits. This directive is under review and is expected to be adopted end 2010. The revised directive may incorporate basic, fundamental research with proper review, however only if implemented and enforced properly.

There is also currently no requirement for GA animals to be cryopreserved and stored within central archive facilities or depositories. Such methods can reduce repetition and duplication of work by providing a central resource for use by the wider scientific community and protect against adverse events such as environmental disasters or genetic drift. They also reduce the need for live transportation of GA animals, with associated welfare problems, because frozen gametes or embryos could be sent instead.

RECOMBINANT DNA AND GENE CLONING

Recombinant DNA is DNA that has been created artificially. DNA from two or more sources is incorporated into a single recombinant molecule.

MAKING RECOMBINANT DNA (RDNA)

- Treat DNA from both sources with the same restriction endonuclease (BamHI in this case).
- BamHI cuts the same site on both molecules
  5' GGATCC 3'
  3' CCTAGG 5'
- The ends of the cut have an overhanging piece of single-stranded DNA.
- These are called "sticky ends" because they are able to base pair with any DNA molecule containing the complementary sticky end.
- In this case, both DNA preparations have complementary sticky ends and thus can pair with each other when mixed.

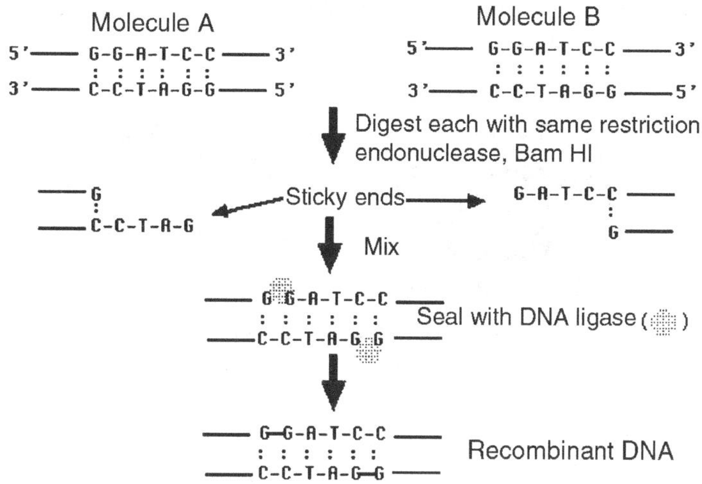

- DNA ligase covalently links the two into a molecule of recombinant DNA.

To be useful, the recombinant molecule must be replicated many times to provide material for analysis, sequencing, etc. Producing many identical copies of the same recombinant molecule is called cloning. Cloning can be done in vitro, by a process called the polymerase chain reaction (PCR). Here, however, we shall examine how cloning is done in vivo.

Cloning in vivo can be done in

- Unicellular microbes like E. coli
- Unicellular eukaryotes like yeast and
- in mammalian cells grown in tissue culture.

In every case, the recombinant DNA must be taken up by the cell in a form in which it can be replicated and expressed. This is achieved by incorporating the DNA in a vector. A number of viruses (both bacterial and of mammalian cells) can serve as vectors. But here let us examine an example of cloning using E. coli as the host and a plasmid as the vector.

PLASMID

A plasmid is an extra-chromosomal DNA molecule separate from the chromosomal DNA which is capable of replicating independently of the chromosomal DNA. In many cases, it is circular and double-stranded. Plasmids usually occur naturally in bacteria, but are sometimes found in eukaryotic organisms (*e.g.*, the 2-micrometre-ring in Saccharomyces cerevisiae). Plasmid size varies from 1 to over 200 kilobase pairs (kbp). The number of identical plasmids within a single cell can be zero, one, or even thousands under some circumstances. Plasmids can be considered to be part of the mobilome, since they are often associated with conjugation, a mechanism of horizontal gene transfer.The term plasmid was first introduced by the American molecular biologist Joshua Lederberg in 1952.

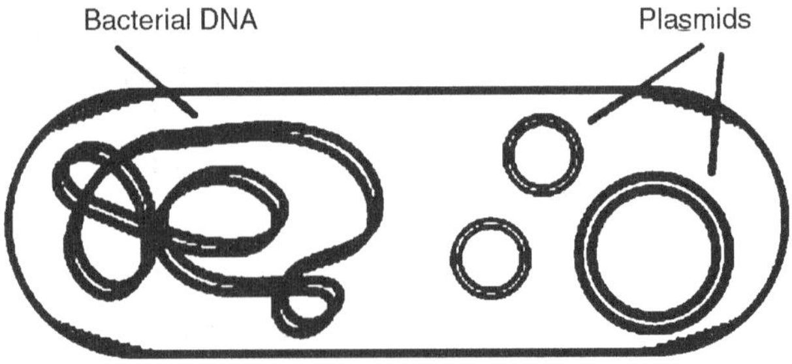

Fig. Plasmids Enclosed Showing Chromosomal DNA and Plasmids.

Plasmids can be considered to be independent life-forms similar to viruses, since both are capable of autonomous replication in suitable (host) environments.

However the plasmid-host relationship tends to be more symbiotic than parasitic (although this can also occur for viruses, for example with Endoviruses) since plasmids can endow their hosts with useful packages of DNA to assist mutual survival in times of severe stress. For example, plasmids can convey antibiotic resistance to host bacteria, who may then survive along with their life-saving guests who are carried along into future host generations.

Vectors

There are two types of plasmid integration into a host bacteria: Non-integrating plasmids replicate as in the top instance; whereas episomes, the lower example, integrate into the host chromosome. Plasmids used in genetic engineering are called vectors. Plasmids serve as important tools in genetics and biotechnology labs, where they are commonly used to multiply (make many copies of) or express particular genes. Many plasmids are commercially available for such uses.

The gene to be replicated is inserted into copies of a plasmid containing genes that make cells resistant to particular antibiotics and a multiple cloning site (MCS, or polylinker), which is a short region containing several commonly used restriction sites allowing the easy insertion of DNA fragments at this location. Next, the plasmids are inserted into bacteria by a process called transformation.

Then, the bacteria are exposed to the particular antibiotics. Only bacteria which take up copies of the plasmid survive the antibiotic, since the plasmid makes them resistant. In particular, the protecting genes are expressed (used to make a protein) and the expressed protein breaks down the antibiotics. In this way the antibiotics act as a filter to select only the modified bacteria. Now these bacteria can be grown in large amounts, harvested and lysed (often using the alkaline lysis method) to isolate the plasmid of interest.

Another major use of plasmids is to make large amounts of proteins. In this case, researchers grow bacteria containing a plasmid harbouring the gene of interest. Just as the bacteria produces proteins to confer its antibiotic resistance, it can also be induced to produce large amounts of proteins from the inserted gene. This is a cheap and easy way of mass-producing a gene or the protein it then codes for, for example, insulin or even antibiotics. However, a plasmid can only contain inserts of about 1-10 kbp. To clone longer lengths of DNA, lambda phage with lysogeny genes deleted, cosmids, bacterial artificial chromosomes or yeast artificial chromosomes could be used.

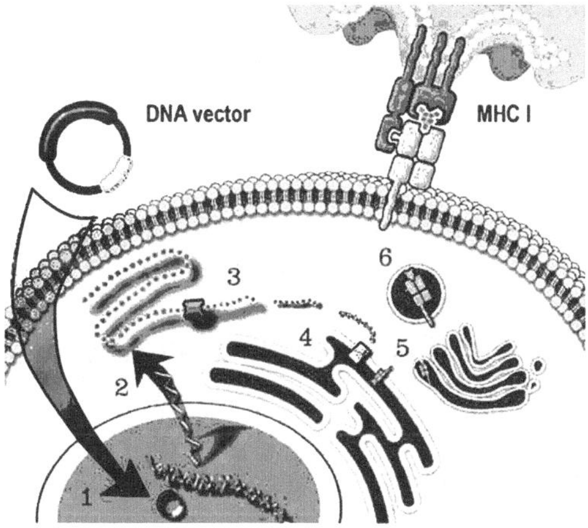

Fig. Plasmid Vector

TYPES

One way of grouping plasmids is by their ability to transfer to other bacteria. Conjugative plasmids contain so-called tra-genes, which perform the complex process of conjugation, the transfer of plasmids to another bacterium. Non-conjugative plasmids are incapable of initiating conjugation, hence they can only be transferred with the assistance of conjugative plasmids, by 'accident'.

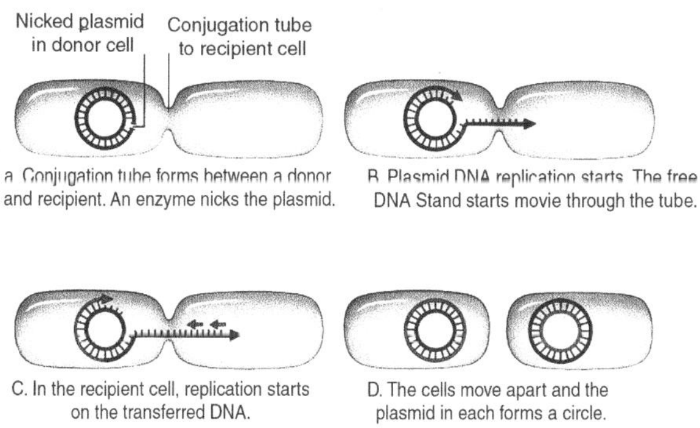

Fig. Bacterial conjugation

An intermediate class of plasmids are mobilizable, and carry only a subset of the genes required for transfer. They can 'parasitize' a conjugative plasmid, transferring at high frequency only in its presence. Plasmids are now being used to manipulate DNA and may possibly be a tool for curing many diseases.

It is possible for plasmids of different types to coexist in a single cell. Seven different plasmids have been found in E. coli. But related plasmids are often incompatible, in the sense that only one of them survives in the cell line, due to the regulation of vital plasmid functions. Therefore, plasmids can be assigned into compatibility groups. Another way to classify plasmids is by function.

There are five main classes:

- Fertility-F-plasmids, which contain tra-genes. They are capable of conjugation.
- Resistance-(R)plasmids, which contain genes that can build a resistance against antibiotics or poisons. Historically known as R-factors, before the nature of plasmids was understood.
- Col-plasmids, which contain genes that code for (determine the production of) bacteriocins, proteins that can kill other bacteria.
- Degradative plasmids, which enable the digestion of unusual substances, *e.g.*, toluene or salicylic acid.
- Virulence plasmids, which turn the bacterium into a pathogen.

Plasmids can belong to more than one of these functional groups. Plasmids that exist only as one or a few copies in each bacterium are, upon cell division, in danger of being lost in one of the segregating bacteria. Such single-copy plasmids have systems which attempt to actively distribute a copy to both daughter cells.

Some plasmids include an addiction system or "postsegregational killing system (PSK)", such as the hok/sok (host killing/suppressor of killing) system of plasmid R1 in Escherichia coli.

They produce both a long-lived poison and a short-lived antidote. Daughter cells that retain a copy of the plasmid survive, while a daughter cell that fails to inherit the plasmid dies or suffers a reduced growth-rate because of the lingering poison from the parent cell.

Plasmid DNA Extraction

As alluded to above, plasmids are often used to purify a specific sequence, since they can easily be purified away from the rest of the genome. For their use as vectors, and for molecular cloning, plasmids often need to be isolated.

There are several methods to isolate plasmid DNA from bacteria, the archetypes of which are the miniprep and the maxiprep/bulkprep. The former can be used to quickly find out whether the plasmid is correct in any of several bacterial clones. The yield is a small amount of impure plasmid DNA, which is sufficient for analysis by restriction digest and for some cloning techniques.

In the latter, much larger volumes of bacterial suspension are grown from which a maxi-prep can be performed. Essentially this is a scaled-up miniprep followed by additional purification. This results in relatively large amounts (several micrograms) of very pure plasmid DNA

DNA LIGASES

DNA ligases are vital enzymes required for important cellular processes such as DNA replication, repair of damaged DNA and recombination. The enzyme mediates the formation of phosphodiester bonds between adjacent 3'-OH and 5'-phosphate termini, thereby joining the nicks in double stranded DNA.

Ligases can be classified into two groups depending on their requirement for ATP or NAD+ as the cofactor. All eukaryotic and virally encoded enzymes are ATP-dependent, whereas most prokaryotic enzymes require NAD^+ for their activity.

ATP-Dependent DNA Ligase

DNA ligase from bacteriophage T7 is a monomer with a molecular weight of 41 kDa. Crystal structures of the apo form of this enzyme and its complex with ATP have been determined at 2.7 Å and 2.6 Å respectively. The structure consists of two distinct domains, a larger N-terminal domain and a C-terminal domain.

The ATP-binding site is situated in the N-terminal domain in a pocket beneath one of the beta-sheets. This pocket is lined by a number of motifs that are conserved across a wide family of nucleotidyltransferases. The structure of the C-terminal domain is remarkably similar to the oligonucleotide binding fold, observed in a number of proteins. The DNA-binding site is proposed to be in a groove running between the two domains.

The structure has provided the basis for biochemical experiments. We have made the two domains separately and looked at their biochemical properties. The larger N-terminal domain is an active ligase but with greatly reduced activity. Both domains are able to bind to DNA. The N-terminal domain binds both single and double stranded DNA, while the smaller C-terminal domain is only able to bind double stranded DNA despite having a fold that is similar to single strand DNA binding proteins. The affinity for nicked DNA comes from a combination of these two DNA-binding affinities at the active site.

Interestingly, the N-terminal domain is poor at the adenylation reaction, but this activity is stimulated greatly by addition of the C-terminal domain. Furthermore, we can demonstrate an association of the domains on gel filtration. These data suggest that a conformational change occurs during the adenylation reaction. This would be similar to that observed directly in our crystal structure of an mRNA capping enzyme, a group of enzymes that are closely related to ligases.

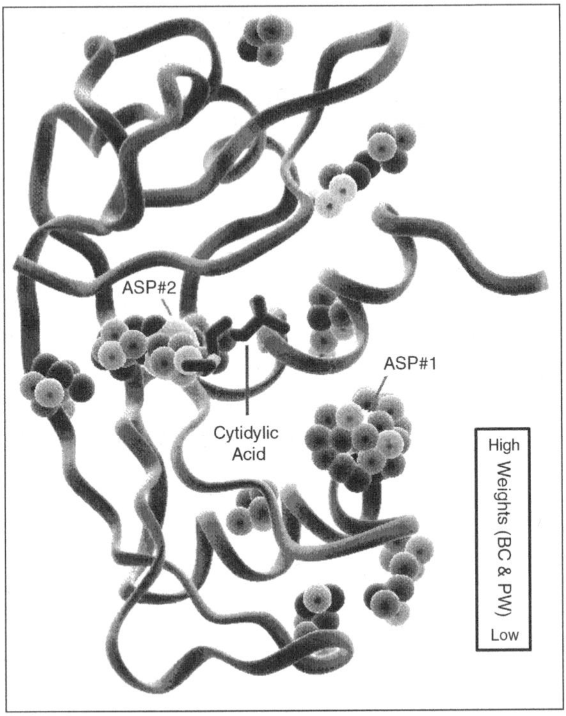

Fig. Structure of RNase A

NAD-Dependent DNA Ligase

An NAD-dependent DNA ligase from *B.stearothermophilus*. The enzyme has been overexpressed and biochemical studies are in progress. Limited proteolysis has been used to investigate the domain structure of the enzyme. There are two domains, with the larger N-terminal domain retaining full self adenylation activity and the smaller C-terminal domain having all of the DNA-binding activity of the full length enzyme.

The situation is therefore different to that we found in the ATP-dependent enzyme, where both domains contribute to the activities of the enzyme. Structural studies are underway with these domains. The structure of the large N-terminal domain has been solved and work is in progress to determine the structure of the small domain.

4

Genes and DNA Recombinant Technology

INTRODUCTION

DNA code is a sequence of chemicals that form information that control how humans are made and how they work. It is a digital code but it is not binary, but quaternary with 4 distinct items. The encoding information in an ordered sequence of 4 different symbols called "bases", typically denoted A, C, G, and T.

- *A*: adenosine
- *C*: cytosine
- *G*: guanine
- *T*: thymine

These 4 substances are the fundamental "bits" of information in the genetic code, and are called "base pairs" because there is actually 2 substances per "bit", as discussed later. Everything else is built on top of this basis of 4 DNA digits.

The entirety of human DNA code, called the "human genome", is about 3 million bases in total. Every human being has 2 copies of this code, one copy from each parent, so a human's cell DNA contains a total of around 6 billion bases.

In computer terms, this is around 6 Gigabytes of symbols, or more like 1 Gigabyte if compacted, since it's about 2 binary bits of information per A/C/G/T base pair. DNA molecules are linear in a twisted double-helix, with a start and an end, and do not contain any cycles.

Chromosomes: These 6 billion odd base pairs are split amongst 46 chromosomes. Each person gets 2 pairs of chromosomes, 23 from each parent, to total 46 chromosomes per human cellhuman cell. A chromosome is the largest form of a DNA molecule, with a large sequence of DNA codes, of differing lengths, usually hundreds of millions of base pairs in each chromosome. chromosomes are independent molecules of DNA, with the typical double-helix, a start and end, but no cycles. chromosomes are physically large enough to be seen on high power microscopes.

Genes: Each chromosome has subsequences of DNA bases that encode particular features, and these are called "genes". Thus genes are not independent molecules, but are abstract sequences within chromosomes. All genes have different lengths. Genes are too small to be physically seen on a microscope, but are analysed using indirect chemical, molecular, and computational methods. The total number of distinct genes in the human genome is believed to be around 30,000 genes according to the Human Genome Project.

So the hierarchy of terminology for genetic components is something like:

- *Base pair*: the smallest element, a single DNA base-4 compound A, C, G, or T.
- *Gene*: a medium-size sequence of around 100,000 DNA base pairs, like a sub-module
- *Chromosome*: a large sequence of hundreds of millions of DNA base pairs, like a computer programme file
- *Human genome*: the entirety of human DNA programme code: 2 pairs of 23 distinct chromosomes, adding to around 6 billion DNA base pairs

Every individual has a unique genetic programme, though all human DNA shares much common code too. A lot of genes and other DNA subsequences are modified or move around within the DNA of a species, such as when they are inherited from parents at conception. DNA does not usually change within a particular individual's body, though this can occur rarely from cell mutations (*e.g.* some cancer cells) and also genetic damage such as from radiation or toxic chemical exposure.

CHROMOSOMES

Each person has 46 chromosomes, in pairs of 2, with 23 from each parent. So there are really 23 distinct chromosomes, and each body cell effectively has 2 different copies of the DNA code, half from each parent.

Each chromosome is distinct and whole. They are ordered, and have clear start and end sequences. In a sense, they are like a file of computer code.

The 23 distinct chromosomes are known and named and have a common structure for each human. The first 22 chromosomes are just named in numbers, simply chromosome 1 through chromosome 22. The name for one of these 22 chromosomes is an "autosome".

The 23rd chromosome is the sex chromosome, which is called either "X" or "Y". Every person has a pair of sex chromosomes, one from each parent. However, unlike the other 22 pairs of chromosomes, a human does not necessarily have 2 similar chromosomes. A male person has a pair of different chromosomes, an X and a Y chromosome, and is usually written as XY. A female has two X chromosomes and is called XX.

***The key issue about chromosomes is to understand their role in reproduction. Firstly, let's make some observations about reproduction*:**

- Children are similar to both parents, with similar traits, but are not identical to either parent.
- Siblings look different, despite sharing the same parents.
- Male and female children occur in about a 50-50 split.

To understand these features, we have to understand how chromosomes are distributed during reproduction. Every person has 46 chromosomes, 23 from the father, 23 from the mother. But the father has 46 chromosomes and so does the mother. Each sperm cell in the father gets 23 chromosomes, and similarly an egg cell gets 23 chromosomes from the mother's set of 46. For each autosome 1..22, the gamete (sperm or egg) gets one of the chromosomes, randomly, without regard to which grandparent the chromosome originally came from. For the sex chromosomes, the egg cell gets one of the mother's X chromosomes, and the sperm gets either the X or Y from the father's chromosomes. Hence, the number of permutations of chromosomes in a father's sperm cell is 2 ^ 23, and similarly the number of egg chromosome permuations is also 2 ^ 23. So even with the same parents, and even with only entire chromosomes inherited, the number of siblings that can be created is about 2 ^ 46.

However, chromosomes are changed during reproduction. They are a natural part of the process. Small or large chunks of chromosome material are swapped during reproductive cell creation. This is called crossover. Thus, the total number of possibilities is even huger than the number purely from simple swapping over.

NON-GENE DNA SEQUENCES

Genes are the best understood subsequence of DNA code. Most genes clearly encode the data sequence representing a particular protein. However, all of the genes together are only a small part of DNA code. Thc 30,000 odd genes in human DNA might only make up 4 per cent of human DNA.

So what is the other DNA code for? These DNA sequences are the least understood of all genetic issues. The main theory is that these DNA sequences are the control mechanisms, that control when particular genes are activated. If the genes are the data sequences for proteins, the remainder must be the real code. This code presumably controls when the genes are activated, so thathuman growthhuman growth follows its normal timetables. It probably also controls how much a gene is activated, controlling how much of each protein is produced by a gene.

DNA AND RNA

There are actually 2 main types of nucleic substances within cell nucleicell nuclei that process information. DNA is the basic form within chromosomes, that is hard-coded into every cell. RNA is a more temporary form that is used to process subsequences of DNA messages. RNA is an intermediate form used to execute the portions of DNA that a cell is using. For example, in the synthesis of proteins, DNA is copied to RNA, which is then used to create proteins: DNA->RNA->Proteins.

The structure of DNA and RNA are very similar. They are both ordered sequences of 4 types of substances: ACGT for DNA, and ACGU for RNA. Thus RNA uses the same three ACGACGsubstances, but uses U (uracil) instead of T (thymine). The molecules uracil and thymine are only slightly different chemically. In DNA, there is pairing between AT and CG, and in RNA, the pairings are AU and CG, but since RNA is not double-stranded, this pairing is much rarer.

Hence, RNA has the 4 substances:

- *A*: adenosine
- *C*: cytosine
- *G*: guanine
- *U*: uracil

Typically, DNA is created from RNA, and this is done by faithfully copying the sequence of base pairs, with the only change converting T to U. Hence, an RNA copy of a DNA sequence encodes the identical information, though it uses a slightly different set of 4 substances.

The differences between DNA and RNA are also many. The underlying sugar molecule that traps the 4 bases is different: deoxyribose in DNA, ribose in RNA. DNA is two strands wrapped in a double-helix, but RNA is a single strand.

CHROMOSOMES AND GENDER

Chromosomes are long, stringy aggregates of genes that carry heredity information. They are composed of DNA and proteins and are located within the nucleus of our cells. Chromosomes determine everything from hair colour and eye colour to gender. Whether you are a male or female depends on the presence or absence of certain chromosomes. Human cells contain 23 pairs of chromosomes for a total of 46.

There are 22 pairs of autosomes and one pair of sex chromosomes. The sex chromosomes are the X chromosome and the Y chromosome. These chromosomes determine gender.

In human sexual reproduction, two distinct gametes fuse to form a zygote. Gametes are reproductive cells produced by a type of cell division called meiosis. They contain only one set of chromosomes and are said to be haploid (one set of 22 autosomes and one sex chromosome). The male gamete, called the spermatozoan, is relatively motile and usually has a flagellum. The female gamete, called the ovum, is non-motile and relatively large in comparison to the male gamete. When the haploid male and female gametes unite in a process called fertilization, they form what is called a zygote. The zygote is diploid, meaning that it contains two sets of chromosomes (two sets of 22 autosomes and two sex chromosomes).

Sex Chromosomes X-Y

The male gametes or sperm cells in humans and other mammals are heterogametic and contain one of two types of sex chromosomes. They are either X or Y. The female gametes or eggs however, contain only the X sex chromosome and are homogametic. The sperm cell determines the sex of an individual in this case. If a sperm cell containing an X chromosome fertilizes an egg, the resulting zygote will be XX or female. If the sperm cell contains a Y chromosome, then the resulting zygote will be XY or male.

Sex Chromosomes X-O

Grasshoppers, roaches, and other insects have a similar system for determining the sex of an individual. Adult males lack a Y sex chromosome and have only an X chromosome. They produce sperm cells that contain either an X chromosome or no sex chromosome, which is designated as O. The females are XX and produce egg cells that contain an X chromosome. If an X sperm cell fertilizes an egg, the resulting zygote will be XX or female. If a sperm cell containing no sex chromosome fertilizes an egg, the resulting zygote will be XO or male.

Sex Chromosomes Z-W

Birds, insects like butterflies, and some species of fish have a different system for determining gender. In these animals it is the female gamete that determines the sex of an individual. Female gametes can either contain a Z chromosome or a W chromosome. Male gametes contain only the Z chromosome. Females of these species are ZW and males are ZZ.

Parthenogenesis

What about animals like most kinds of wasps, bees, and ants that have no sex chromosomes? How is gender determined? In these species, fertilization determines gender. If an egg becomes fertilized it will develop into a female. A non-fertilized egg may develop into a male.

The female is diploid and contains two sets of chromosomes, while the male is haploid. This development of an unfertilized egg into an individual is called parthenogenesis.

STRUCTURAL CHROMOSOME ABERRATIONS

Structural changes in the normal banding patterns of chromosomes occur as the result of one of three biologic processes. The most common cause of structural chromosome aberrations is physical breakage involving one or more chromosomes. Two other less common causes of structural chromosome aberrations are unequal crossing over due to mispairing of homologous chromosomes during meiosis and defective DNA synthesis during chromosome replication. When a chromosome breaks, the broken ends are *sticky* or unstable, and cellular repair mechanisms usually rejoin the two ends promptly (*i.e.* restitutional healing). In such cases, the occurrence of chromosome breakage would not be detected. However, if more than one chromosome break occurs, the repair mechanisms cannot differentiate between sticky ends and the wrong ends may join (*i.e.* non-restitutional healing). Chromosome breakage occurs spontaneously or is induced by exposure to ionizing radiation or mutagenic chemicals. The rate of spontaneous, non-restitutional chromosome breakage exceeds 1 in 1000 gametes, which is 100 times greater than the spontaneous mutation rate for single gene loci.

Understanding the nature and behaviour of structural chromosome aberrations is of considerable clinical importance. Prospective parents carrying structurally rearranged chromosomes are at significant reproductive risk for genetically unbalanced conceptions. Structural chromosome aberrations also have been the primary means for developing a physical map of the human genome and the first step in the eventual cloning of disease-causing genes.

There are five distinct classes of structural chromosome aberrations. They are translocation, deletion, duplication, inversion, and isochromosome.

TRANSLOCATIONS

Translocations are interchromosomal exchanges brought about by breakage and transfer of segments of chromosomes to different locations. There are three basic types of translocations: *reciprocal*; *centric fusion* or *Robertsonian*; and*insertional*. In a reciprocal translocation, segments distal to breaks in two chromosomes are exchanged. The exchange may involve the long or the short arms of any pair of chromosomes, autosomal or sex (or both), homologous or non-homologous. Such an exchange usually does not result in any loss of genetic material, and the individual is clinically normal and called a *balanced carrier*. Nevertheless, a balanced carrier is at significant reproductive risk for producing chromosomally unbalanced gametes that adversely affect conception and embryonic development.

Figure shows a reciprocal exchange between the long arm of chromosome 4 and the long arm of chromosome 13 occurring in a balanced carrier. During gametogenesis, homologous segments of chromosomes normally synapse (*i.e.*pair together), and in the case of reciprocal translocation between chromosome 4 and 13, this would produce an association of four chromosomes in the form of a cross-shaped *quadrivalent*. At anaphase of meiosis I, depending on the type of chromosome segregation, 12 different gametes involving chromosomes 4 and 13 are possible, six from 2:2 segregation (two chromosomes into each daughter cell), and six from 3:1 segregation. With the former combination, three types of segregation are possible: alternate, producing one chromosomally normal gamete and the other that is a chromosomally balanced gamete; adjacent I, resulting in gametes that carry duplications and deficiencies for segments of chromosomes 4 and 13; and adjacent II, also producing gametes with deficiencies and duplications of segments of chromosomes 4 and 13. With 3:1 segregation, six other chromosome combinations are possible, but the genetic imbalance causes early pregnancy failure, and in most cases, the pregnancy goes unrecognized. Because centromeres from homologous chromosomes regularly segregate from one another during meiosis I of gametogenesis, 2:2 segregation is more likely, particularly alternate and adjacent I forms. Although in theory gametes with twelve different chromosome complements are possible, in practice the actual number is limited by the pattern of chromosome segregation and their effect on embryonic viability. The actual risk of unbalanced offspring is always lower than the theoretical risk.

A *de novo,* apparently balanced translocation detected in the course of prenatal genetic diagnostic studies cannot ensure a normal phenotypic outcome. Balanced translocations have been observed in liveborn infants with congenital malformations and developmental disabilities whose parental chromosomes are normal. A survey of centers performing prenatal diagnosis showed that the risk of an adverse pregnancy outcome with a *de novo* translocation detected by second-trimester amniocentesis was high, exceeding 9 per cent.

Table. Minimal Risk of Chromosomally Unbalanced Conception Detected by Prenatal Diagnosis for Carriers of Balanced Structural Rearrangements

Rearrangement	**Carrier**	**Conception (%)**
Centric fusion 13;14	Father	1
Centric fusion 13;14	Mother	1
Centric fusion 14;21	Father	1
Centric fusion 14;21	Mother	11
Centric fusion 21;22	Father	5
Centric fusion 21;22	Mother	10
Centric fusion 21;21	Father	100
Centric fusion 21;21	Mother	100
Reciprocal translocation	Father	12
Reciprocal translocation	Mother	12
Pericentric inversion*	Father	4
Pericentric inversion*	Mother	8

* Does not include the common chromosome 9 pericentric inversion

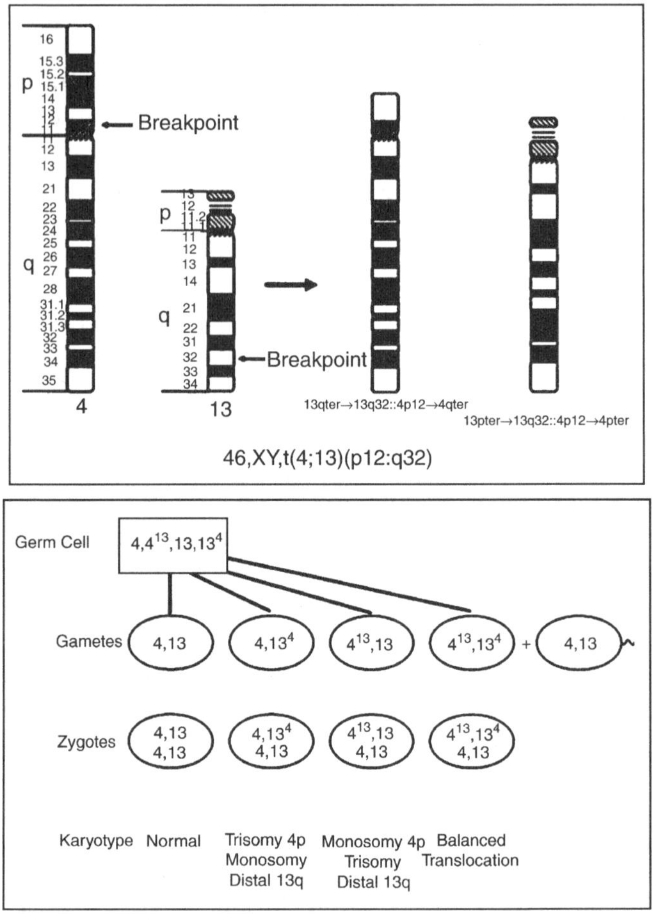

Fig. Meiotic Segregation (2:2) in a Reciprocal Translocation Involving Chromosomes 4 and 13.

Although the specific risk of mental retardation and birth defects cannot be predicted with certainty, empirical data indicate that 1 of 10 sporadic, apparently balanced translocations may be associated with loss of normal gene function and abnormal embryonic development.

The recurrence risk (*i.e.* the chance of a second conception with the same chromosome rearrangement) is low, although rare instances of gonadal mosaicism for structural aberrations have been observed. Prenatal diagnostic studies in subsequent pregnancies are warranted in cases of *de novo* translocations, despite the low risk of recurrence.

Centric fusion, or a Robertsonian translocation, results as a consequence of chromosome breakage at or near the centromeres of two acrocentric chromosomes (13, 14, 15, 21, and 22). The breaks usually occur just above the centromeres, resulting in a single metacentric chromosome with two centromeres (*i.e.* dicentric) and a satellite fragment with no centromere (*i.e.* acentric). Any acentric fragment invariably is lost during cell division. Alternatively, Robertsonian translocations may arise in some cases as a consequence of crossing over between homologous segments of two acrocentric chromosomes during meiosis I. In either case, the formation of a Robertsonian translocation is associated with reduction of chromosome number by one, with the diploid number being 2N = 45. The genetic material contained on the short arms of acrocentric chromosomes appears to have no clinical effects. Despite their loss, individuals who carry *de novo* Robertsonian translocations are phenotypically normal. There is, however, an increased reproductive risk of forming unbalanced gametes during gametogenesis.

Centric fusion of chromosomes 13 and 14 is the most common structural chromosome aberration in humans, occurring with a frequency of 1 in 1000 individuals. The next most common Robertsonian translocation involves chromosome 21 with chromosomes 13, 14, 15, 21, and 22. As shown in Figure, a balanced carrier of a 14;21 centric fusion translocation has 45 chromosomes and, although healthy, is at significant reproductive risk.

When homologous chromosome segments synapse during meiosis, a trivalent is formed of the two normal chromosomes, 14 and 21, with the 14;21 translocation chromosome. At anaphase I, the three chromosomes can segregate in three ways, producing gametes of six different constitutions. Alternate segregation is associated with one chromosomally normal gamete and one balanced gamete. Adjacent I segregation results in one gamete carrying a duplicated segment of chromosome 14 and one deficient in the same segment; because both are unbalanced, embryonic and/or fetal viability are likely to be compromised, and spontaneous abortion is likely. Adjacent II segregation produces one gamete with duplication of the long arm of chromosome 21 and one gamete with complete deficiency of chromosome 21.

Although liveborn infants with only one 21 chromosome (from the other parent) has not been observed, the gamete with duplication of the long arm of 21 results in a conception with Down syndrome. Although six different gametic combinations are theoretically possible, only three types of conceptions are observed: conceptions with a normal complement of 46 chromosomes; a balanced 14;21 translocation carrier with 45 chromosomes and a normal pregnancy outcome expected; and an unbalanced chromosome constitution because of the presence of the 14;21 translocation and two chromosomes 21, resulting in Down syndrome.

These three types of conceptions do not occur with equal frequency in the liveborn population, partly because of selection factors favoring the gametes carrying the 14;21 translocation and partly because of the sex of the carrier parent. Empiric data obtained in familial forms of 14;21 translocations indicate that approximately 20 per cent of pregnancies end in spontaneous abortion; as many as 50 per cent of all liveborn infants are translocation carriers like one of the parents. If the maternal parent is the carrier, the risk of a Down syndrome liveborn infant is 11 per cent, whereas if the paternal parent is the carrier, the risk of Down syndrome is less than 5 per cent.

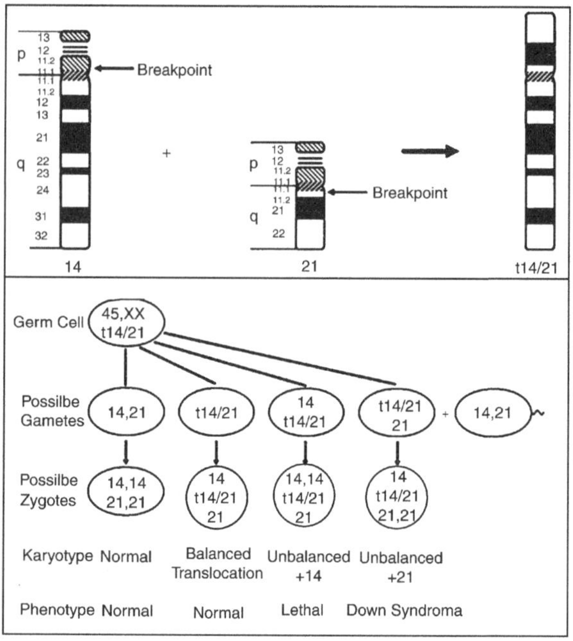

Fig. Meiotic Segregation of Robertsonian (Centric Fusion) Translocation Involving Chromosomes 14 and 21.

In approximately 4.5 per cent of all newborn cases, Down syndrome occurs as a consequence of an unbalanced Robertsonian translocation involving chromosome 21. Chromosome studies in such newborns, however, show that only one half of the cases are familial. The remaining cases arise *de novo* in the meiotic cell division producing the gamete that participated in fertilization. Down syndrome as a consequence of a Robertsonian 21;21 translocation presents a unique situation. In approximately 95 per cent of cases, the 21;21 translocation observed in a newborn with Down syndrome has arisen *de novo*. The remaining cases most likely are caused by gonadal mosaicism, because the origin of a balanced carrier for a 21;21 Robertsonian translocation would require two biologic errors (formation of the translocation in one parental gamete and loss of a normal chromosome 21 in the other parental gamete).

If centric fusion of the two 21 chromosomes occurred during gametogenesis, the resulting conception could not be balanced but rather would be affected with Down syndrome because of the normal chromosome contributed by the other parent's gamete. If centric fusion of the two 21 chromosomes occurred postzygotically, presumably there would be cells with 46 chromosomes and cells with 45 chromosomes carrying the translocation, with one unlikely exception: formation occurring before the first cell division of the zygote. What is unique to carriers of 21;21 translocations is that all of their gametes result in chromosomally unbalanced conceptions. Approximately one half of the conceptions are affected with Down syndrome, based on the contribution of the 21;21 translocation from one parent and a normal 21 chromosome from the other parent.

The chromosome constitution of the remaining conceptions has only one chromosome 21 (monosomy 21), and will not be viable. If all of the cells of a potential parent carry the 21;21 translocation in a balanced state, prenatal diagnosis would be of no value, because all pregnancies are chromosomally unbalanced and developmentally compromised. When Down syndrome occurred secondary to a Robertsonian 21;21 translocation and parental karyotypes were normal, there were no recurrences of Down syndrome.

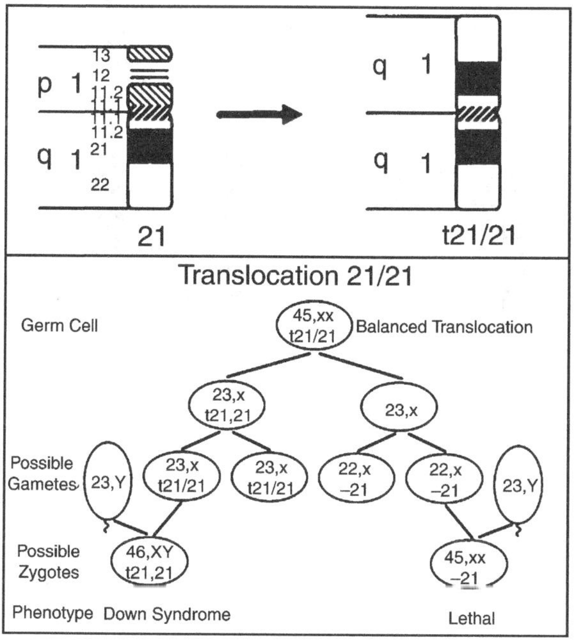

Fig. Meiotic Segregation of t(21;21) in a Balanced Translocation Carrier.

DELETION

A deletion is the loss of any part of a chromosome. It can occur in several ways. A terminal deletion occurs when a single break in a chromosome produces centric and acentric chromosomes. The latter chromosome, lacking a centromere, is lost in the next cell generation, whereas when the centric chromosome duplicates, the broken ends of the chromatids can fuse together, forming a dicentric chromosome. More commonly, an interstitial deletion arises from the breakage at two points along a chromosome. If the chromosome segment between the two sites of breakage does not include the centromere, the acentric fragment so formed will be lost. If breakage occurs in both chromosome arms, the terminal, acentric ends are lost, and the two proximal ends may unite to form a ring chromosome. A ring chromosome with a centromere is usually able to complete mitosis successfully, although if sister-strand crossing over occurs, a ring chromosome that is twice the size of the original ring chromosome, with two centromeres, is formed. Deletion of segments of chromosomes can occur as a consequence of adjacent I and II disjunction involving a parental translocation.

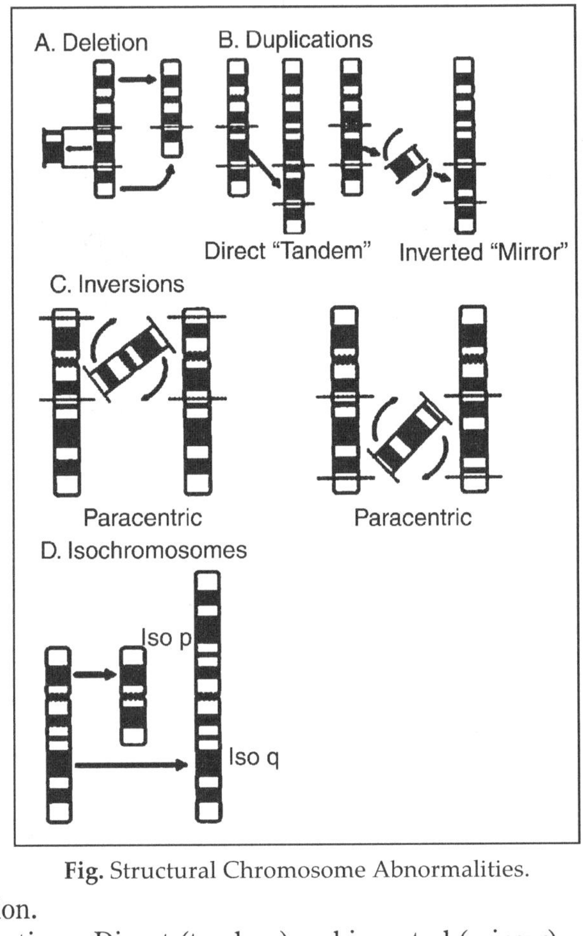

Fig. Structural Chromosome Abnormalities.

A. Deletion.
B. Duplications: Direct (tandem) and inverted (mirror).
C. Inversions: pericentric and paracentric.
D. Isochromosomes.

The smallest loss of chromosomal material detected microscopically is approximately 400,000 base pairs. Conceptions with visible deletions can be monosomic for a large number of genes. With autosomal deletions that are viable, developmental disabilities with mental retardation are to be expected. Specific syndromes associated with deletions include the cri-du-chat syndrome with visible loss of genetic material from the short arm of chromosome 5; the

Wolf-Hirschhorn syndrome with deletion in the short arm of chromosome 4; and the Prader-Willi syndrome, with paternally derived deletion of 15q11. Part of a group of syndromes designated as *contiguous gene syndromes* includes conditions that occasionally have been associated with small but cytogenetically detectable deletions or duplications. Included in this group are deletions of 8q24 in Langer-Giedion syndrome, deletions of 11p13 associated with aniridia and Wilms tumor, deletions of 13q14 associated with retinoblastoma, deletions of 17p13 in the Miller-Dieker syndrome, and deletions of 22q11 in DiGeorge syndrome. When cytogenetic analysis does not show a visible deletion, molecular analysis can demonstrate the loss of specific gene sequences representing submicroscopic deletions of the band or bands previously associated with the specific syndrome. The ability to apply molecular techniques for the cloning of the gene responsible for Duchenne and Becker muscular dystrophies, for example, first required determining its physical location, which was accomplished through *deletion mapping.*

DUPLICATION

Duplication of any segment of a chromosome may arise as the result of unequal crossing over during meiosis or from meiotic events in a parent with a translocation, inversion, or isochromosome. In general, duplications as a form of chromosomal imbalance appear to be better tolerated clinically than deletions. At the molecular level, duplications in the form of repeats of specific nucleotide sequences appear to play a primary role in such gene disorders as the fragile-X syndrome, myotonic dystrophy, and the Kennedy syndrome and may be an important feature in many other human genes. For example, the fragile-X syndrome is caused by a heritable unstable DNA sequence because of changes in copy number of a trinucleotide repeat, CGG.

The overall incidence of visible deletions, duplications, or a combination of the two in newborns is estimated at 1 case 1 per 2000. For any chromosomal duplication or deletion detected prenatally or postnatally, parental chromosomes must be examined to exclude a balanced structural rearrangement. If parental karyotypes are normal, the recurrence risk is not increased above the general population incidence, with the rare exception of gonadal mosaicism.

INVERSIONS

Inversions arise from two chromosome breaks occurring in the same chromosome, followed by a rotation of 180 degrees of the segment between the breaks. If both breaks occur in the same arm so that the centromere is not included, the inversion is called *paracentric.* If the breaks are on either side of the centromere, the inversion is called *pericentric.* The change in gene order has not been associated with any clinical abnormalities, but inversions place their carriers at reproductive risk as a consequence of their structure and behaviour during meiosis. Inversions may interfere with the pairing of homologous chromosomes and thereby reduce crossing over within the inverted segment, but for synapsis to occur, one member must form a loop in the region of the inversion. If crossing over occurs within the loop of a paracentric inversion, a dicentric chromatid and an acentric fragment are formed, both of which are unstable and unlikely to result in abnormal offspring. If a single crossover occurs within the loop of a pericentric inversion, the resulting two chromatids have complementing deletions and duplications, and abnormal offspring may result.

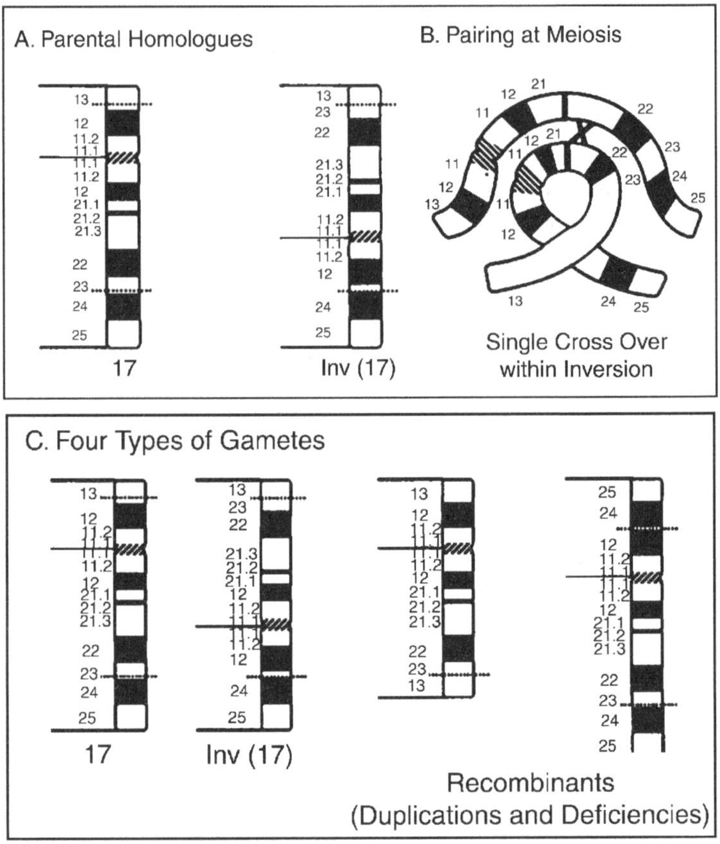

Fig. Products of Meiotic Crossing Over in a Pericentric Inversion.

A, Parental homologues.

B. Loop formation in meiotic pairing of normal and inverted chromosomes. Only two of the four chromatids are shown.

C. Exchange products, both with duplications and deletions.

The closer both breakpoints are to the ends of the chromosomes, the greater is the chance that the pregnancy will come to term. Paracentric inversions are not an indication for prenatal diagnosis, because if crossing over does occur within the inversion loop, the degree of genetic imbalance is incompatible with viability in all but the most exceptional instances. Carriers of pericentric inversions, however, are at risk for viable, chromosomally unbalanced offspring, especially when a large inversion is involved. Empirically, this risk is 8 per cent for maternal carriers and 4 per cent for paternal carriers. One exception is worthy of mention: small pericentric inversions of chromosome 9. These common inversions, which are present in 1 per cent of the general population, have not been associated with abnormal offspring as a consequence of crossing over within the inverted segment.

GENES: PROTEIN DATA SEQUENCES IN THE DNA CODE

Some parts of DNA sequences are known to be purely data. These are the "genes". The best understood aspect of DNA coding is the encoding of amino acid information in genes that is used by the body to synthesize proteins. These are data blocks that represent protein structures. All proteins are substances made up of only 20 basic building blocks called amino acids. Proteins are ordered sequences of these 20 amino acids. Another terminology is that an amino acid is a "peptide" and a protein is a sequence of many peptides called a "polypeptide".

So how does DNA encode the structure of a protein? It uses triplets of base pairs. There are 4x4x4=64 possible combinations in a base pair triplet, and only 20 amino acids. Some extra codes are used as start and stop signal markers at each end of the data sequence. Other triplets are mapped so that more than one triplet can represent a particular amino acid.

However, the representation is unique across all DNA mapping base pair triplets to the 20 amino acids:

- *Phenylalanine (Phe)*: UUU, UUC
- *Leucine (Leu)*: UUA, UUG
- *Isoleucine (Ile)*: AUU, AUC, AUA
- *Methionine (Met)*: AUG
- *Valine (Val)*: GUU, GUC, GUA, GUG
- *Serine (Ser)*: UCU, UCC, UCA, UCG, AGU, ACG
- *Proline (Pro)*: CCU, CCC, CCA, CCG
- *Threonine (Thr)*: ACU, ACC, ACA, ACG
- *Alanine (Ala)*: GCU, GCC, GCA, GCG
- *Tyrosine (Tyr)*: UAU, UAC
- *Histidine (His)*: CAU, CAC
- *Glutamine (Gln)*: CAA, CAG
- *Asparagine (Asn)*: AAU, AAC
- *Lysine (Lys)*: AAA, AAG
- *Aspartic acid (Asp)*: GAU, GAC
- *Glutamic acid (Glu)*: GAA, GAG
- *Cysteine (Cys)*: UGU, UGC
- *Tryptophan (Trp)*: UGG

- *Arginine (Arg)*: CGU, CGC, CGA, CGG, AGA, AGG
- *Glycine (Gly)*: GGU, GGC, GGA, GGG

In addition, the following triplet codes are special:

- *STOP*: UAA, UAG, UGA
- *START*: AUG (same code as the Methionine amino acid)

Clearly, there are not unique 1-1 mappings of triplets to amino acids. However, although there is redundancy, it is not ambiguous. Any triplet can represent only 1 amino acid.

Why this redundancy? Perhaps there is some meaning to it? Perhaps simply a primitive form of error prevention? Perhaps it is simply an accident of nature that occurs because 3 digits were needed, since 2 DNA digits could only encode 4x4=16 codes, which is not enough to represent the 20 amino acids and start/stop codes.

This DNA encoding appears to be almost the same for all genetics on the planet. A few species of single-celled protists have slightly different codes. The DNA data sequences are of varying length depending on the size of the protein. Proteins can range from tiny proteins with about 50 amino acids to huge proteins with 5,000 amino acids. The DNA start and stop sequences are not the same as the RNA start and stop triplets. DNA has a promoter sequence to show where RNA should start to be copied, and a terminator sequence to tell RNA where to stop. The RNA then uses only a single triplet as the start and stop markers. The DNA promoter and terminator sequences are more complex.

Introns: Surprisingly, not all of the DNA code is useful. Certain sequences called "introns" are simply occurred. These are like comments in protein coding sequences. They are transcribed to mRNA properly, but then they are excised from the mRNA to produce the final mRNA. The resulting mRNA is the same order and codes as the original mRNA, but with the introns sequences removed.

RNA DATA SEQUENCES IN DNA

Proteins are not the only substances that are synthesized directly from data within the DNA. Some forms of RNA are specialized, and also have their formula encoded directly in digital DNA formulae.

Not all types of RNA are temporary intermediate forms with their form depending on whatever DNA they are copying. There are certain forms of RNA that have a particular form that is the same across all individuals.

Some of these special-purpose RNA forms are:

- tRNA: transfer RNA
- rRNA: ribosome RNA

There are exactly 20 forms of tRNA, one each to transfer a particular amino acid. tRNA molecules contain about 75-80 bases. tRNA recognizes one of the 64 triplets, and matches it to one of the 20 amino acids. Since there are 20 tRNA types, and not 64, each tRNA molecule has to recognize more than one triplet ordering as a match.

The DNA code contains multiple repetitionsrepetitions of the codes for tRNA and rRNA. About 280 copies are spread over 5 chromosomes. Presumably, this allows each cell to make multiple copies of tRNA and rRNA molecules at once from its single copy of the DNA.

EXECUTING THE DNA PROGRAMME: PARALLEL EXECUTION

Every cell has a full copy of the entire DNA, complete with around 6 billion DNA base pairs jammed into the cell's nucleus. Whenever cells divide to replicate, they duplicate the entire DNA code so that each cell retains a full DNA copy.

The only cells that do not have the entire DNA code are reproductive sperm or egg cells that have only 23 chromosomes each, and thus only about a half copy of DNA.

CONCEPT OF CELLS

Cells are the building blocks of life, and all living things—from bacteria to human beings—are composed of them. The number of cells varies greatly from organism to organism: A bacterium has just one cell; an average-sized adult human has between 60 trillion and 100 trillion. Most cells are too small to be seen with the naked eye and must be viewed under a microscope. In the human body, almost all cells vary between [1] D25,000 of an inch and [1] D125,000 of an inch in diametre, although there is a nerve cell in the upper leg that, while extremely thin, is several feet long. Other cells are larger. A hen's egg, for example, is a single cell, and the largest cell of any organism on Earth is the ostrich egg, which weighs about a pound. Even the smallest cell contains a complete copy of the genetic information that gives an individual organism the traits that make it what it is and not some other thing. Cells have what are called life cycles. They are created, live, reproduce, and die. It is during the process of reproduction, or cell division, that genetic information is passed along from generation to generation.

SIMPLE AND COMPLEX CELLS

There are two types of cells, prokaryotes and eukaryotes. (The words are pronounced "pro-carry-oats" and "you-carry-oats.") Prokaryotes are the kind of cells of which bacteria and other single-celled organisms are composed. More complex forms of life, including plants and animals, are made of many eukaryotes.

The two types of cells are distinguished from each other by their internal structure. Both are enclosed within an external membrane that separates them from their environment. All the internal components of a prokaryote float within this membrane. Eukaryotes, on the other hand, have a number of additional internal membranes that divide them into compartments.

The principal compartments of a eukaryotic cell are the nucleus and the cytoplasm. To use an analogy from the world of business, the nucleus is the cell's executive office and the cytoplasm is its factory. The nucleus generates instructions for a number of cellular processes that are, in turn, carried out in the cytoplasm. The nucleus of a eukaryotic cell is itself a complex structure. It is composed of DNA (deoxyribonucleic acid), which contains genetic information and

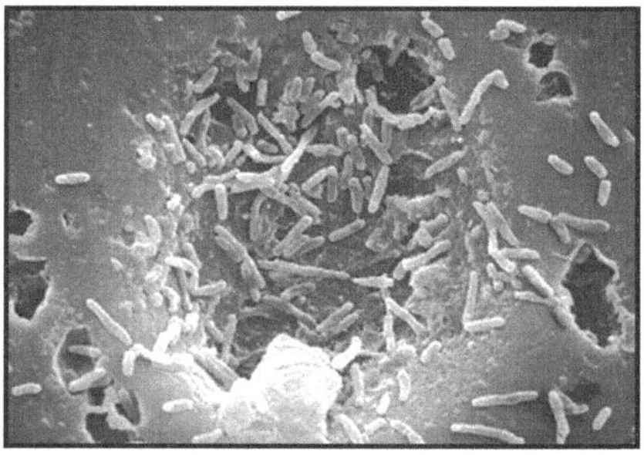

Fig. Single-celled Bacteria Infect a host. These Organisms are far Simpler Creatures than their Multicelled Counterparts.

combines with proteins to produce a substance called chromatin.

"Throughout most of the life cycle of the cell, chromatin exists as exceedingly long, thin, entangled threads that cannot be clearly distinguished by any microscope," says geneticist William K. Purves.

However, when the nucleus is about to divide, the chromatin condenses and coils tightly to form a precise number of readily visible objects called chromosomes. Each chromosome contains one long molecule of DNA. The chromosomes are the bearers of hereditary instructions; their DNA carries the information required to perform the functions of the cell and endow the cell's descendants with the same instructions.

The number of chromosomes in each cell varies from organism to organism, but they always come in matched pairs called homologues. Human beings have 23 pairs of chromosomes for a total of 46. Chimpanzees, humanity's closest relative in the animal kingdom, have 24 pairs; dogs have 39; cats, 17; ferns, 256.

Thus, in humans, the genetic information contained in DNA is divided among 23 pairs of homologous (matching) chromosomes in such a way that the nucleus of each individual cell contains a complete copy of the organism's entire genetic code. Each chromosome is a long molecule which is further divided into subsections of genetic information. These subsections are the factors that Mendel discovered—what today we call genes.

CHROMOSOMES, DNA, AND GENES

Although the chromosomes contain a copy of an individual's genetic makeup—all the genes necessary to produce all the individual's traits—the two members of each matching set of chromosomes are not exactly identical to each other.

This is because one set of chromosomes is inherited from the maternal:

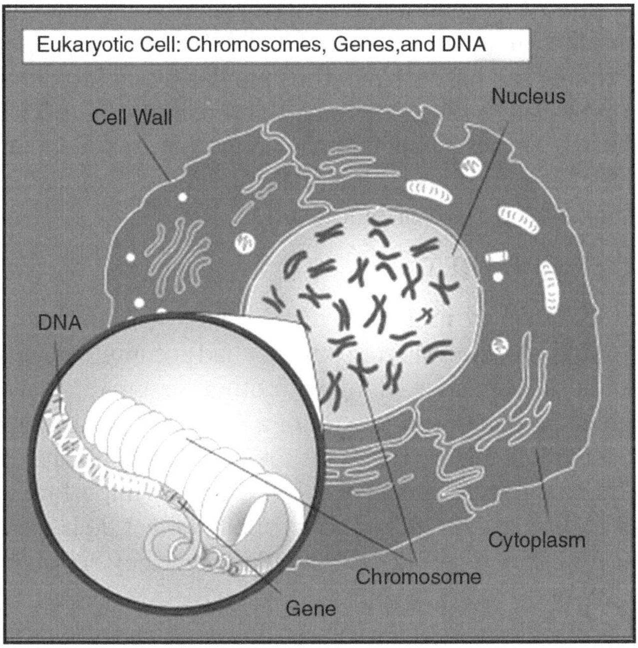

parent and its matching partner from the paternal parent. Thus, it is possible for homologous chromosomes to contain different alleles for the same gene. In Mendel's experiments with peas, with respect to height some of his plants contained two dominant alleles for tallness, some contained two recessive alleles for shortness, and some contained both a dominant and recessive allele.

This fact is crucial for an understanding of genetic inheritance. It explains why children are not identical to one or the other of their parents. It also explains why parents can pass on the genes for recessive traits, traits that they do not themselves exhibit, to their progeny and why a specific trait may lie dormant for generations before making its appearance in a family tree. That matching homologous chromosomes can contain different alleles for the same gene accounts for the pervasive fact of genetic diversity—how from a relatively limited number of genes a virtually unlimited number of unique individuals can be produced.

Science writer Laura Gould sums up the role of the cell nucleus and its contents in the production of the traits that distinguish one individual from another: Chromosomes are thread-like structures in the nucleus of almost every cell [some red blood cells don't have a nucleus]; they are made in part of DNA. They come in matching pairs, one member of the pair providing genetic information from the mother, the other from the father.... Genes are just little pieces of chromosomes: tiny segments of DNA.... Each gene has a fixed location on its chromosome and helps to specify a trait.

Genes determine traits because they function as a code that tells structures in the cytoplasm how to function. The cytoplasm is found outside the cell's nucleus. A computer analogy is useful: The DNA contained in genes performs like software, telling the hardware in the cytoplasm what to do. Specifically, it sends a message through the membrane that encloses the nucleus to entities called ribosomes in the cytoplasm to manufacture one or several of a wide range of proteins. It is these proteins that actually do the work of making peas tall or short, or humans brown-eyed or blue-eyed.

THE GENETIC ALPHABET

DNA was discovered in 1869 by a German scientist named Friedrich Miescher. He found it while studying pus that had accumulated on the bandages of wounded soldiers. Miescher, along with other scientists, learned that DNA was a large molecule composed mostly of a type of sugar called deoxyribose, which is related to table sugar. They also found traces of phosphate, a chemical derived from the element phosphorous. But the most important discovery was that DNA also contained four substances called nucleotide bases. These bases are adenosine, cytosine, guanine, and thymine, and they are abbreviated A, C, G, and T.

Miescher suspected that these bases combined to form chemical messages, and in so doing he came close to discovering the genetic code that governs all life. In fact, later research has shown that the bases that compose DNA function exactly like an alphabet that encodes meaningful expressions. In the same way that the twenty-six letters of the English alphabet can be combined to form an enormous number of intelligible words, phrases, and sentences, the four letters of the genetic alphabet—A, C, G, and T—combine with each other to create chemical messages that are then transmitted to the ribosomes and other parts of the cell.

However, for a language to work as a method of communication, the various letters have to be associated with each other according to a set of rules. In human languages, these rules are called grammar and syntax. The genetic code also has a set of rules, but it took scientists a long time to discover exactly what it was. The first clue came in the early part of the twentieth century when they found that in any DNA molecule the number of As must equal the number of Ts and the number of Cs must equal the number of Gs, but the number of A-T, C-G combinations does not have to be equal.

The importance of this piece of information, however, was not understood for almost fifty years until scientists developed a complete description of how the various components of a DNA molecule fit together. This feat was accomplished by a process called X-ray crystallography, in which a substance is combined with salt and allowed to form crystals. When these crystals are viewed under a powerful electron microscope, the structure of molecules becomes apparent. However, electron microscopes do not produce precise visual images of what they are focused on. Instead, they generate data that have to be interpreted.

The Structure of DNA

In the 1940s, two British scientists, Rosalind Franklin and Maurice Wilkins, applied the techniques of X-ray crystallography to DNA. Two other scientists, American geneticist James Watson and British biophysicist Francis Crick, became aware of their work and began to construct a physical model of the DNA molecule. The 1953 paper in which Watson and Crick published the results of their painstaking research has been recognized as one of the most revolutionary and influential documents in the history of science.

Watson, Crick, and Wilkins each received a Nobel Prize for their work. Unfortunately, Franklin died before her invaluable contribution could be recognized in this way. Recently, historians of science have finally begun to recognize that without Franklin's groundbreaking work, the discovery of the structure of the DNA molecule would have been delayed by years, if not decades.

Watson and Crick concluded that the DNA molecule was shaped like a double helix, two strands spiraling around each other. Former New York Times science editor Boyce Rensberger explains: A helix is the shape of a corkscrew. A double helix is the shape of two corkscrews, one intertwined with the other and curving parallel to it, like the railings of a spiral staircase. Another way to think of the double helix is to imagine a twisted rope ladder with rigid rungs, each rope forming a helix. The easiest way to think about DNA is to start by splitting the two helices [the plural form of the word helix] apart. Think of it as sawing down through the middles of the wooden rungs of a rope ladder. The result is two single ropes with half-rungs hanging off each rope.

Watson and Crick's discovery of the double helix structure of DNA unlocked the secret of the rules that govern the genetic code. In every DNA molecule each A is associated with a T and each C with a G."In all DNA the rope part of the ladder is the same up and down its length-a monotonous alteration of sugar and phosphate molecules," Rensberger continues."The half-rungs, which stick out from the sugars, are the interesting parts.

They serve as the four letters of the genetic alphabet.... As in the English language, the sequence of bases along one strand of DNA-the sequence of As, Ts, Cs, and Gs-spells out the genetic message.... Opposite every T would be an A (and vice versa) and opposite every G would be a C (and vice versa)." 11 Extending the comparison between the genetic code and human language, a gene, as a sequence of As, Ts, Cs, and Gs along the entwined strands of DNA that make up a chromosome, becomes the equivalent of a sentence.

The question then arises: How can just four letters create code for the thousands of proteins that the body produces and the correspondingly large number of traits they govern? The answer lies in the astonishingly large number of A-T, C-G combinations, or base pairs as they are called, that make up a gene. If all the DNA compacted into a cell were stretched out, it would be about seven feet long. Thus, says R. Scott Hawley,"The average human DNA molecule is 65 million base pairs in length for a total DNA content of six trillion base pairs." So, although the four genetic letters can be put together in only a few unique ways, the fact that each pairing on the 6 million rungs of the DNA ladder can be different from creature to creature or plant to plant accounts for the complexity of life.

GENES AND CHROMOSOMES

Linkage occurs when genes are on the same chromosome. Remember that sex-linked genes are on the X chromosome, one of the sex chromosomes. Linkage groups are invariably the same number as the pairs of homologous chromosomes an organism possesses.Recombination occurs when crossing-over has broken linkage groups, as in the case of the genes for wing size and body colour that Morgan studied. Chromosome mapping was originally based on the frequencies of recombination between alleles.

Since mutations can be induced (by radiation or chemicals), Morgan and his coworkers were able to cause new alleles to form by subjecting fruit flies to mutagens (agents of mutation, or mutation generators). Genes are located on specific regions of a certain chromosome, termed the gene locus (plural: loci). A gene therefore is a specific segment of the DNA molecule.

Alfred Sturtevant, while an undergraduate student in Morgan's lab, postulated that crossing-over would be less common between genes adjacent to each other on the same chromosome and that it should be possible to plot the sequence of genes along a fruit fly chromosome by using crossing-over frequencies. Distances on gene maps are expressed in map units (one map unit = 1 recombinant per 100 fertilized eggs; or a 1 per cent chance of recombination).

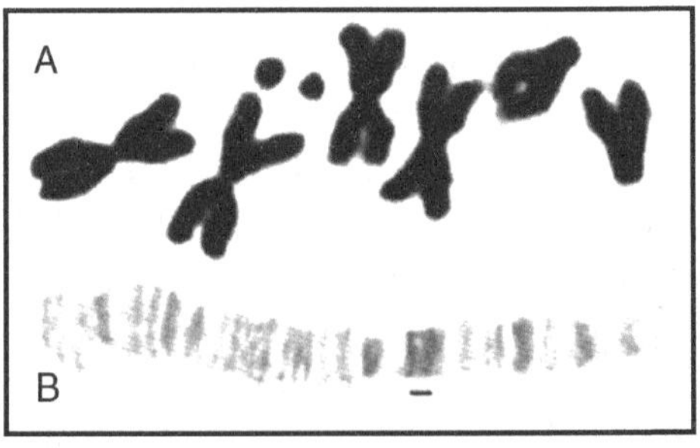

Fig. A Illustrates the Fruit fly Chromosomes at Metaphase; figure B shows a Polytene Chromosome.

The map for *Drosophila melanogaster* chromosomes is well known. Note that eye colour and aristae length are far apart, as indicated by the occurrence of more recombinants (crossing-overs) between them, while wing length is closer to eye shape (as indicated by the low frequency of recombination between these two features).

CHROMOSOME ABNORMALITIES

Chromosome abnormalities include inversion, insertion, duplication, and deletion. These are types of mutations. Since DNA is information, and information typically has a beginning point, an inversion would produce an inactive or altered protein. Likewise deletion or duplication will alter the gene product.

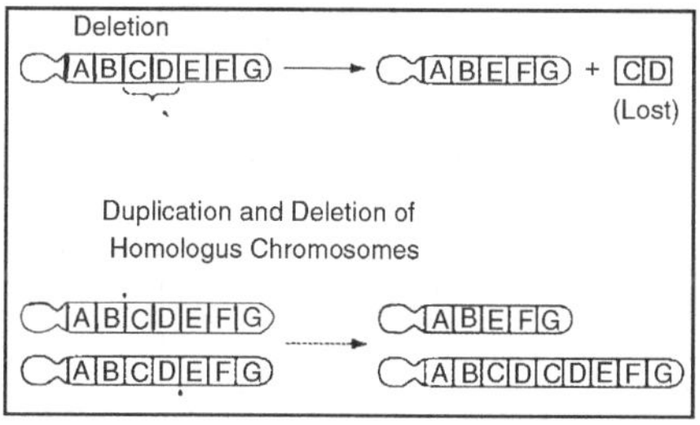

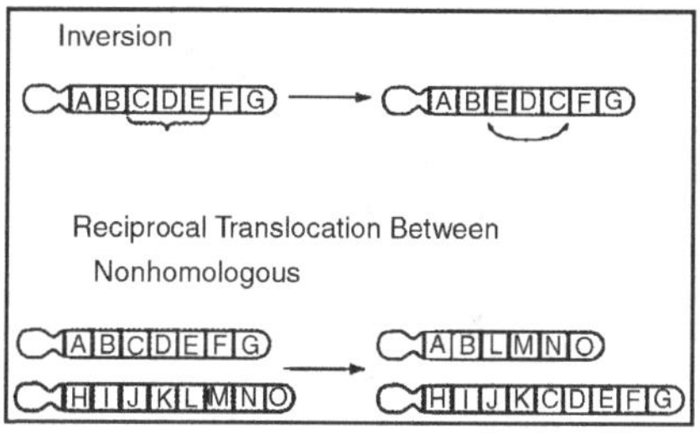

DNA AND RNA

The sequences of DNA that make up genes communicate instructions to the ribosomes to manufacture proteins that work together to produce an organism's traits. However, genes do not perform this function directly.

First, a process called transcription must occur. In transcription, a single gene, which could contain thousands of base pairs, unravels from the chromosome on which it is located. The DNA that constitutes that gene then splits into its two complementary strands. A special type of protein called an enzyme moves along one of the strands letter by letter and creates a corresponding strand of a substance called ribonucleic acid (RNA).

RNA is similar to DNA in that it that it has four nucleotide bases. Three of these bases-A, C, and G-are the same in both, but in place of thymine (T), RNA has a base called uracil, which is abbreviated with the letter U. As the enzyme creeps along the gene's DNA, it transcribes each base it encounters into a corresponding base on the newly emerging strand of RNA. Thus, a C on the

DNA strand becomes a G on the RNA strand, a G becomes a C, a T becomes an A, but the As are transcribed not into Ts but into Us. For example, a strand of DNA that reads ACGGCAT would be transcribed as UGCCGUA.

Once the strand of RNA has been completely transcribed, it travels through the membrane that encloses the cell's nucleus into the cytoplasm. There, it attaches itself to a ribosome, providing the instructions needed to manufacture a protein. This process is called translation, and it works like this.

The thousands of proteins that any organism:

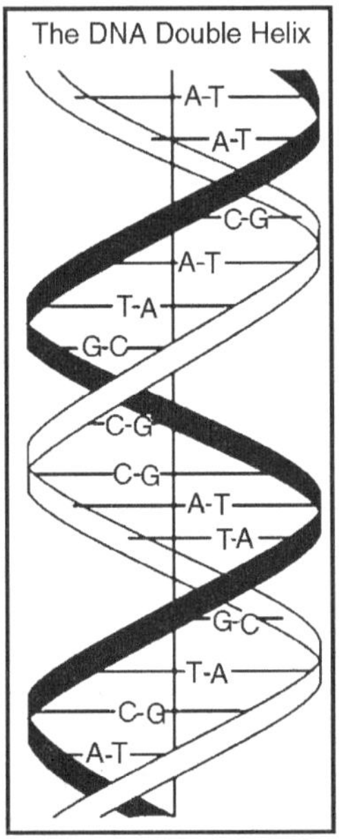

contains are made up of various combinations of twenty substances called amino acids. Each three-letter sequence of RNA tells the ribosome to make one of these amino acids. The process continues in sequence, each three letters of RNA bringing into existence a new amino acid, which attaches itself to those already made. At the conclusion of the process, a complete protein has been manufactured and is ready to do its part in producing the traits that characterize the organism in which all this activity has been taking place. The three-letter sequence of RNA that codes for an amino acid is called a condon. Together, the condons form a set of instructions—the genetic code mentioned earlier. This code is the basis for all the forms that life can take. It bridges the gap between the hereditary information that genes contain and the biochemical processes that give each individual organism the traits that define it.

MITOSIS

The way genes encode instructions for the manufacture of proteins is similar to the mechanism they use to pass from generation to generation. A key part of the life cycle of every cell takes place when the cell reproduces, or makes a copy of itself. Eukaryotic cells, those that make up complex organisms like plants and animals, come in two types. These types are somatic cells and sex cells, also called gametes. Somatic cells combine with each other to make up a body's tissues and organs.

Sex cells combine with the sex cells of another organism to produce offspring. The two sorts of eukaryotes reproduce in different ways. When somatic cells divide, they include a complete copy of all the genetic information contained in the original cell. That is, both corresponding sets of chromosomes are replicated in the new cell. On the other hand, when sex cells divide, only one of the two sets of chromosomes is reproduced. When the sex cell combines with the opposite sex cell of another organism—a sperm cell with an egg cell—the new cell produced by this union will then contain two matching sets of chromosomes, but one will have come from the father and the other will have come from the mother.

The reproduction of somatic cells is called mitosis; that of sex cells is termed meiosis. The purpose of mitosis is growth, so that organs and other body parts can form completely as an organism progresses from infancy to adulthood. Mitosis also creates new cells to replace those that die off at the end of their life cycles. The purpose of meiosis is to create an entirely new organism.

Compared to meiosis, mitosis is a relatively straightforward process. First, the chromosomes become thicker and double into cross-shaped forms, each limb of the cross containing a complete copy of the DNA in all of the organism's genes. Biochemists Paul Berg and Maxine Singer describe how the phenomenon continues: "All of the chromosomes eventually line up in the central plane of the cell and then divide into two groups; the two groups then move to opposite ends of the cell. Two cells are formed when a membrane grows and separates the two ends of the original cell. Each of the new cells (referred to as daughter cells) has a full set of chromosome pairs."

MEIOSIS

Because sex cells combine to form a new organism, they cannot each have a full complement of matching chromosomes. If they did, the offspring resulting from the union of two sex cells would have twice the number of chromosomes-and twice the amount of genetic information-as either of the two parent cells. In humans, for example, each child would have not the required forty-six chromosomes but two times that number, or ninety-two. Human offspring with more or less than the necessary forty-six chromosomes usually do not survive; consequently the human race would have died out after the first generation. Therefore, sex cells divide twice.

The first division is like mitosis, except that during the stage when the number of chromosomes doubles, individual genes often jump from one chromosome to the other. This is possible because similar genes occur at the same location on each of the chromosomes.

The process, known as crossing over or recombination, plays a: Key role in genetic engineering and also promotes genetic diversity.

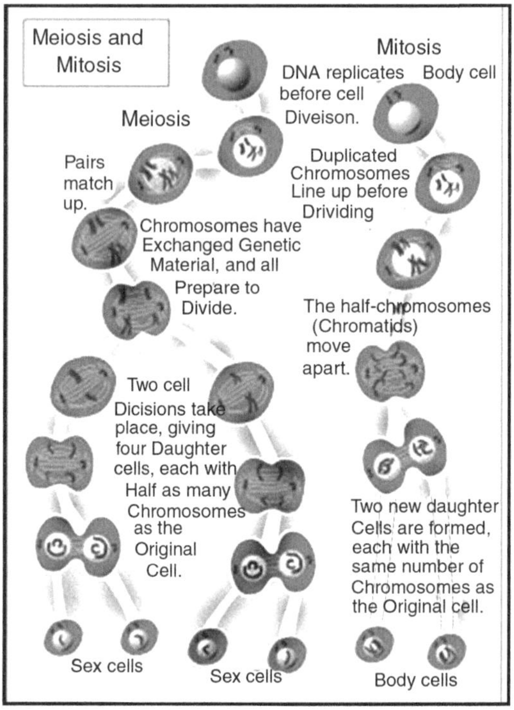

Fig. Source: Colin Tudge, The Engineer in the Garden: Genes and Genetics: From the Idea of Heredity to the Creation of Life. London:Jonathan Cape, 1993.

Since one member of each pair of chromosomes has been inherited from the mother and the other from the father and since each gene has two forms (alleles), recombination creates a novel arrangement of genetic information to be passed on to future generations.

Once the chromosomes have doubled and two new cells have been formed, a further division takes place to guarantee that each sex cell produced has only one set of chromosomes (in humans, twenty-three rather than forty-six)."These two paired sets of recombined chromosomes now separate in the phase known as meiosis-II, or reduction division," says Colin Tudge."Without further doubling of chromosome material, each pair of chromosomes from each set separates from its homologue [matching partner]. Meiosis results in four cells, each... containing just one set of chromosomes, and each of those chromosomes is a unique new entity, combining genetic material from both parents." 14

When each individual sex cell meets up with its partner of the opposite sex during reproduction, a new cell is formed that contains the chromosomes, and the recombined genes on those chromosomes, from each of the parent cells. Thus a new creature is created. It is similar to its parents because it has the same genes as they did, but it is also different because each of those genes may contain alleles different from the ones that constituted the genetic makeup of the parents. As this process continues from generation to generation, the individuals produced tend to differ to an ever-greater degree from the original parent pair. These differences enable subsequent generations to adapt to changing environment and form the genetic basis of evolution, explaining how species have changed, and new species have arisen, during the course of the history of life on Earth.

SEX DETERMINATION

There is one exception to the rule that all chromosomes come in matching pairs, and that exception determines whether newly conceived organisms develop into males or females. The pair of dissimilar chromosomes have been designated X and Y. It is important to remember that although they have different names, they do form a pair in the same way their more alike counterparts do. The Y chromosome is shorter than the X. That is just another way of saying that it has fewer genes, or less genetic material. But it does contain one gene that the X does not, and that is the TDF gene-the gene that causes maleness.

Every normal human female has two X chromosomes; every normal male has both an X and a Y. (There are variations to this rule in some other types of organisms.) When eggs, the female sex cells, are formed during the first stage of meiosis, each one gets two copies of the X chromosome. When sperm cells are formed, each gets an X and a Y. When these cells divide again during the second stage of meiosis to create a cell with only one copy of each chromosome, the egg must have an X and only an X because that is all it had to start with. A sperm cell can have either an X or a Y.

When egg and sperm combine during the conception of a new individual, the cell that is produced must therefore inherit one X chromosome from its mother. From its father, however, it can inherit either an X or a Y. Whether a sperm cell has an X or a Y chromosome is random, accounting for the roughly equal number of males and females born.

For the first six weeks of life, the human embryo, whether it is to become a male or female, develops in the same way. Sometime during the seventh or eighth week of pregnancy, the TDF gene provided by the Y chromosome, if there is one, becomes active. During this period of activity, which lasts only during this phase of the embryo's growth, the gene sends chemical instructions to other genes, telling them to produce the proteins that will cause certain cells to mature into male testicles. The testicles produce hormones which create other male sexual characteristics. If only X chromosomes are present, this process does not occur and the embryo develops into a female.

DNA TECHNOLOGY AND LAW OF GENES

Besides blazing new paths in health science, genetics is transforming the way in which criminal investigations and trials are conducted. The importance of DNA evidence rests on a single but crucial fact: Every individual's DNA is unique (except in the case of identical twins, who have identical genomes). A person's DNA profile can be used like a fingerprint to link suspects to crime scenes and victims. DNA profiling-also called DNA fingerprinting or DNA typing-has been responsible for overturning verdicts and saving innocent people from execution. The process is complicated and controversial. Nevertheless, it has earned itself a place in police probes and courtroom proceedings, and the science behind it is compelling enough that it is likely to play a major role in the future.

BATTLE OVER DNA

Proponents of DNA analysis had to fight against strenuous opposition before the judicial system consented to recognize its validity in courtroom proceedings. "In its early days, DNA was highly controversial and not always accepted by the courts," says James F. Crow of the Genetics Department at the University of Wisconsin in Madison:

The arguments and uncertainties led to a study by the National Academy of Sciences. Publication of this report started with a bang. A badly mangled report by the usually reliable *New York Times* implied that there should be a moratorium on DNA use.

This was followed the next day by a front retraction. The [National Academy of Sciences] report had made no such recommendation, although it raised a number of questions about the techniques and their interpretation.

The NSF conducted further studies and concluded that DNA was, in fact, a valid tool in criminal investigation. "We strongly believe that DNA testing properly performed belongs in the courtroom," say Howard Coleman and Eric Swenson in a guide to the subject prepared for journalists.

We share this view with the U.S. Congress Office of Technology Assessment, the National Academy of Sciences and the vast majority of scientists in America. The continuing controversy over the admissibility of DNA evidence makes little sense except to the very few people who benefit from continued confusion.

The DNA war is a clash of the different approaches of law and science. It would be difficult to think of two disciplines that have more disparate approaches to seeking the truth. Scientific truth evolves by the building of consensus through peer review and the replicating of experiments. For the law, truth is less absolute and more relative to the case at hand. Legal truth is achieved through adversarial argument and judgement. The DNA war is a dramatic, hard-fought conflict between these two worlds.

DNA evidence can be gathered from a wide variety of sources, including blood, semen, hair, fingernails and toenails, saliva, urine, skin, and other body tissues. The DNA of every human being is 99.9 per cent

DNA evidence can be gathered by taking samples of a person's saliva, hair, or fingernails: Similar to that of every other human being; however, since there are more than 3 billion base pairs to consider, that 0.1 per cent difference is equivalent to 3 million ways in which each genome can be uniquely different. These differences manifest themselves as different arrangements of As, Ts, Cs, and Gs, called DNA polymorphisms. These lie in vast stretches of DNA known as non-coding, DNA, or the DNA that lies between genes on each chromosome.

Non-coding DNA, or junk DNA as it is also called, is characterized by sequences of letters (As, Ts, Cs, and Gs) that repeat in a regular manner up to thirty times in a row. These patterns are called variable number tandem repeats (VNTRs), and it is extremely unlikely that any two individuals will have similar patterns of repeating junk DNA provided that a sufficiently large number of patterns are observed. The key circumstance in DNA evidence is that the number of VNTRs at specific locations on the chromosomes differs between individuals. The fundamental technique used to determine a DNA profile is called restriction fragment length polymorphisms (RFLPs). This complicated phrase simply means that investigators determine the number of VNTR repeats at a series of distractive locations in order to come up with an individual's profile. An alternative method is to use a polymerase chain reaction (PCR). RFLP testing requires more DNA, from 20 to 50 billionths of a gram, to be accurate. PCR requires only 2 billionths of a gram.

DNA PROFILES

To create a DNA profile, the following steps must be followed. First, DNA is taken from the crime scene. The DNA can come from any body fluid or tissue left behind. The DNA must be cleaned up to make sure that it has not been contaminated by dust, soil, carpet fibres, or other extraneous substances. Every cell will contain a complete copy of the subject's entire genome, far too much genetic material to be analysed all at once. Therefore, the DNA is cut into manageable pieces using restriction enzymes that were developed as part of the tool kit of genetic engineering. The choice of which restriction enzymes to use is important. They should not cut the DNA within the VNTR locations that are being examined. Ideally, they should do their work just outside a VNTR region.

Once the DNA has been reduced to fragments, the pieces, which range in size from one hundred to ten thousand base pairs, are sorted according to length. The process used to do this is called gel electrophoresis. The DNA fragments are placed on a flat tray in a chemical gel called agarose. The tray is then placed in an electric field. Since the DNA has a negative charge, it will be pulled towards the positive electrode, and larger pieces will progress more slowly than smaller ones. The relative size of each fragment of DNA is determined by how far it has moved through the agarose in a given period of time.

To prevent the fragments of DNA from decomposing, they are uncoiled from their spiral (double helix) form and affixed to a nylon membrane that has been treated with preservatives. The DNA pieces are then attached to radioactive probes that identify each VNTR location. The radioactive probe can be photographed on X-ray film to provide a visual image of the microscopic letters in the VNTR sequence. These show up on the film as a series of dark bands that appear similar to the bar codes on consumer items. To obtain a complete genetic profile, this procedure has to be repeated for a number of VNTR locations on different chromosomes. If the profile is based on thirteen different locations, the standard used by the Federal Bureau of Investigation (FBI) and many other law enforcement organizations, the chances that the DNA sample can be linked to the wrong individual are approximately one in 100 billion.

However, crime scenes can be very messy places. DNA from the victim, or victims if there are more than one, frequently gets mixed with that of the suspect or suspects. The precision of DNA profiling is often complicated by these and other circumstances. "The interpretation of the data... requires that inferences based on the science of statistics, population genetics and probability theory be applied to

A labouratory worker studies a DNA profile. DNA testing has been a vital tool in criminal investigations since it was first used in 1987. the measurements of physical reality that the [DNA tests] reveal," say Coleman and Swenson. "Forensic scientists use these mathematical concepts to calculate and report an estimate of how frequently the genetic profile they have observed might be found in major population groups."

DNA'S COURTROOM DEBUT

DNA evidence was first used in a courtroom in the mid-1980s. In one early case, it was employed to exonerate one suspect and convict another. Police in the East Midlands area of England were struggling to solve the brutal rape and murder of two schoolgirls. Their prime suspect was a kitchen worker at a nearby mental hospital. The police brought samples of the worker's blood and samples of semen taken from the crime scenes to geneticist Alec Jeffreys at the University of Leicester for analysis. Jeffreys was able to conclude that the same person had committed both crimes, but that the perpetrator was not the kitchen worker.

Police were convinced that the killer was a local resident, so they requested that every male in town between the ages of seventeen and thirty-four submit a blood sample for testing. Within a few months more than 4,583 DNA tests had been performed and none of them matched the DNA retrieved from the crime scene semen. The authorities were about to fall back on more conventional investigative methods when they were informed by a bakery manager that one of her employees, a man named Colin Pitchfork, had persuaded a coworker to be tested in his place.

When officers confronted Pitchfork, he confessed to the rapes and murders. When the police lab tested his DNA, they found it matched perfectly the samples taken from the sites where the two girls were killed.

The date was September 1987, and DNA testing has been a vital tool in criminal investigations ever since. DNA testing spread from England to America, where it quickly became a major weapon in the war against crime. The mass media hailed it as a virtually infallible way to determine guilt or innocence. Even judges were overwhelmed by its apparent ability to cut through the often confusing and inconclusive nature of other forms of evidence. One judge, Joseph Harris, of Albany, New York, proclaimed that DNA evidence was the "greatest single advance in the search for truth since the advent of cross-examination."

At first, defence attorneys were powerless to challenge the apparent certitude of DNA evidence. In more than a hundred cases, genetics played the decisive role in getting a conviction. Some legal experts felt that the phenomenon was eroding a defendant's right to a fair trial. "In rape cases, when the semen has been matched with the defendant's and the chance that it came from another person is 33 billion to one, you don't need a jury," complained lawyer Robert Bowser. But chinks soon began to appear in the seemingly invincible DNA armor—not because the science was faulty, but because the tests were extremely complicated and the labouratories where they were conducted were prone to error.

DNA CHALLENGED

The first case on which DNA evidence was successfully disputed took place in New York in 1987. A woman named Vilma Ponce, who was seven months pregnant, was found dead with more than sixty stab wounds on her body. Her two-year-old daughter lay murdered in a similarly brutal way in the bathroom of Ponce's apartment. The principal suspect was José Castro, a janitor in a nearby apartment building. Detectives found a blood stain on his watch. DNA tests conducted by an independent labouratory indicated that the blood on Castro's watch matched Ponce's blood. The odds against the blood being from someone else were put at one in 189 million.

Castro's defence lawyer challenged the test results and produced witnesses who testified that the lab had not followed accepted procedures. The DNA evidence was thrown out of court and Castro appeared to be on his way to acquittal when he suddenly confessed to the crime. The DNA evidence was correct, but it was not allowed to be presented to the jury because the labouratory's procedures for handling such evidence were deemed to be too lax. The scientific basis of applying genetics to criminal proceedings has never been questioned; what has neutralized much such evidence has been sloppiness on the part of lab technicians and police investigators in handling the samples upon which the DNA tests were conducted. The most famous trial in which DNA evidence played a role was that of former football star O.J. Simpson, who was accused in 1994 of murdering his ex-wife, Nicole Brown Simpson, and her friend Ron Goldman. The prosecution produced DNA evidence that appeared overwhelmingly to link Simpson to the crime scene. Yet the jury refused to convict him.

His defence attorneys did not challenge the validity of DNA testing in general. In fact, they distanced themselves from the labouratory procedures that were done. What they did attack—successfully—was how police investigators handled the blood samples once they had been collected. By focusing on breaks in the chain of evidence, how samples were passed from one investigator to another, how long and under what circumstances they were stored, and a number of other irregularities, the defence team managed to convince the jury that the possibility existed that the samples had been tampered with. It was enough to create "reasonable doubt" as to Simpson's guilt in the minds of the jurors and led to his acquittal. Again, the scientific basis of DNA analysis was never in question, but science alone was not the deciding factor in the outcome of the trial.

A forensic scientist places blood samples into evidence envelopes. Careless handling of samples and other lab mistakes can make DNA data unusable in court.

PATERNITY TESTING

Determining guilt or innocence in criminal investigations is not the only application of DNA technology to legal matters. The technique is used even more frequently in paternity testing, with approximately two hundred thousand such tests being performed every year. It has all but supplanted conventional blood tests as a convenient, accurate way to determine the identity of a child's father. While blood tests can conclusively prove that a man is not the father of a child, they cannot determine beyond doubt that he is.

DNA tests, on the other hand, provide almost 100 per cent assurance in either case. Moreover, blood tests must be performed by licensed doctors, whereas DNA tests can be done by a wide range of qualified technicians. In many cases, all that is required is a cotton swab rubbed on the inside of the cheek. In addition, blood tests could not be performed on babies until they were six months old. DNA samples can be taken immediately after birth—or while the baby is still in the womb. Finally, DNA can be obtained from the remains of a deceased father to establish whether or not a child is entitled to a share of his estate. Paternity testing is possible because every individual contains two copies, or alleles, of every gene, one from the mother and the other from the father. If a child has an allele that did not come from the mother, then it must have come from the biological father. Conversely, if the allele in question is shown not to have come from the suspected father, paternity is disproved.

"Most paternity testing is done for financial reasons, *i.e.* to establish legal responsibility and provide for child support," say Coleman and Swenson.

Even more important are the emotional and social issues. When testing can demonstrate conclusively to a man that he is the father of a child, then he is more likely to provide not only financial support, but emotional support as well. He may bond with the child and take an active part in its life.... It wasn't very long ago that obtaining a paternity test was an uncertain, expensive and inconvenient process.... DNA testing has made the process convenient and the result conclusive. In all but the rarest of instances, the DNA test results provide a level of certainty so high that paternity will, for all practical purposes, be proven or disproved.

STREAMLINING THE JUSTICE SYSTEM

DNA paternity tests can also be used to prove childhood sexual abuse. Often in these cases, the allegation is not made until years after the event, when physical evidence that a crime has taken place is no longer available. If the abuse resulted in pregnancy, the tests can easily and definitively determine paternity and hence identify the perpetrator.

Barry Scheck, one of O.J. Simpson's defence attorneys, is the cofounder of the Innocence Project, an organization that has been fighting to get false convictions overturned on the basis of DNA evidence. He believes that genetic testing will streamline the criminal justice system by enabling police to determine at a preliminary stage whether a suspect was at the scene of the crime independently of often unreliable eyewitness testimony. "One of the things that we have to recognize about DNA testing is that we can use it... in an earlier stage of the investigation than it is being used today," Scheck says. "We can take smaller amounts of sample and test them at the beginning of the investigation.... We can now use DNA to get suspects and to eliminate the wrong people very, very early in the investigation."

Scheck insists that the key to the success of DNA testing in the courts is the proper handling of samples by the police and by the labouratories where the tests are carried out.

"DNA testing is truly revolutionary," he says. "It's transforming the way we do business in the criminal justice system. And it's only beginning. That's one reason why it's very important to make sure that the testing itself is done correctly, is done with the highest standards of quality assurance so that we don't make mistakes."

Scheck believes that DNA testing will be able to exonerate many innocent people who are now in prison after having been falsely convicted on the basis of other types of evidence. It can also shed light on troubling, unsolved cases and prevent further crime by getting dangerous individuals off the streets. "We can go back to these old cases and see whether or not the system made mistakes," he says. "We can look at old, unsolved cases and find the real perpetrator."

Finally, DNA testing, because it is relatively cheap, can help accused people with few resources at their disposal mount an effective defence. Scheck cites the Simpson case, in which, despite the Hall-of-Fame football player's wealth and the top-quality lawyers it enabled him to hire, the prosecution still had a preponderance of resources on its side. "The prosecution had 47 prosecutors working on the case. FBI, Interpol, the Los Angeles Police Department, the crime lab had tremendous resources, much greater than ours, but the point was... we were able to demonstrate that a lot of things went wrong."

MOLECULAR GENETICS AND DNA TECHNOLOGY

While the period from the early 1900s to World War II has been considered the "golden age" of genetics, scientists still had not determined that DNA, and not protein, was the hereditary material. However, during this time a great many genetic discoveries were made and the link between genetics and evolution was made.

Friedrich Meischer in 1869 isolated DNA from fish sperm and the pus of open wounds. Since it came from nuclei, Meischer named this new chemical, nuclein. Subsequently the name was changed to nucleic acid and lastly to deoxyribonucleic acid (DNA). Robert Feulgen, in 1914, discovered that fuchsin dye stained DNA. DNA was then found in the nucleus of all eukaryotic cells.

During the 1920s, biochemist P.A. Levene analysed the components of the DNA molecule. He found it contained four nitrogenous bases:cytosine, thymine, adenine, and guanine; deoxyribose sugar; and a phosphate group. He concluded that the basic unit (nucleotide) was composed of a base attached to a sugar and that the phosphate also attached to the sugar.

He (unfortunately) also erroneously concluded that the proportions of bases were equal and that there was a tetranucleotide that was the repeating structure of the molecule. The nucleotide, however, remains as the fundemantal unit (monomer) of the nucleic acid polymer. There are four nucleotides: those with cytosine (C), those with guanine (G), those with adenine (A), and those with thymine (T).

Molecular structure of three nirogenous bases. In this diagram there are three phosphates instead of the single phosphate found in the normal nucleotide. Images from Purves et al

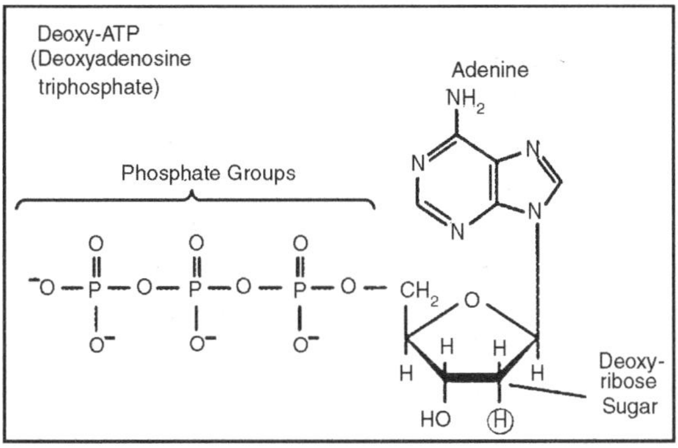

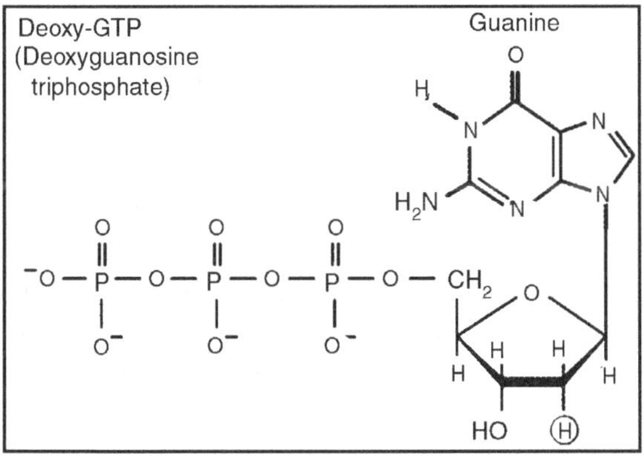

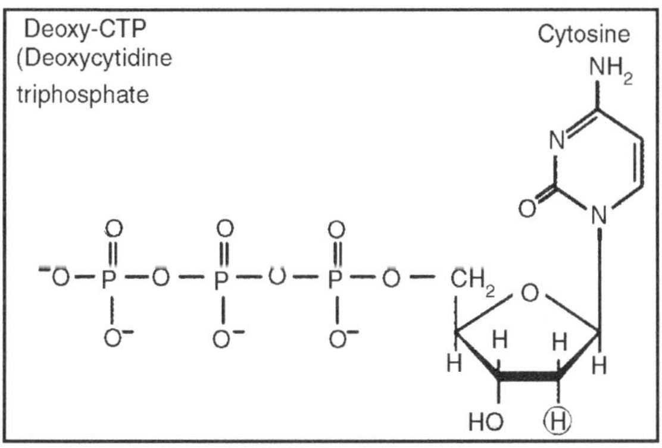

During the early 1900s, the study of genetics began in earnest: the link between Mendel's work and that of cell biologists resulted in the chromosomal theory of inheritance; Garrod proposed the link between genes and "inborn errors of metabolism"; and the question was formed: what is a gene?

The answer came from the study of a deadly infectious disease: pneumonia. During the 1920s Frederick Griffith studied the difference between a disease-causing strain of the pneumonia causing bacteria (*Streptococcus peumoniae*) and a strain that did not cause pneumonia.

The pneumonia-causing strain (the S strain) was surrounded by a capsule. The other strain (the R strain) did not have a capsule and also did not cause pneumonia. Frederick Griffith was able to induce a non-pathogenic strain of the bacterium *Streptococcus pneumoniae* to become pathogenic.

Griffith referred to a transforming factor that caused the non-pathogenic bacteria to become pathogenic. Griffith injected the different strains of bacteria into mice. The S strain killed the mice; the R strain did not.

He further noted that if heat killed S strain was injected into a mouse, it did not cause pneumonia. When he combined heat-killed S with Live R and injected the mixture into a mouse (remember neither alone will kill the mouse) that the mouse developed pneumonia and died. Bacteria recovered from the mouse had a capsule and killed other mice when injected into them!

Hypotheses:

- The dead S strain had been reanimated/resurrected.
- The Live R had been transformed into Live S by some "transforming factor".

Further experiments led Griffith to conclude that number 2 was correct.

In 1944, Oswald Avery, Colin MacLeod, and Maclyn McCarty revisited Griffith's experiment and concluded the transforming factor was DNA. Their evidence was strong but not totally conclusive. The then-current favourite for the hereditary material was protein; DNA was not considered by many scientists to be a strong candidate.

The breakthrough in the quest to determine the hereditary material came from the work of Max Delbruck and Salvador Luria in the 1940s. Bacteriophage are a type of virus that attacks bacteria, the viruses that Delbruck and Luria worked with were those attacking*Escherichia coli*, a bacterium found in human intestines.

Bacteriophages consist of protein coats covering DNA. Bacteriophages infect a cell by injecting DNA into the host cell. This viral DNA then "disappears" while taking over the bacterial machinery and beginning to make new virus instead of new bacteria.

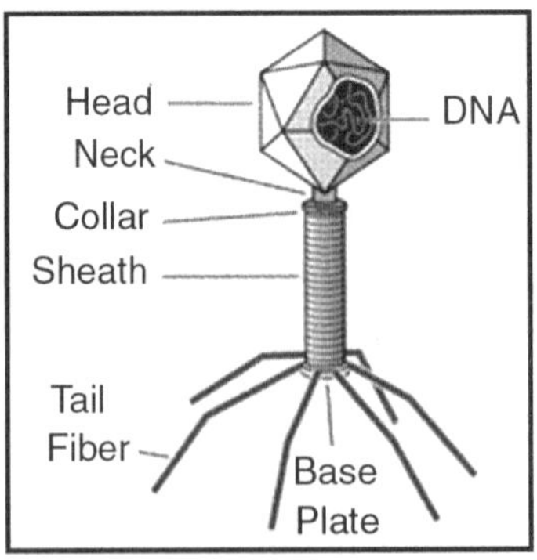

Fig. Structure of a Bacteriophage Virus. Image from Purves et al

After 25 minutes the host cell bursts, releasing hundreds of new bacteriophage. Phages have DNA and protein, making them ideal to resolve the nature of the hereditary material. In 1952, Alfred D. Hershey and Martha Chase conducted a series of experiments to determine whether protein or DNA was the hereditary material.

By labeling the DNA and protein with different (and mutually exclusive) radioisotopes, they would be able to determine which chemical (DNA or protein) was getting into the bacteria. Such material must be the hereditary material (Griffith's transforming agent). Since DNA contains Phosphorous (P) but no Sulfur (S), they tagged the DNA with radioactive Phosphorous-32. Conversely, protein lacks P but does have S, thus it could be tagged with radioactive Sulfur-35. Hershey and Chase found that the radioactive S remained outside the cell while the radioactive P was found inside the cell, indicating that DNA was the physical carrier of heredity.

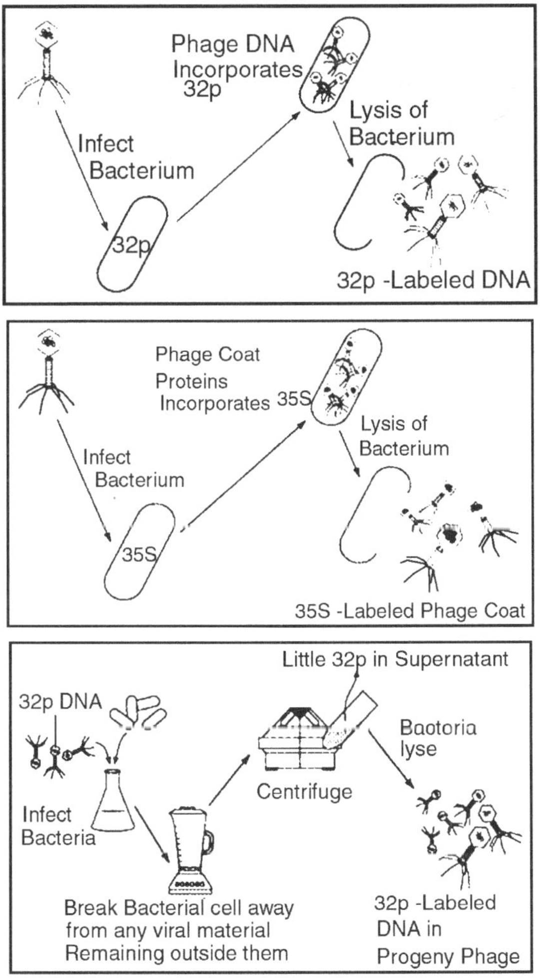

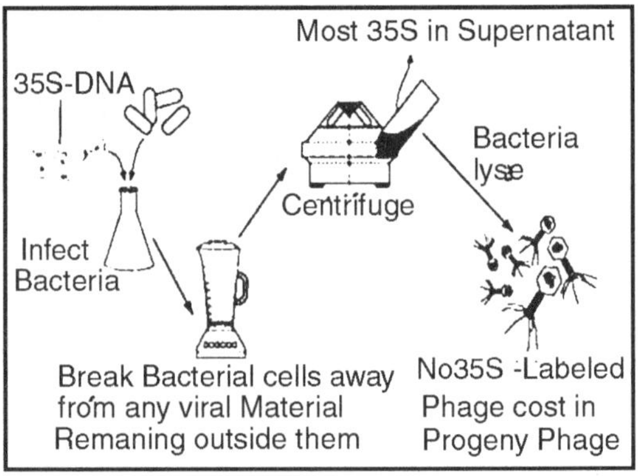

Diagrams illlustrating the Hershey and Chase experiment that supported DNA as the hereditary material while it also showed protein was NOT the hereditary material. Images from Purves et al.,

THE STRUCTURE OF DNA

Erwin Chargaff analysed the nitrogenous bases in many different forms of life, concluding that the amount of purines does not always equal the amount of pyrimidines (as proposed by Levene). DNA had been proven as the genetic material by the Hershey-Chase experiments, but how DNA served as genes was not yet certain.

DNA must carry information from parent cell to daughter cell. It must contain information for replicating itself. It must be chemically stable, relatively unchanging. However, it must be capable of mutational change. Without mutations there would be no process of evolution.

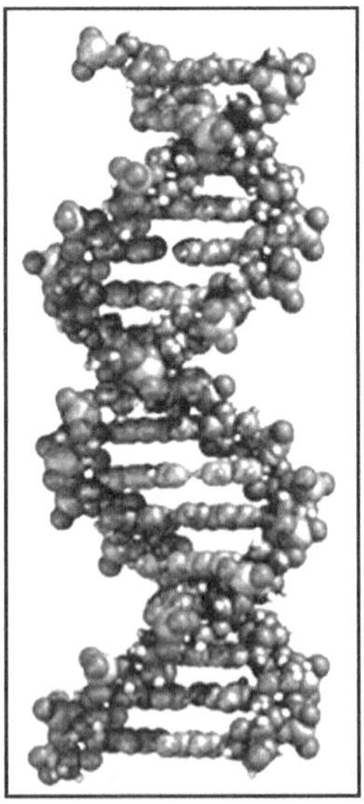

Fig. Ball and Stick Model of DNA. Image from Purves et al.

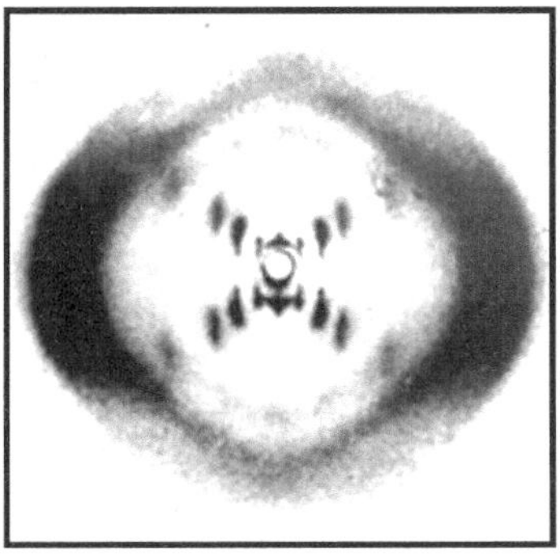

Fig. X-ray Diffraction Photograph of the DNA Double Helix. Image from the Internet.

Many scientists were interested in deciphering the structure of DNA, among them were Francis Crick, James Watson, Rosalind Franklin, and Maurice Wilkens. Watson and Crick gathered all available data in an attempt to develop a model of DNA structure. Franklin took X-ray diffraction photomicrographs of crystalline DNA extract, the key to the puzzle. The data known at the time was that DNA was a long molecule, proteins were helically coiled (as determined by the work of Linus Pauling), Chargaff's base data, and the x-ray diffraction data of Franklin and Wilkens.

DNA is a double helix, with bases to the centre (like rungs on a ladder) and sugar-phosphate units along the sides of the helix (like the sides of a twisted ladder). The strands are complementary (deduced by Watson and Crick from Chargaff's data, A pairs with T and C pairs with G, the pairs held together by hydrogen bonds).

Notice that a double-ringed purine is always bonded to a single ring pyrimidine. Purines are Adenine (A) and Guanine (G). We have encountered Adenosine triphosphate (ATP) before, although in that case the sugar was ribose, whereas in DNA it is deoxyribose. Pyrimidines are Cytosine (C) and Thymine (T). The bases are complementary, with A on one side of the molecule you only get T on the other side, similarly with G and C. If we know the base sequence of one strand we know its complement.

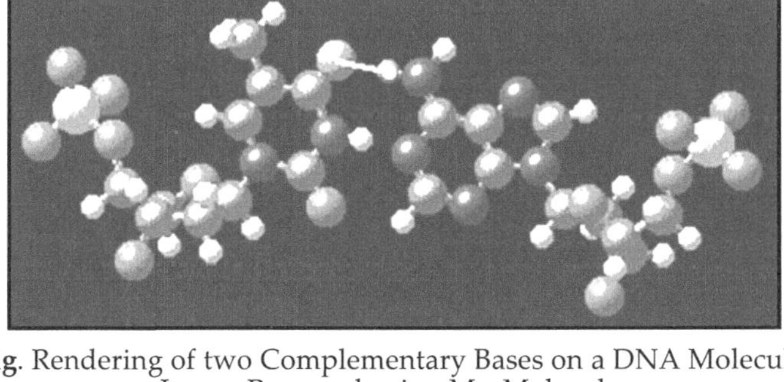

Fig. Rendering of two Complementary Bases on a DNA Molecule. Image Prepared using MacMolecule.

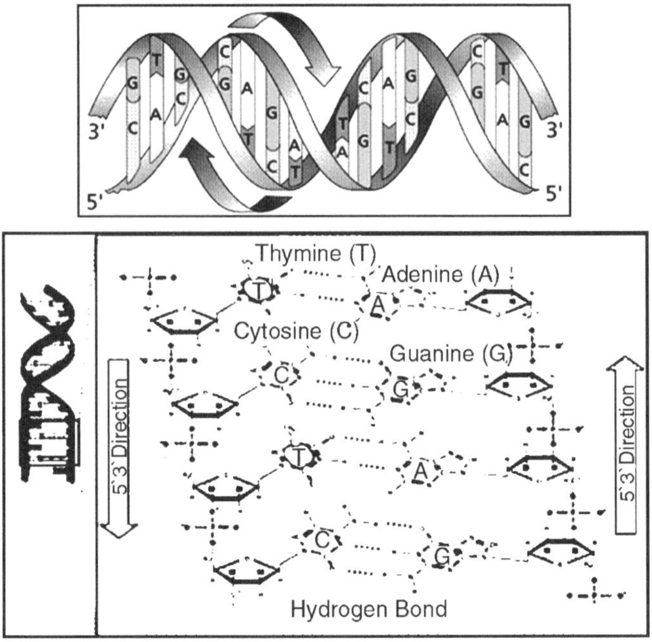

Fig. The Ribbon Model of DNA. Image from Purves et al.

DNA REPLICATION

DNA was proven as the hereditary material and Watson et al. had deciphered its structure. What remained was to determine how DNA copied its information and how that was expressed in the phenotype. Matthew Meselson and Franklin W. Stahl designed an experiment to determine the method of DNA replication. Three models of replication were considered likely.

1. Conservative replication would somehow produce an entirely new DNA strand during replication.

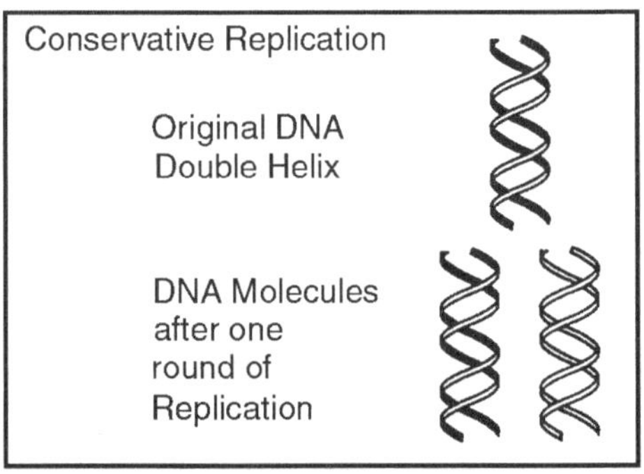

Fig. Conservative Model of DNA Replication. Image from Purves et al

2. Semiconservative replication would produce two DNA molecules, each of which was composed of one-half of the parental DNA along with

an entirely new complementary strand. In other words the new DNA would consist of one new and one old strand of DNA. The existing strands would serve as complementary templates for the new strand.

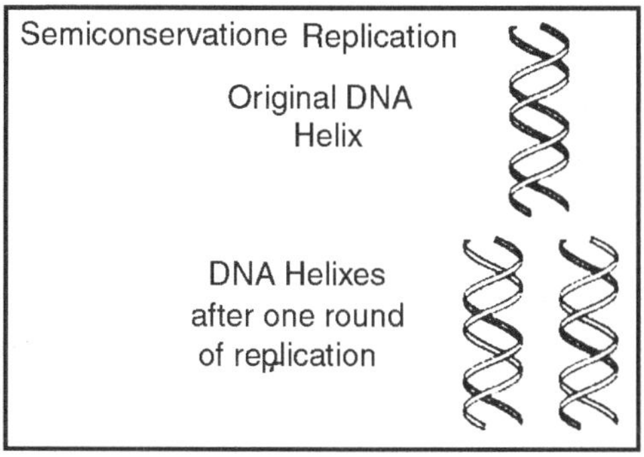

Fig. The Semiconservative Model of DNA Structure. Image from Purves et al

3. Dispersive replication involved the breaking of the parental strands during replication, and somehow, a reassembly of molecules that were a mix of old and new fragments on each strand of DNA.

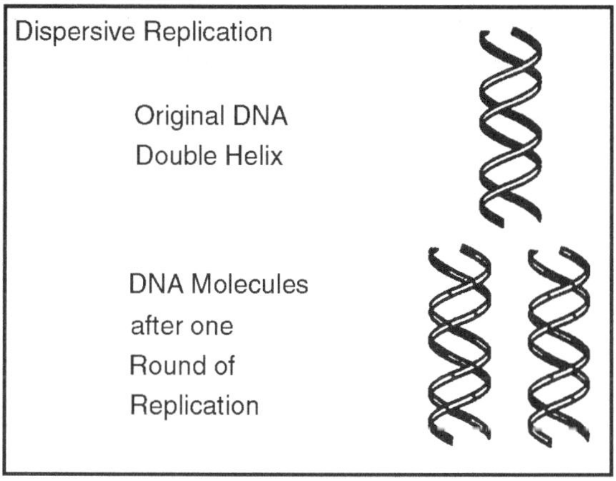

Fig. The Dispersive Replication model of DNA Replication. Image from Purves et al.

The Meselson-Stahl experiment involved the growth of *E. coli* bacteria on a growth medium containing heavy nitrogen (Nitrogen-15 as opposed to the more common, but lighter molecular weight isotope, Nitrogen-14). The first generation of bacteria was grown on a medium where the sole source of N was Nitrogen-15. The bacteria were then transferred to a medium with light (Nitrogen-14) medium. Watson and Crick had predicted that DNA replication was semi-conservative. If it was, then the DNA produced by bacteria grown on light medium would be intermediate between heavy and light. It was.

DNA replication involves a great many building blocks, enzymes and a great deal of ATP energy (remember that after the S phase of the cell cycle cells have a G phase to regenerate energy for cell division). Only occurring in a cell once per (cell) generation, DNA replication in humans occurs at a rate of 50 nucleotides per second, 500/second in prokaryotes.

Nucleotides have to be assembled and available in the nucleus, along with energy to make bonds between nucleotides. DNA polymerases unzip the helix by breaking the H-bonds between bases. Once the polymerases have opened the molecule, an area known as the replication bubble forms (always initiated at a certain set of nucleotides, the origin of replication). New nucleotides are placed in the fork and link to the corresponding parental nucleotide already there (A with T, C with G). Prokaryotes open a single replication bubble, while eukaryotes have multiple bubbles. The entire length of the DNA molecule is replicated as the bubbles meet.

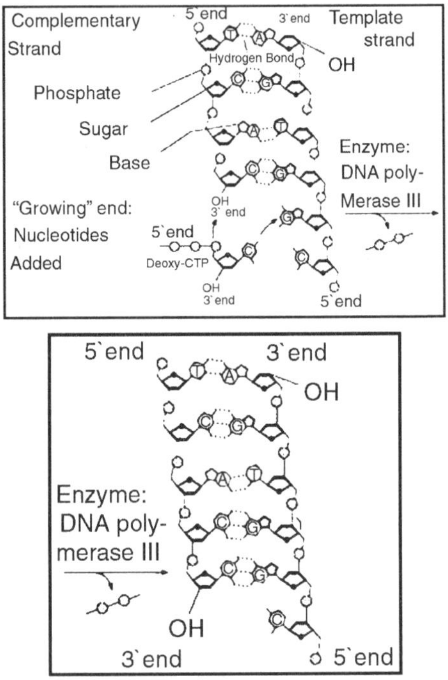

Fig. The Roles of DNA Polymerases in DNA Replication. Image from Purves et al.

Since the DNA strands are antiparallel, and replication proceeds in thje 5' to 3' direction on EACH strand, one strand will form a continuous copy, while the other will form a series of short Okazaki fragments.

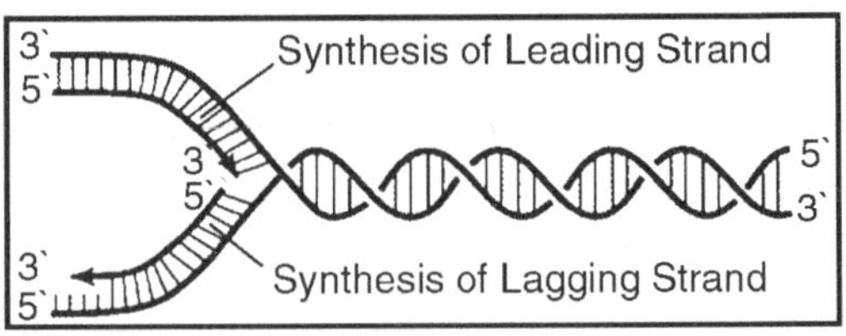

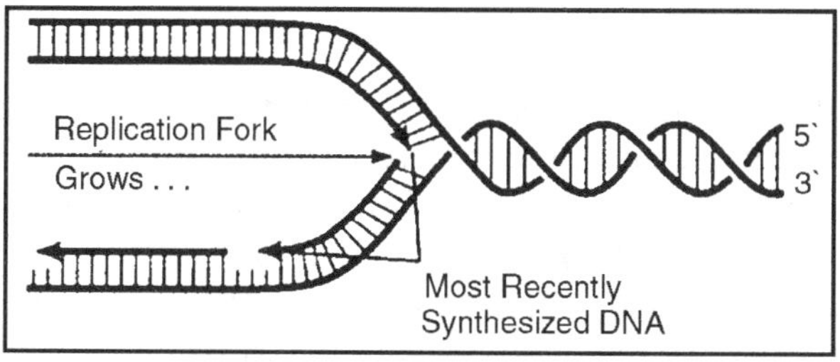

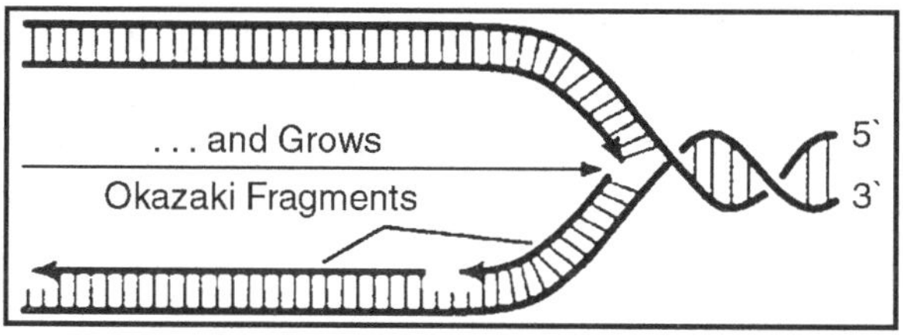

DNA AND CHROMOSOMES

DeoxyriboNucleic Acid is commonly known as DNA. DNA at its most basic level is in essence the blueprint of an organism's genetic make-up. The shape of DNA, discovered by Francis Crickand James Watson, is a double helix. The composition of DNA is a series of nitrogenous bases (known as base pairs), and sugar and phosphate strands. Within the ladder model of DNA (at left), the sugar and phosphate strands compose the sides of the DNA model, or molecule, while the actual rungs of the ladder are made up of the 4 nitrogen bases. For a molecule of DNA to maintain its shape there are a number of rules between pair bonds, and the strands of phosphates.

They are as follows:

- The four pair bonds, Cytosine, Guanine, Adenine, and Thymine, do not entirely intermix. Only Cytosine can attach with Guanine, and only Adenine can attach to Thymine. Or only C's can attach to G's, and only T's to A's.
- The Phosphate and Sugar strands (edges of the ladder), are arranged as follows, S - P - S - P - S - P, etc.

In the composition of DNA, codons are a grouping of three base pair sequences, which determine the production of different types of amino acids. There are 20 different types of amino acids, which a cell uses to produce proteins, and which are central to the functioning of cells within an organism. Even the most minute differentiation in the amount and type of amino acid, can have drastic effects on a cell's composition and function. DNA replicates itself by splitting itself in two, or unzipping itself like a zipper. Once detached, the split side of the ladder connects with free floating nucleotides (C, T, A, and G base pairs). Connections between Nucleotides are governed by the base pair rule(*i.e.* C's can only connect with G's, and T's can only connect with A's). When replication is complete, two identical strands of DNA are produced.

Chromosomes

Cells, whether through meiosis or mitosis, share genetic information through chromosomes.Chromosomes are the package for DNA and genes. A living organism has roughly 5 feet or 1.5 metres of compacted DNA within it's nucleus. To contain the DNA within the nucleus of a cell, the genetic information is wound up into strands like coiled rope. These strands take the shape of X's. Each place on a chromosome holds genetic information which pertains to the expression of a trait. The genetic information a chromosome holds, or one piece of genetic information a chromosome can hold, is known as a gene. An allele is a point or place on a chromosome. The locus is a place on a specific chromosome where a gene is found.

5

Gene Expression and DNA Structure

DNA CARRIES GENETIC

The material of heredity should have certain properties:

- It must be very stable so that genetic information can be stored in it and transmitted countless times to subsequent generations.
- It must be capable of precise copying or replication so that its information is not lost or altered.
- Although stable, it must also be subject to change in order to account, in the short term, for the appearance of mutant forms and, in the long term, for evolution.

The first evidence that deoxyribonucleic acid, or DNA, might be the material of heredity came from investigations on *Streptococcus pneumoniae,* one of the types of bacteria that cause pneumonia. In 1928, Frederick Griffith, an English microbiologist, was comparing the properties of two strains of pneumococcus bacteria.

One strain, Type S (S for smooth colonial morphology), is virulent because it is enclosed within a slippery polysaccharide coat, or capsule, that protects it from the immune system of its host. The other strain, Type R (R for rough-looking colonies), lacks an enzyme for the biosynthesis of the polysaccharide coat and is not virulent because it cannot resist attack by the host's immune system.

When Griffith injected Type S bacteria into mice, the blood became filled with S bacteria and the mice died. Heat-killed Type S bacteria had no effect on the mice, but if mice were injected with non-virulent Type R bacteria that had been mixed with heat-killed Type S bacteria, the mice died and virulent Type S bacteria could be recovered from their blood.

Somehow, the heat-killed Type S bacteria had transformed the non-virulent R Type into the virulent S Type.

In 1931, M.H. Dawson and R. H. P. Sia showed that extracts of heat-killed Type S cells could transform non-pathogenic R cells into genetically stable, pathogenic S cells.

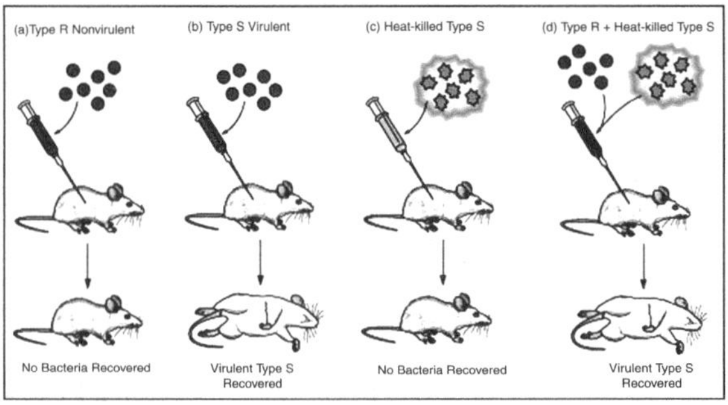

Fig. Griffith Experiment on Pneumococcal Transformation:

a. Mice are resistant to Type R *Streptococcus pneumoniae* bacteria, but
b. Are killed by injection with virulent Type S *S. pneumoniae* bacteria.
c. Injection with heat-killed virulent bacteria does not kill mice, but
d. If heat-killed Type S bacteria are mixed with non-virulent Type R bacteria, they have the capacity to *transform* non-virulent Type R bacteria into the virulent Type S form.

THE "TRANSFORMING PRINCIPLE"

In 1944, Oswald T. Avery and his associates Colin M. MacLeod and Maclyn McCarty at the Rockefeller Institute made the discovery that the substance active in transforming Type R bacteria to virulence was, in fact, DNA. This finding was surprising and not immediately accepted because most scientists at the time thought that proteins, substances chemically more complex and diverse than nucleic acids, were the genetic material.

Avery, MacLeod, and McCarty showed that highly purified preparations of "transforming principle" contained no detectable protein and were unaffected by trypsin or chymotrypsin (two proteolytic enzymes) or by pancreatic RNase (which hydrolyzes RNA).

However, the transforming substance was readily inactivated by treatment with pancreatic DNase, an enzyme that specifically degrades DNA. Thus, DNA must have been the agent carrying the information that transforms R bacteria to virulence. Because transformation was stably inherited, DNA merited strong consideration as the actual material of heredity.

GENE INTERACTIONS

FINDING THE GENES

Between 1884 (the year Mendel died) and 1888 details of mitosis and meiosis were reported, the cell nucleus was identified as the location of the genetic material, and "qualities" were even proposed to be transmitted on chromosomes to daughter cells at mitosis. In 1903 Walter Sutton and Theodore Boveri formally proposed that chromosomes contain the genes. The

Chromosome Theory of Inheritance is one of the foundations of genetics and explains the physical reality of Mendel's principles of inheritance.

Fig. Fruit Fly, *Drosophila Melanogaster*.

The location of many genes (Mendel's factors) was determined by Thomas Hunt Morgan and his coworkers in the early 1900's. Morgan's experimental organism was the fruit fly (*Drosophila melanogaster*). Fruit flies are ideal organisms for genetics, having a small size, ease of care, susceptibility to mutations, and short (7-9 day) generation time. The role of chromosomes in determination of sex was deduced by Morgan from work on fruit flies.

During Metaphase I, homologous chromosomes will line up. A karyotype can be made by cutting and arranging photomicrographs of the homologous chromosomes thus revealed at Metaphase I. Two types of chromosome pairs occur. Autosomes resemble each other in size and placement of the centromere, for example pairs of chromosome 21 are the same size, while pairs of chromosome 9 are of a different size from pair 21. Sex chromosomes may differ in their size, depending on the species of the organism they are from.

In humans and *Drosophila*, males have a smaller sex chromosome, termed the Y, and a larger one, termed the X. Males are thus XY, and are termed heterogametic. Females are XX, and are termed homogametic. In grasshoppers, which Sutton studied in discovering chromosomes, there is no Y, only the X chromosome in males. Females are XX, while males are denoted as XO. Other organisms (notably birds, moths and butterflies) have males homogametic and females heterogametic. Males (if heterogametic) contribute either an X or Y to the offspring, while females contribute either X. The male thus determines the sex of the offspring. Remember that in meiosis, each chromosome is replicated and one copy sent to each gamete.

Morgan discovered a mutant eye colour and attempted to use this mutant as a recessive to duplicate Mendel's results. He failed, instead of achieving a 3:1 F2 ratio the ratio was closer to 4:1 (red to white). Most mutations are usually recessive, thus the appearance of the white mutant presented Morgan a chance to test Mendel's ratios on animals. The F1 generation also had no white eyed females. Morgan hypothesized that the gene for eye colour was only on the X chromosome, specifically in that region of the X that had no corresponding region on the Y. White eyed fruit flies were also more likely to die prior to adulthood, thus explaining the altered ratios. Normally eyes are red, but a variant (white) eyed was detected and used in genetic study. Cross a homozygous white eyed male with a homozygous red eyed female, and all the offspring have red eyes. Red is dominant over white. However, cross a homozygous white eyed female with a red eyed male, and the unexpected results show all the males have white eyes and all the females red eyes. This can be explained if the eye colour gene is on the X chromosome.

Explanation

If the gene for eye colour is on the X chromosome, the red eyed male in the second cross will pass his red eyed X to only his daughters, who in turn received only a recessive white-carrying X from their mother. Thus all females had red eyes like their father. Since the male fruit fly passes only the Y to his sons, their eye colour is determined entirely by the single X chromosome they receive from their mother (in this case white). Thus all the males in the second cross were white eyed.

These experiments introduced the concept of sex-linkage, the occurrence of genes on that part of the X that lack a corresponding location on the Y. Sex-linked recessives (such as white eyes in fruit flies, hemophilia, baldness, and colourblindness in humans) occur more commonly in males, since there is no chance of them being heterozygous. Such a condition is termed homozygous.

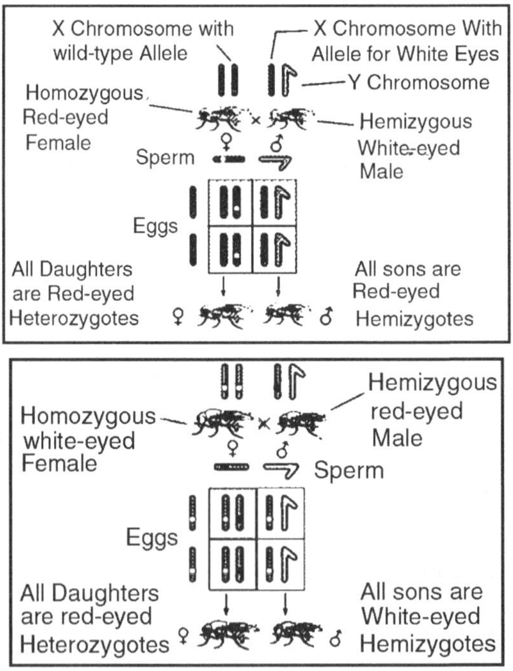

Fig. Inheritance of Eye Colour in Fruit flies.

Characteristics of X-linked Traits

- Phenotypic expression more common in males
- Sons cannot inherit the trait from their fathers, but daughters can.

Sons inherit their Y chromosome from their father.

Only a few genes have been identified on the Y chromosome, among them the testis-determining factor (TDF) that promotes development of the male phenotype.

Barr bodies are interpreted as inactivated X chromosomes in mammalian females. Since females have two X chromosomes, the Lyon hypothesis suggests that one or the other X is inactivated in each somatic (non-reproductive) cell during embryonic development. Cells mitotically produced from these embryonic cells likewise have the same inactivated X chromosome.

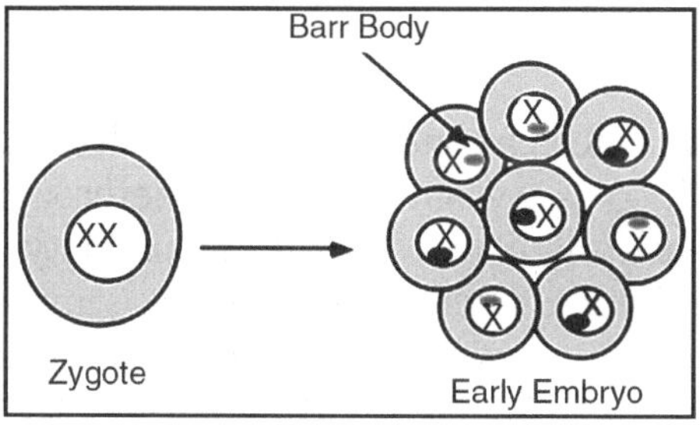

Fig. The Role of Deactivated X Chromosomes in Mammalian Female Development.).

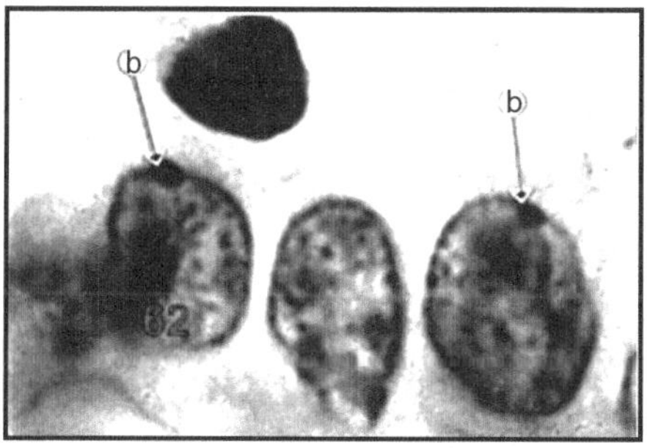

Fig. Barr Bodies (b) in Mammalian Cells.

Calico cats (sometimes called tortoiseshell) are almost always female since the calico trait is caused by some areas of the cat's fur expressing one allele and others expressing the other colour. Can there be a male tortoiseshell cat? How would such a cat get its genes? Remember that fur colour in cats is a sex-linked feature. Would the male calico be fertile or sterile?

THE MODERN VIEW OF THE GENE

While Mendel discussed traits, we now know that genes are segments of the DNA that code for specific proteins. These proteins are responsible for the expression of the phenotype. The basic principles of segregation and independent assortment as worked out by Mendel are applicable even for sex-linked traits.

Codominant alleles

Codominant alleles occur when rather than expressing an intermediate phenotype, the heterozygotes express both homozygous phenotypes. An example is in human ABO blood types, the heterozygote AB type manufactures antibodies to both A and B types. Blood Type A people manufacture only anti-B antibodies, while type B people make only anti-A antibodies. Codominant alleles are both expressed. Heterozygotes for codominant alleles fully express both alleles. Blood type AB individuals produce both A and B antigens. Since neither A nor B is dominant over the other and they are both dominant over O they are said to be codominant.

Incomplete Dominance

Incomplete dominance is a condition when neither allele is dominant over the other. The condition is recognized by the heterozygotes expressing an intermediate phenotype relative to the parental phenotypes. If a red flowered plant is crossed with a white flowered one, the progeny will all be pink. When pink is crossed with pink, the progeny are 1 red, 2 pink, and 1 white.

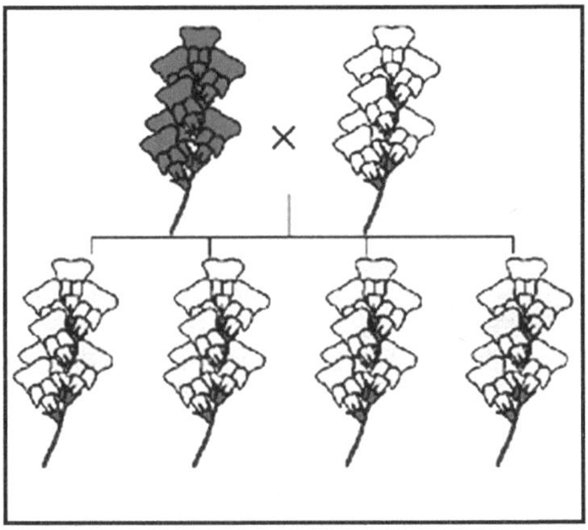

Fig. Inheritance of Flower Colour in Snapdragons

Flower colour in snapdragons is an example of this pattern. Cross a true-breeding red strain with a true-breeding white strain and the F_1 are all pink (heterozygotes). Self-fertilize the F1 and you get an F_2 ratio of 1 red: 2 pink: 1 white. This would not happen if true blending had occurred (blending cannot explain traits such as red or white skipping a generation and pink flowers crossed with pink flowers should produce ONLY pink flowers).

Multiple Alleles

Many genes have more than two alleles (even though any one diploid individual can only have at most two alleles for any gene), such as the ABO blood groups in humans, which are an example of multiple alleles. Multiple alleles result from different mutations of the same gene. Coat colour in rabbits is determined by four alleles. Human ABO blood types are determined by alleles A, B, and O. A and B are codominants which are both dominant over O. The only possible genotype for a type O person is OO. Type A people have either AA or AO genotypes. Type B people have either BB or BO genotypes. Type AB have only the AB (heterozygous) genotype. The A and B alleles of gene I produce slightly different glycoproteins (antigens) that are on the surface of each cell. Homozygous A individuals have only the A antigen, homozygous B individuals have only the B antigen, homozygous O individuals produce neither antigen, while a fourth phenotype (AB) produces both A and B antigens.

INTERACTIONS AMONG GENES

While one gene may make only one protein, the effects of those proteins usually interact (for example widow's peak may be masked by expression of the baldness gene). Novel phenotypes often result from the interactions of two genes, as in the case of the comb in chickens. The single comb is produced only by the rrpp genotype. Rose comb (b) results from R_pp. (_ can be either R or r). Pea comb (c) results from rrP_. Walnut comb, a novel phenotype, is produced when the genotype has at least one dominant of each gene (R_P_).

Epistasis

Epitasis is the term applied when one gene interferes with the expression of another (as in the baldness/widow's peak mentioned earlier). Bateson reported a different phenotypic ratio in sweet pea than could be explained by simple Mendelian inheritance. This ratio is 9:7 instead of the 9:3:3:1 one would expect of a dihybrid cross between heterozygotes. Of the two genes (C and P), when either is homozygous recessive (cc or pp) that gene is epistatic to (or hides) the other. To get purple flowers one must have both C and P alleles present.

ENVIRONMENT AND GENE EXPRESSION

Phenotypes are always affected by their environment. In buttercup (*Ranunculus peltatus*), leaves below water-level are finely divided and those above water-level are broad, floating, photosynthetic leaf-like leaves. Siamese cats are darker on their extremities, due to temperature effects on phenotypic expression.

Expression of phenotype is a result of interaction between genes and environment. Siamese cats and Himalayan rabbits both animals have dark coloured fur on their extremities. This is caused by an allele that controls pigment production being able only to function at the lower temperatures of those extremities. Environment determines the phenotypic pattern of expression.

POLYGENIC INHERITANCE

Polygenic inheritance is a pattern responsible for many features that seem simple on the surface. Many traits such as height, shape, weight, colour, and metabolic rate are governed by the cumulative effects of many genes. Polygenic traits are not expressed as absolute or discrete characters, as was the case with Mendel's pea plant traits.

Instead, polygenic traits are recognizable by their expression as a gradation of small differences (a continuous variation).

The results form a bell shaped curve, with a mean value and extremes in either direction.

Height in humans is a polygenic trait, as is colour in wheat kernels. Height in humans is NOT discontinuous.

If you line up the entire class a continuum of variation is evident, with an average height and extremes in variation (very short [vertically challenged?] and very tall [vertically enhanced]). Traits showing continuous variation are usually controlled by the additive effects of two or more separate gene pairs.

This is an example of polygenic inheritance. The inheritance of EACH gene follows Mendelian rules.

Usually polygenic traits are distinguished by:

- Traits are usually quantified by measurement rather than counting.
- Two or more gene pairs contribute to the phenotype.
- Phenotypic expression of polygenic traits varies over a wide range.

Human polygenic traits include:

- Height
- SLE (Lupus)
- Weight
- Eye Colour
- Intelligence
- Skin Colour
- Many forms of behaviour

The role of many genes (polygenic inheritance) in production of a continuum of phenotypes.

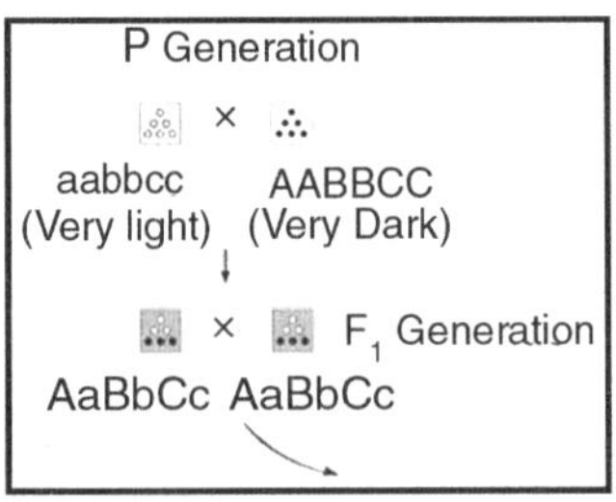

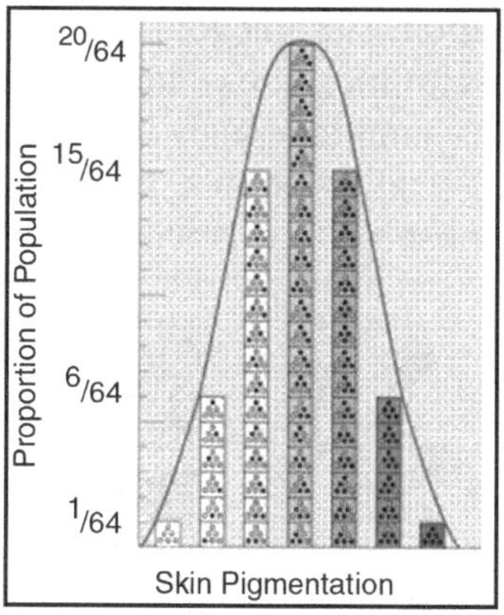

PLEIOTROPY

Pleiotropy is the effect of a single gene on more than one characteristic. An example is the "frizzle-trait" in chickens. The primary result of this gene is the production of defective feathers. Secondary results are both good and bad; good include increased adaptation to warm temperatures, bad include increased metabolic rate, decreased egg-laying, changes in heart, kidney and spleen. Cats that are white with blue eyes are often deaf, white cats with a blue and an yellow-orange eye are deaf on the side with the blue eye. Sickle-cell anemia is a human disease originating in warm lowland tropical areas where malaria is common. Sickle-celled individuals suffer from a number of problems, all of which are pleiotropic effects of the sickle-cell allele.

MOLECULAR MECHANISM OF RECOMBINATION

Genetic recombination is the natural process by which genetic information is rearranged to form new associations. Such recombination is a powerful genetic and evolutionary force that reshapes the genomes of all organisms. At the molecular level, *genetic recombination is the exchange (or incorporation) of one DNA sequence with (or into) another.*

For example, homologous recombination involves an exchange of DNA sequences between homologous chromosomes, resulting in the arrangement of genes into new combinations. The process underlying homologous recombination is termed general recombination because the enzymatic machinery that mediates the exchange can use essentially any pair of homologous DNA sequences as substrates.

Homologous recombination occurs during the production of gametes (*meiosis*) in diploid organisms. In higher animals, that is, those with immune systems, recombination also occurs in the DNA of somatic cells responsible for expressing proteins of the immune response, such as the immunoglobulins. This somatic recombination rearranges the immunoglobulin genes, dramatically increasing the potential diversity of immunoglobulins available from a fixed amount of genetic information. Homologous recombination can also occur in bacteria. Indeed, even viral chromosomes undergo recombination. For example, if two mutant viral particles simultaneously infect a host cell, a recombination event between the two viral genomes can lead to the formation of a recombination virus chromosome that is wild-type.

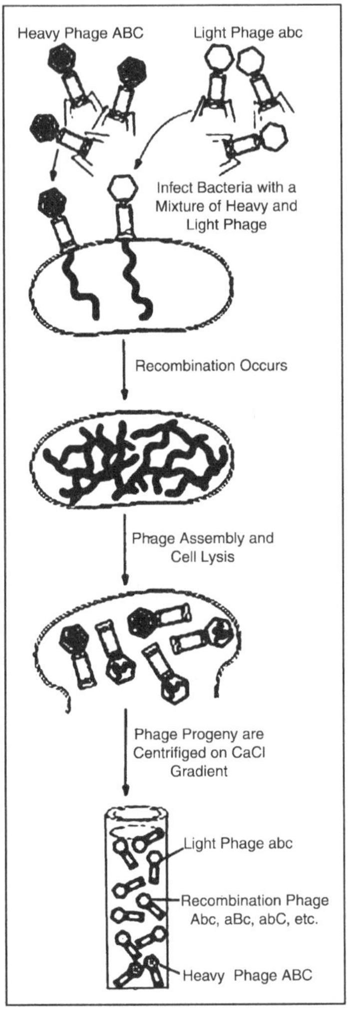

Fig. Meselson and Weigle's Experiment Demonstrated that a Physical Exchange of Chromosome Parts Actually Occurs During Eecombination. Density-labeled, "Heavy" Phage, Symbolized as ABC Phage in the Diagram, was Used to Coinfect Bacteria Along with "Light" Phage, the abc Phage. The Progeny from the Infection were Collected and Subjected to CsCl Density Gradient Centrigugation. Parental-type ABC and abc Phage were Well-separated in the Gradient, but Recombinant Phage (ABc,Abc,aBc,aBC, and so on) were Distributed Diffusely between the Two Parental Bands because they Contained Chromosomes Constituted from Fragments of Both "Heavy" and "Light" DNA. These Recombinant Chromosomes Formed by Breakage and Reunion of Parental "Heavy" and "Light" Chromosomes.

Bacteriophage genomes can insert into bacterial chromosomes by a form of recombination, but because the integration of the bacteriophage DNA into the host DNA occurs only at a unique site on the host chromosome and involves specific DNA sequences on both phage DNA and bacterial DNA, the process is called site-specific recombination. Only a short length of homology (often less than 15 bp) is necessary for site-specific recombination events, and the enzymes involved act only on these sequences. In transposition, yet another type of recombination, particular DNA sequences known as *transposons* are inserted somewhat independently of any sequence homology on the DNA into which the insertion occurs. However, transposons themselves carry a specific sequence essential to insertion. Transposition serves as a mechanism by which genetic material may be moved from one chromosomal location to another. A fourth, rare form of recombination, illegitimate recombination, occurs between non-homologous DNA independently of any unique sequence element.

GENERAL RECOMBINATION

Recombination occurs by the breakage and reunion of DNA strands, so that a physical exchange of parts takes place. Matthew Meselson and J. J. Weigle demonstrated in 1961 that this happens by coinfecting E. *coli* with two genetically distinct bacteriophage l strains, one of which had been density-labeled by growth in ^{13}C- and ^{15}N-containing media. The phage progeny were recovered and separated by CsCl density gradient centrifugation. Phage particles that displayed recombinant genotypes were distributed throughout the gradient while parental (non-recombinant) genotypes were found within discrete "heavy" and "light" bands in the density gradient. The results showed that recombinant phage contained DNA derived in varying proportions from both parents. The obvious explanation is that these recombinant DNAs arose via the breakage and rejoining of DNA molecules.

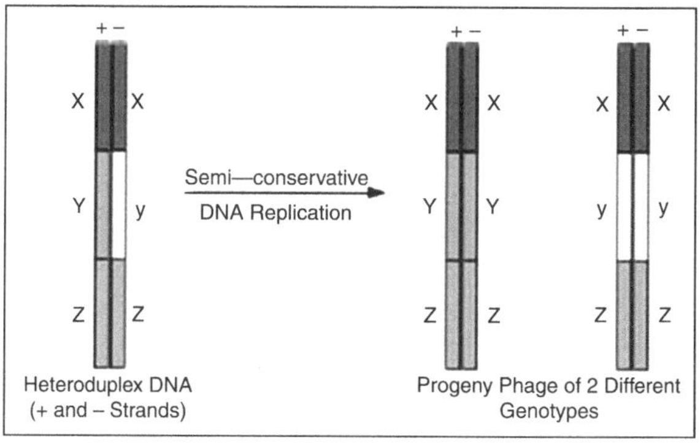

Fig. The Generation of Progeny Bacteriophage of Two Different Genotypes from a Single Phage Particle Carrying a Heteroduplex DNA Region within its Chromosome. The Heteroduplex DNA is Composed of One Strand that is Genotypically XYZ (the + strand), and the Other Strand that is Genotypically XyZ (the - strand). That is, the Genotype of the Two Parental Strands for Gene Y is Different (One is Y, the Other y).

A second important observation made during this type of experiment was that some of the plaques formed by the phage progeny contained phage of two different genotypes, even though each plaque was caused by a single phage infecting one bacterium.

Therefore, some infecting phage chromosomes must have contained a region of heteroduplex DNA, duplex DNA in which a part of each strand is contributed by a different parent.

THE HOLLIDAY MODEL

In 1964, Robin Holliday proposed a model for homologous recombination that has proven influential. The two homologous DNA duplexes are first juxtaposed so that their sequences are aligned. This process of chromosome pairing is called synapsis. Holliday suggested that recombination begins by introduction of single-stranded nicks at homologous sites on the two paired chromosomes.

The two duplexes partially unwind, and the free, single-stranded end of one duplex begins to base-pair with its nearly complementary, single-stranded.

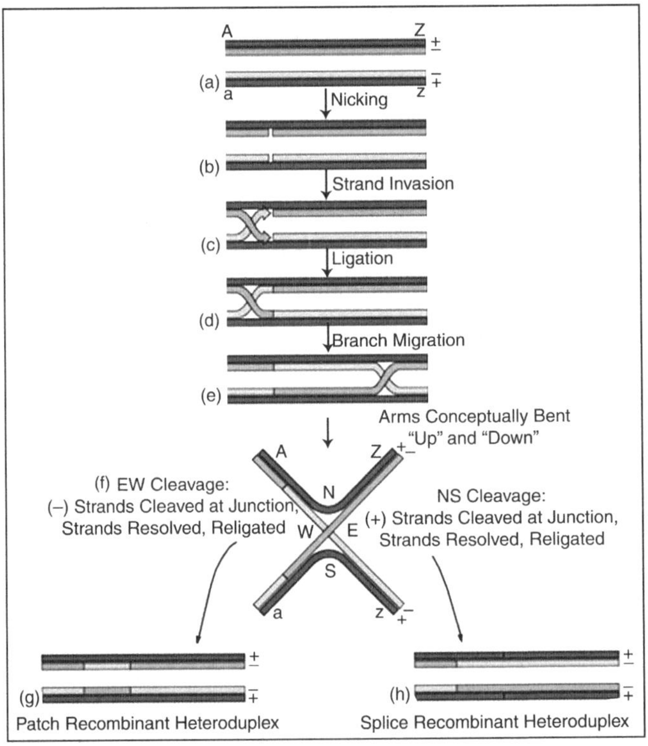

Fig. The Holliday Model for Hologuous Recomibination. The + Signs and - Signs Label Strands of Like Polarity. For example, Assume that the Two Strands Running 5' ® 3' as Read Left to Right are Labeled

Region along the intact strand in the other duplex, and vice versa. This strand invasion is followed by ligation of the free ends from different duplexes to create a cross-stranded intermediate known as a Holliday junction. The cross-stranded junction can now migrate in either direction (branch migration) by unwinding and rewinding of the two duplexes.

Branch migration results in strand exchange; heteroduplex regions of varying length are possible. In order for the joint molecule formed by strand exchange to be resolved into two DNA duplex molecules, another pair of nicks must be introduced.

Resolution can be represented best if the duplexes are drawn with the chromosome arms bent "up" or "down" to give a planar representation. Nicks then take place, either at E and W, that is, in the - strands that were originally nicked or at N and S, that is, the + strands (the strands not previously nicked). Duplex resolution is most easily kept straight by remembering that + strands are complementary to - strands and any resultant duplex must have one of each.

Nicks made in the strands originally nicked lead to DNA duplexes in which one strand of each remains intact. Although these duplexes contain heteroduplex regions, they are not recombinant for the markers (*AZ, az*) that flank the heteroduplex region; such heteroduplexes are called patch recombinants. Nicks introduced into the two strands not previously nicked yield DNA molecules that are both heteroduplex and recombinant for the markers *A/a* and *Z/z*; these heteroduplexes are termed splice recombinants.

Although this Holliday model explains the outcome of recombination, it provides no mechanistic explanation for the strand exchange reactions and other molecular details of the process.

THE ENZYMOLOGY OF GENERAL RECOMBINATION

To illustrate recombination mechanisms, we focus on general recombination as it occurs in *E. coli*. The principal players in the process are the RecBCD enzyme complex, which initiates recombination; the RecA protein, which binds single-stranded DNA, forming a nucleoprotein filament capable of strand invasion and homologous pairing; and the RuvA, RuvB, and RuvC proteins, which drive branch migration and process the Holliday junction into recombinant products.

Eukaryotic homologs of these prokaryotic recombination proteins have been identified, indicating that the fundamental process of general recombination is conserved across all organisms.

a. RecBCD binds to a duplex DNA end and its helicase activity begins to unwind the DNA double helix. "Rabbit ears" of ssDNA loop out from RecBCD because the rate of DNA unwinding exceeds the rate of ssDNA release by RecBCD.

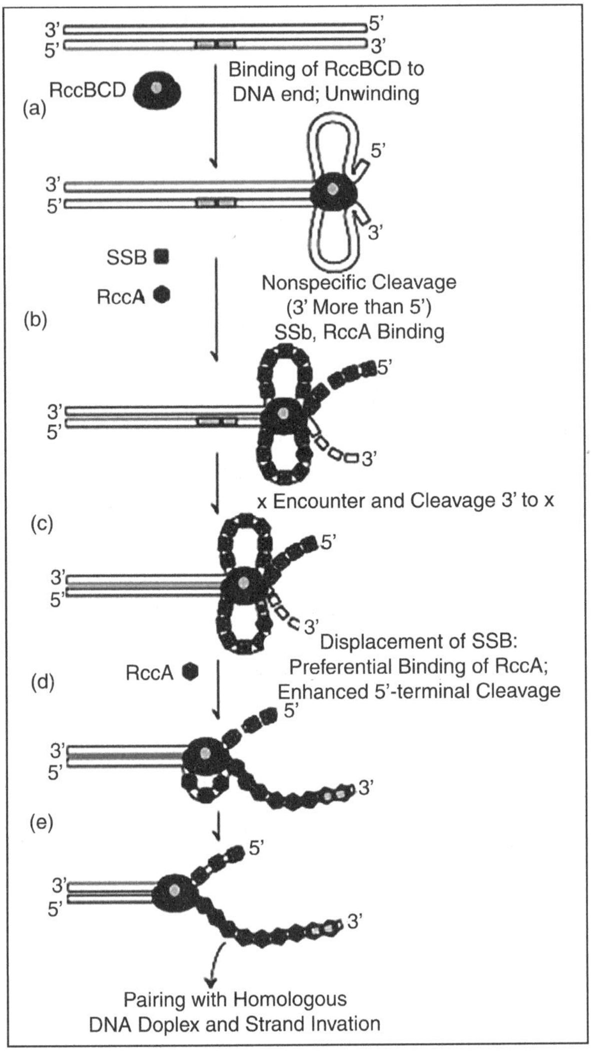

Fig. Model of RecBCD-dependent Initiation of Recombination.

b. As it unwinds the DNA, SSB (and some RecA) bind to the single-stranded regions; the RecBCD endonuclease activity randomly cleaves the ssDNA, showing a greater tendency to cut the 3'-terminal strand rather than the 5'-terminal strand.

c. When RecBCD encounters a properly oriented c site, the 3'-terminal strand is cleaved just below the 39-end of c.

d. RecBCD now directs the binding of RecA to the 3'-terminal strand, as RecBCD endonuclease activity now acts more often on the 5'-terminal strand.

e. A nucleoprotein filament consisting of RecA-coated 3'-strand ssDNA is formed.

The RecBCD Enzyme Complex

The proteins RecB (140 kD; 1180 amino acids), RecC (130 kD; 1122 amino acids), and RecD (67 kD; 608 amino acids) form a multifunctional enzyme complex having both helicase and nuclease activity. The RecBCD complex initiates recombination by attaching to the end of a DNA duplex (or at a double-stranded break in the DNA) and using its ATP-dependent helicase function to unwind the dsDNA. As RecBCD progresses along unwinding the duplex, its nuclease activity cleaves both of the newly formed single strands (although the strand that provided the 3'-end at the RecBCD entry site is cut more frequently than the 5'-terminal strand.

Single-stranded DNA-binding protein (SSB) (and some RecA protein) readily binds to the emerging single strands. Sooner or later, RecBCD encounters a particular nucleotide sequence, a so-called *Chi* (or c) site, characterized by the sequence 5'-GCTGGTGG-3'. These c sites are recombinational "hot spots"; 1009 c sites have been identified in the E. *coli* genome (on average, about one every 4.5 kb of DNA). When a c sequence is encountered by a RecBCD complex approaching its 3'-side (the..G-3'-side), RecBCD cleaves the c-bearing DNA strand four to six bases to the 3' side of c. Interaction of RecBCD with the c site causes the D subunit of RecBCD to become irreversibly altered such that the RecBCD complex no longer expresses nuclease activity against the 3'-terminal strand, but nuclease activity against the 5'-terminal strand increases.

Resuming its helicase function, RecBCD unwinds the dsDNA, and collectively these processes generate a ssDNA tail bearing a site at its 3'-terminal end. This ssDNA may reach serveral kilbases in length. At the same time, the c-altered RecBDC complex now promotes preferential binding of RecA protein, instead of SSB, to the 3'-terminal strand to form a Nucleoprotein filament active in pairing and strand invasion with a homologous region in another dsDNA molecule.

The RecA Protein

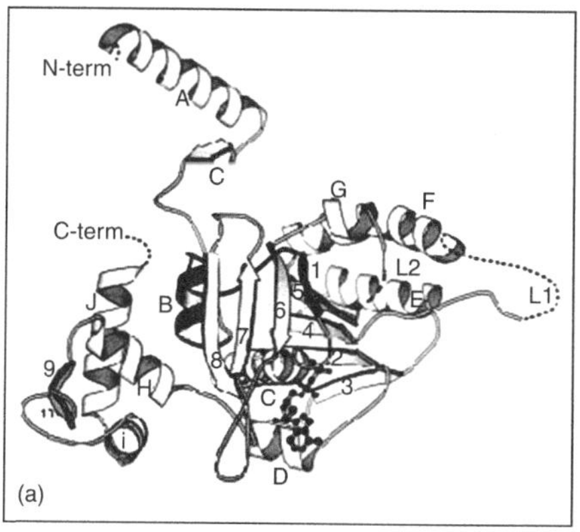

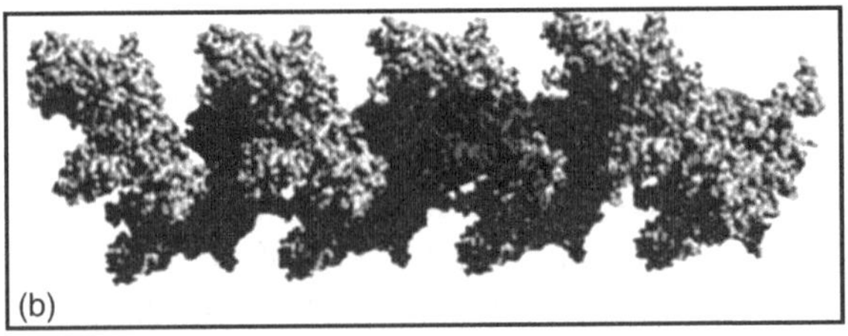

Fig. The Structure of RecA protein.

a. Ribbon diagram of the RecA monomer. Note the ADP bound at the site near helices C and D.
b. RecA filament. Four turns of a helical filament that has six RecA monomers per turn. A RecA monomer is highlighted in red

The RecA protein, or recombinase, is a multifunctional 352-residue (38 kD) enzyme that acts in general recombination to catalyze the ATP-dependent DNA strand exchange reaction, leading to formation of a Holliday junction. RecA protein crystallizes in the absence of DNA to form a helical filament having six monomers per turn. This filament has a deep spiral groove large enough to accommodate three strands of DNA. RecA binds single-stranded DNA with a stoichiometry of one RecA per three nucleotides, and the resultant nucleoprotein filament has a helical pitch of 8.5 to 10 nm and about six RecA monomers per turn. The DNA in both RecA:ssDNA and RecA:dsDNA filaments is extended 150 per cent relative to B-form DNA. The nucleoprotein filament formed by binding of RecA protein to the 3'-terminal ssDNA has affinity for other DNA molecules. In fact, binding of multiple DNA strands is the hallmark of RecA function.

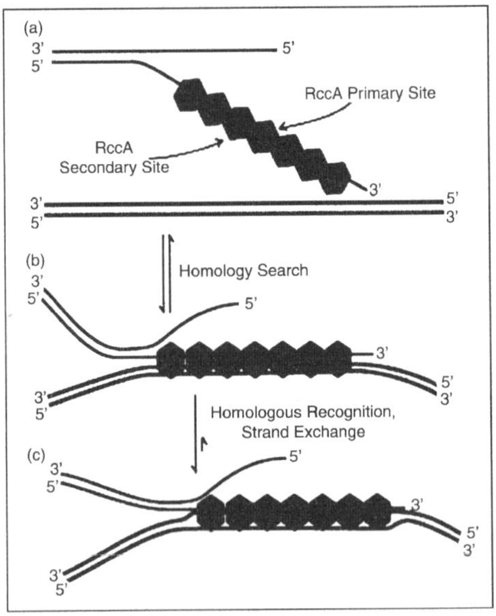

Fig. Model for the Strand exchange Function of RecA, as Based on the Relative DNA Affinities of the Primary and Secondary DNA-binding Sites on RecA.

a. ssDNA is bound in the primary DNA-binding site of RecA.
b. dsDNA is bound weakly in the RecA secondary DNA-binding site, and RecA scans this dsDNA for homology.
c. Homology recognition leads to DNA strand exchange as the RecA-bound ssDNA forms a heteroduplex with a newly found complementary strand; this heteroduplex fills RecA's primary DNA-binding site.

In recombination, RecA uses its so-called high-affinity primary DNA-binding site to bind ssDNA. This complex then interacts with other DNA molecules through a secondary DNA-binding site within RecA. This secondary site has higher affinity for ssDNA than dsDNA. The relative affinity of this secondary site suggests a mechanism for RecA in DNA strand exchange during recombination: a recA:ssDNA nucleoprotein complex transiently binds dsDNA in its secondary site and scans along the minor groove of the dsDNA, searching it for sequence homology with its bound ssDNA.

When homology is found, a hybrid DNA duplex is formed between the ssDNA in the primary site and the complementary strand found in the scanned dsDNA. Formation of this hybrid duplex displaces the strand of the scanned dsDNA that is most like the ssDNA brought in by RecA. This process is the essence of DNA strand exchange. The displaced DNA strand is bound with higher affinity in the RecA secondary site than the dsDNA that was occupying this site. High-affinity binding of ssDNA in the secondary site stabilizes the strand-exchange complex and ensures proper heteroduplex formation by RecA in homologously aligned DNA strands.

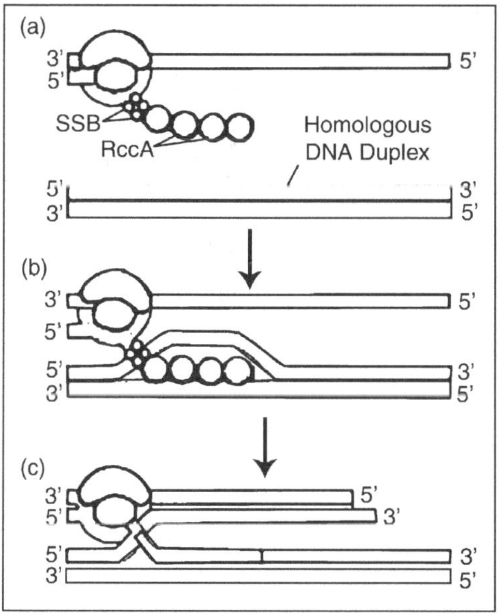

Fig. Model for Homologous Recombination as Promoted by RecA Enzyme.

a. RecA protein (and SSB) aid strand invasion of the 3'-ssDNA into a homologous DNA duplex,
b. Forming A D-loop.
c. The D-loop strand that has been displaced by strand invasion pairs with its complementary strand in the origninal duplex to form a Holliday junction as strand invasion continues.

Procession of base unpairing of dsDNA and re-pairing into hybrid strands along the DNA duplex initiatesbranch migration. Branch migration drives the displacement of the homologous DNA strand from the DNA duplex and its replacement with the ssDNA strand, a process known as single-strand assimilation (or single-strand uptake). Strand assimilation does not occur if there is no homology between the ssDNA and the invaded DNA duplex. The DNA strand displaced by the invading 3'-terminal ssDNA is free to anneal with the 5'-terminal strand in the original DNA, a step that is also mediated by RecA protein and SSB.

The result is a Holliday junction, the classic intermediate in genetic recombination. Proteins that assist RecA in the formation of a Holliday junction include RecF, RecO, and RecR.

Resolution of the Holliday Junction by the RuvA, RuvB, and RuvC Proteins

The Holliday junction is then processed into recombination products by RuvA, RuvB, and RuvC. Specifically, RuvA (203 amino acids) and RuvB (336 amino acids) work together as a Holliday junction-specific helicase complex that dissociates the RecA filament and catalyzes branch migration. An RuvA tetramer fits precisely within the junction point, which has a square-planar geometry, and this RuvA tetramer targets the assembly of RuvB around opposite arms of the DNA junction.

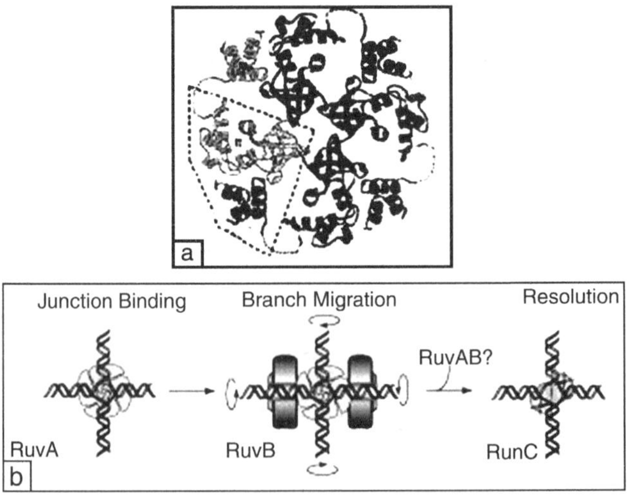

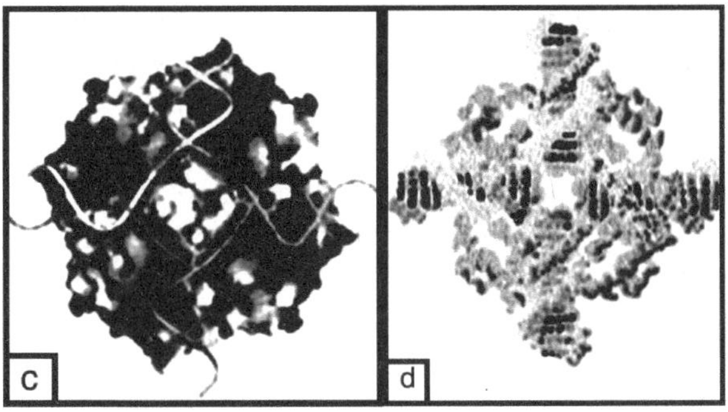

Fig. Model for the Resolution of a Holliday Junction in E. *coli* by the RuvA, RuvB, and RuvC proteins.

a. Ribbon diagram of the RuvA tetramer. RuvA monomers have an overall L shape (one of them is outlined by the dashed white line); four of them form a tetramer with fourfold rotational symmetry, a structure reminiscent of a four-petaled flower.
b. Model for RuvA/RuvB action. Structure of a multisubunit complex that promotes DNA branch migration.

The RuvB protein binds to form two oppositely oriented, hexameric [$(RuvB)_6$] ring structures encircling the dsDNAs, one on each side of the Holliday junction. Rotation of the dsDNAs by the RuvB hexameric rings pulls the dsDNAs through $(RuvB)_6$ and unwinds the DNA strands across the "spool" of RuvA, which threads the separated single strands into newly forming hybrid (recombinant) duplexes. The RuvA tetramer is a disklike structure, one face of which has an overall positive charge, with the exception of four negatively charged central pins, each contributed by an RuvA monomer. These four pins fit neatly into the hole at the center of the Holliday junction. The negatively charged sugar-phosphate backbones of the four DNA duplexes of the Holliday junction are threaded along grooves in the positively charged RuvA face, with the negatively charged central pins appropriately situated to transiently separate the dsDNA molecules into their component single strands through repulsive electrostatic interactions with the phosphate backbones of the DNA. The separated strands of each parental duplex are then channeled into grooves in the RuvA face, where they are led into hydrogen-bonding interactions with bases contributed by strands of the other parental DNA to form the two daughter hybrid duplexes flowing out from the RuvAB complex. Figure illustrates a model for the RuvA tetramer with the square-planar Holliday junction.

Depending on how the strands in the Holliday junction are cleaved and resolved, patch or splice recombinant duplexes result. RuvC (173 amino acids) is an endonuclease that resolves Holliday junctions into heteroduplex recombinant products (RuvC resolvase). An RuvC dimer binds at the Holliday junction and cuts pairs of DNA strands of similar polarity; whether a patch or a splice recombinant results depends on which DNA pair is cleaved.

Recombination is a fundamental process that is involved not only in generating genetic diversity, but is also involved in DNA repair and chromosome segregation during cell division. Hexameric ring helicases such as RuvB are DNA-driving molecular motors; similar motors act during DNA replication to propel strand separation and initiate DNA synthesis. Thus, the RuvABC system for processing Holliday junctions may represent a general paradigm for DNA manipulation in all cells.

TRANSPOSONS

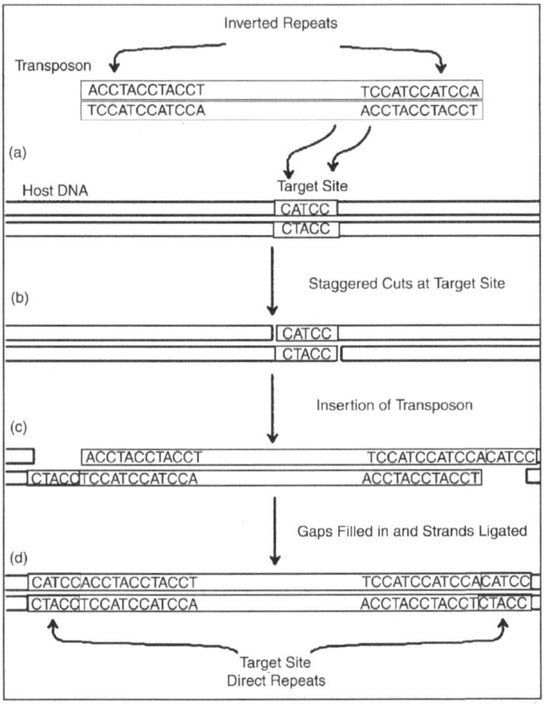

Fig. The Archetypal Transposon has Inverted Nucleotide-sequence Repeats at its Termini, Represented here as the 12-bp Sequence ACGTACGTACGT

a. It acts at a target sequence (shown here as the sequence CATGC) within host DNA by creating a staggered cut
b. Whose protruding single-stranded ends are then ligated to the transposon
c. The gaps at the target site are then filled in, and the filled-in strands are ligated
d. Transponson insertion thus generates direct repeats of the target site in the host DNA, and these direct repeats flank the inserted transponson.

In 1950, Barbara McClintock reported the results of her studies on an activator gene in maize (*Zea mays* or, as it's usually called, *corn*) that was recognizable principally by its ability to cause mutations in a second gene. Activator genes were thus an internal source of mutation. A most puzzling property was their ability to move relatively freely about the genome. As we have seen, scientists had labored to establish that chromosomes consisted of genes arrayed in a fixed order, so most geneticists viewed as incredible this idea of genes moving around. The recognition that McClintock so richly deserved for her explanation of this novel phenomenon had to await verification by molecular biologists. In 1983, Barbara McClintock was finally awarded the Nobel Prize in physiology or medicine. By this time, it was appreciated that many organisms, from bacteria to humans, possessed similar *"jumping genes"* able to move from one site to another in the genome. This mobility led to their designation as mobile elements, transposable elements, or, simply, transposons.

Transposons are segments of DNA that are moved enzymatically from place to place in the genome. That is, their location within the DNA is unstable. Transposons range in size from several hundred bp to more than 8 kbp. Transposons contain a gene encoding an enzyme necessary for insertion into a chromosome and for the remobilization of the transposon to different locations. These movements are termedtransposition events. The smallest transposons are called insertion sequences, or ISs, signifying their ability to insert apparently at random in the genome.

Insertion into a new site causes mutation because it disrupts the DNA sequence at that site. Insertion occurs at sites that show little homology to the insertion sequence or transposon. Although certain transposons (such as E. *coli* transposon Tn 7) may undergo transposition once per cell generation, most transposition events are infrequent, taking place only once every 10^4 to 10^7generations. Larger and more complex transposons also carry genes that are not involved in the enzymology of insertion and excision of the transposon, such as genes conferring resistance to antibiotics. Episomes, plasmids that can reversibly integrate into bacterial genomes, contain transposons.

HEREDITARY MOLECULE OF BACTERIOPHAGE

Further proof that DNA is the material of heredity came from the study of bacteriophage. In 1952, Alfred Hershey and Martha Chase devised an elegant experiment to trace the fates of the two major components of bacteriophage¾coat protein and DNA¾following infection. They took advantage of the fact that nucleic acids lack sulfur and proteins lack phosphorus to uniquely label bacteriophage DNA with ^{32}P and bacteriophage protein with ^{35}S.

Bacterio-phage labeled with either isotope were obtained from cultures of *bacteriophage T2*grown on *Escherichia coli* in medium containing radioactive ^{32}P-labeled inorganic phosphate or radioactive ^{35}S-labeled methionine.

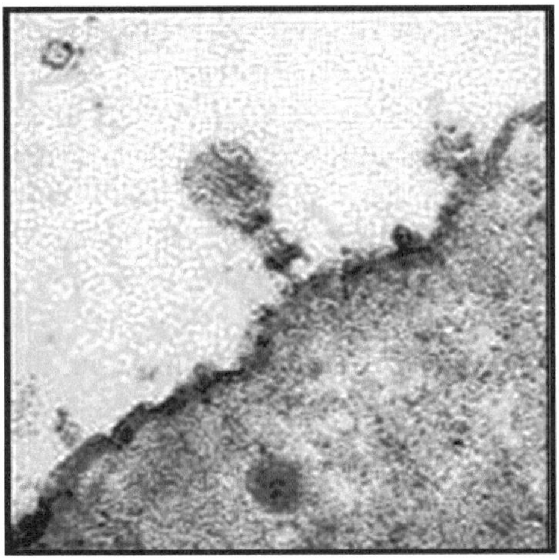

Fig. Electron Micrograph of Bacteriophage Particle Attached to a Bacterial Cell. A Single T4 Bacteriophage Weighs 5 ® 10^{-13} g and Consists of 60 per cent DNA and 40 per cent Protein. Its Volume is about 1/1000 the Volume of an *E. coli* Cell. T4 Phage Heads are 100 nm ´ 65 nm Icosahedra Attached to Tails 100 nm Long by 25 nm in Diameter.

Phage infection of bacteria involves attachment of the bacteriophage to the bacterial cell at specific attachment sites. The phage DNA enters the bacterial cell, leaving its protein coat behind on the surface of the bacterium. Hershey and Chase mixed labeled bacteriophage T2 with unlabeled E. *coli* cells, permitting sufficient time for the phage to attach. Then they vigorously agitated the culture in a blender to shear the phage coats from the bacterial surface.

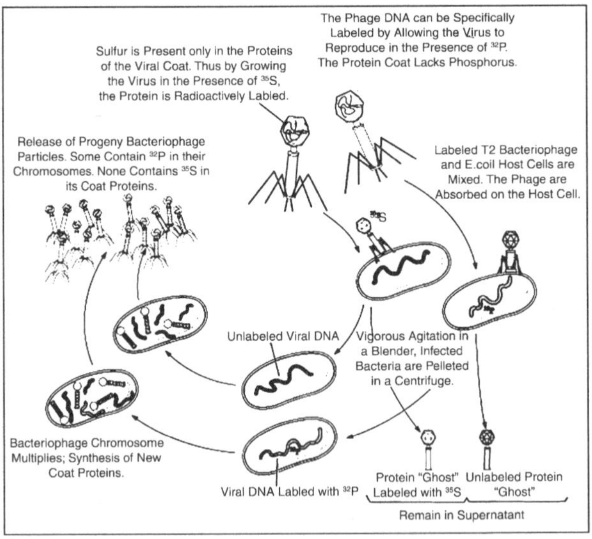

Fig. The Hershey and Chase Experiement Demonstrated that the DNA Component of Bacteriophage T2 Carried the Requisite Genetic Information for Bacteriophage Reproduction.

Following centrifugation of the culture, infected bacteria could be recovered in the pellet, whereas the phage coats containing most of the ^{35}S label remained suspended in the supernatant. In contrast, when E. *coli* cells were infected with ^{32}P-labeled T2 phage, the bacterial pellet contained most of the ^{32}P. Furthermore, upon lysis, 30 per cent of the original ^{32}P but only 1 per cent of the ^{35}S was recovered in the bacteriophage progeny produced by the infection. Hershey and Chase surmised that the bacteriophage DNA was sufficient for bacteriophage reproduction. That is, DNA must be the material of heredity.

GENETIC INFORMATION IN BACTERIA

Bacteria are very useful organisms for genetic analysis: Under optimal conditions of growth and reproduction, some bacteria (such as E. *coli*) divide every 20 minutes, the progeny of each division being a new generation. A genetic experiment can be completed with bacteria in hours, whereas an analogous experiment with a multicellular organism would take months or years because the generation times of such organisms are months or even years in duration. Further, a single milliliter of bacterial culture can contain enormous numbers of bacteria¾as many as 10^{10} ¾all derived from a single parental bacterium:

A single bacterium growing with a generation time of 20 min can give rise to 10^{10} progeny in less than 11 hr. N, the number of cells after n number of generations, is given by $N = 2^n$. For $N = 10^{10} = 2^n$, $n = 33.22$ ($2^{32.22} = 10^{10}$). At 0.33 hr per generation, 33.22 generations (the time to accumulate 10^{10} cells from a single bacterium) occur in about 11 hours. Because of these vast numbers, very rare genetic events can be observed. That is, a one-in-a-million occurrence could be present in thousands of bacteria in a culture. In addition, because bacteria are haploid organisms (organisms with only one chromosome or one set of chromosomes), each cell contains but one set of genetic instructions.

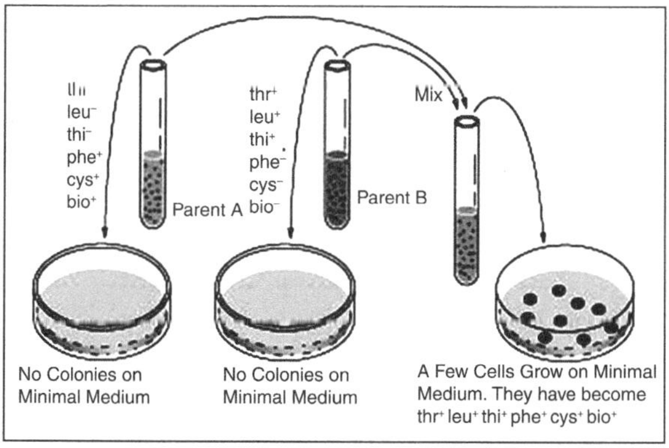

Fig. The Use of Nutritional Mutants to Demonstrate Sexuality in Bacteria. The Genetic Markers are *thr⁻*, *leu⁻*, *thi⁻* (Inability to Grow in the Absence of Threonine, Leucine, and Thiamine, Respectively) on one Chromosome (Parent A), and *phe⁻*, *cys⁻*, *bio⁻* (Defects in Genes for Phenylalanine, Cystine, and Biotin Synthesis, Respectively) on the Other Chromosome (Parent B).

Consequently, any mutation in a gene is not masked or corrected by a second, normal copy of the gene, as it usually is in diploid organisms (organisms having two, essentially duplicate, sets of chromosomes). In haploid organisms like bacteria, the phenotype, or perceptible characteristics of the organism, reflects its genotype, or genetic composition. In contrast, diploid organisms may exhibit a wild-type, or normal, phenotype for any trait, even though their genotype might contain one mutant copy and one wild-type copy of the gene responsible for the trait.

MAPPING THE STRUCTURE OF BACTERIAL CHROMOSOMES

In 1946, Joshua Lederberg and Edward Tatum discovered that genetic information could be transferred between bacteria. They used two strains of E. *coli* that differed in their growth requirements due to mutations each carried. One strain (thr^-, leu^-, thi^-) required threonine, leucine, and thiamine to grow; the other (phe^-, cys^-, bio^-) required phenylalanine, cystine, and biotin. These two strains were mixed together and spread on the surface of a petri plate of minimal medium lacking any of the required supplements. After a day, a very small number of bacterial colonies were observed to be growing. Somehow, these growing bacteria had acquired functional (wild-type) copies of each of the mutant genes.

This remarkable result suggested strongly that the chromosomes of the two different cell types were brought together in a process akin to sexual exchange. In order for the progeny cells (which contain but one chromosome) to have acquired genetic information from the parental strains, genetic recombination must have occurred. This represents, in the words of Lederberg and Tatum, "the assortment of genes in new combinations." Apparently, at some point in time, parental DNA molecules must have aligned along regions of homology (sequence similarity), and segments from one of these molecules must have been interchanged with similar segments from the other parents so that some DNA molecules (chromosomes) now carried wild-type thr^+ leu^+ thi^+ phe^+ cys^+ bio^+ genes.

Lederberg and Tatum speculated that, in order for the various genes to have had the opportunity to recombine, the cells of one strain must have interacted with the cells of the other.

SEXUAL CONJUGATION IN BACTERIA

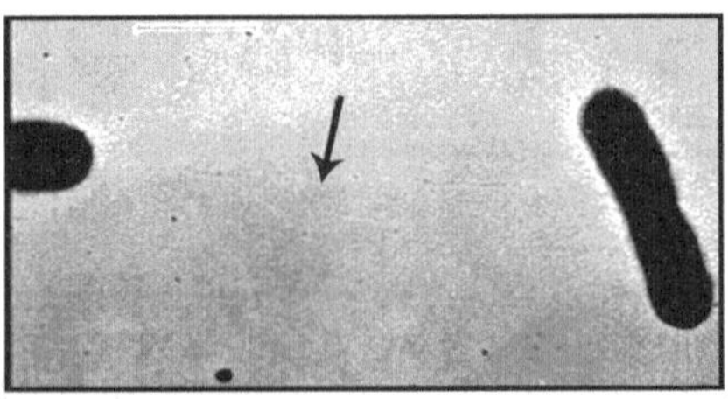

Fig. Electron Micrograph of Two *E. coli* Cells, one F^+, the Other F^-, Joined in Sexual Conjugation. The Pilus Joining them in Indicated by the Arrow.

The transfer of DNA between bacteria takes place via a process known as sexual conjugation, a phenomenon unsuspected prior to the Lederberg-Tatum experiment. Bacterial cells sometimes contain, in addition to their chromosome, extrachromosomal DNA molecules called plasmids.

Plasmids represent "extra" or auxiliary genetic information. Bacterial cells are capable of conjugation if they possess a particular plasmid called the F factor (F for fertility). Such F^+, or *donor, cells* have thin, hollow tubes projecting from their surface known as sex pili or F pili (singular = *pilus*). One or more pili can bind to specific receptors on the surface of cells that lack an F factor. The pilus provides a connection between the two cells. Upon conjugation, a single strand of the F factor is passed to the F^- cell, where its complementary strand is synthesized.The recipient F^- cell thus becomes F^+ by virtue of now having a double-stranded F factor plasmid. The F factor plasmid consists of about 94,000 base pairs; about one-third of this DNA is devoted to about 25 genes that function specifically in the transfer of genetic material from F^+ to F^- cells. Among these genes are those necessary for the formation of pili. In reality, the F factor is an infectious agent.

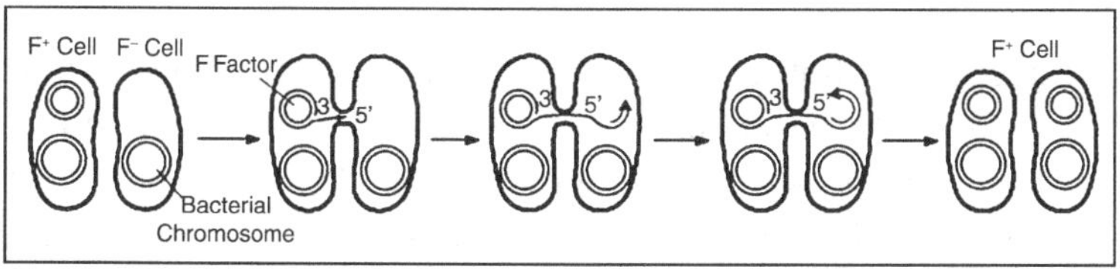

Fig. The Transfer of F Factor from an F^+ to an F^- Cell. A Single Strand of the F Factor is Nicked and Transferred into the Recipient F^- Cell. The Complementary Strand is then Synthesized within the Recipient F^- Cell to Create a New Double-stranded F Factor, Transforming the F^- Cell into an F^+ One.

HIGH FREQUENCY OF RECOMBINATION

In rare instances, the F factor will integrate into the bacterial host chromosome. (Plasmids capable of chromosomal integration are termed episomes.) Cells harboring F factor integrated into the chromosome show a much higher frequency of recombination of chromosomal genes upon conjugation, or "mating," with F^- cells and so are referred to as *Hfr* cells, for "high frequency of recombination." In *Hfr* cells, the conjugal process determined by the F factor operates as it does when the F factor is acting autonomously.

That is, a single strand is passed to the recipient F^- cell, where its complementary strand is synthesized. However, because of its integrated position within the *Hfr* chromosome, the F factor carries along genes adjacent to it on the chromosome. If conjugation continues long enough, a single-stranded copy of the entire host chromosome is passed to the F^- cell. However, conjugation rarely persists the 100 minutes or more required for complete transfer, so usually only part of the *Hfr* chromosome is transferred.

Genes from the *Hfr* chromosomes are transferred into the F^- cell in a fixed order. Therefore, the order of the genes along the *Hfr*chromosome must be fixed, and this order can be mapped by the technique of interrupted mating. Further, because the F factor is integrated at different sites in different *Hfr* strains of E. *coli*, genes difficult to map because they were transferred very late (and hence rarely) in one *Hfr* strain are readily mapped in another. The genetic map obtained by the interrupted mating method reveals a circular arrangement of genes, consistent with the circular organization of the E. *coli* chromosome. Other bacterial chromosomes show a similar circular organization.

IMMUNOGLOBULIN GENES

The immunoglobulin genes are a highly evolved system for maximizing protein diversity from a finite amount of genetic information. This diversity is essential for gaining immunity to the great variety of infectious organisms and foreign substances that cause disease.

THE IMMUNE RESPONSE

Only vertebrates show an immune response. If a foreign substance, called an antigen, gains entry to the bloodstream of a vertebrate, the animal responds via a protective system called the immune response. The immune response involves production of proteins capable of recognizing and destroying the antigen. This response is mounted by certain white blood cells¾the B and T cell lymphocytes and themacrophages. B cells are so named because they mature in the bone marrow; T cells mature in the thymus gland. Each of these cell types is capable of gene rearrangement as a mechanism for producing proteins essential to the immune response. Antibodies, which can recognize and bind antigens, are immunoglobulin proteins secreted from B cells. Because antigens can be almost anything, the immune response must have an incredible repertoire of structural recognition. Thus, vertebrates must have the potential to produce immunoglobulins of great diversity in order to recognize virtually any antigen.

The Immunoglobulin G Molecule

Immunoglobulin G (IgG or g-globulin) is the major class of antibody molecules found circulating in the bloodstream. IgG is a very abundant protein, amounting to 12 mg per mL of serum. It is a 150-kD a_2b_2-type tetramer. The a or *H* (for *heavy*) chain is 50 kD; the b or *L*(for *light*) chain is 25 kD. A preparation of IgG from serum is heterogeneous in terms of the amino acid sequences represented in its L and H chains. However, the IgG L and H chains produced from any given B lymphocyte are homogeneous in amino acid sequence. L chains consist of 214 amino acid residues and are organized into two roughly equal segments, the V_L and C_L regions. The V_L designation reflects the fact that L chains isolated from serum IgG show variations in amino acid sequence over the first 108 residues, V_L symbolizing this "variable" region of the L polypeptide. The amino acid sequence for residues 109 to 214 of the L polypeptide is constant, as represented by its designation as the "constant light," or C_L, region. The heavy, or H, chains consist of 446 amino acid residues.

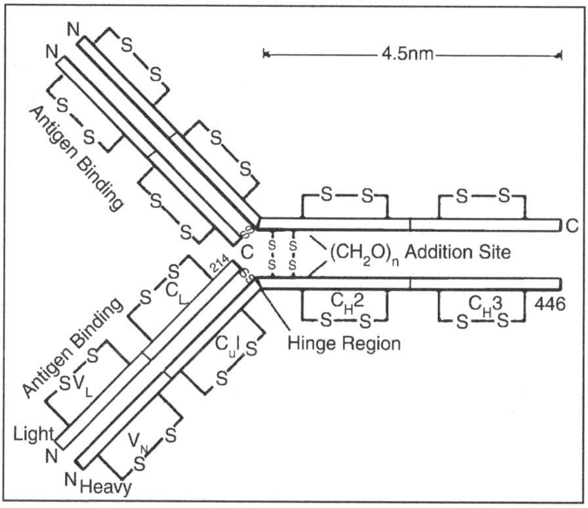

Fig. The Organization of the IgG Molecule. Two Identical L Chains are Joined with Two Identical H Chains. Each L Chain is Held to an H Chain Via an Interchain Disulfide Bond. The Variable Regions of the Four Polypeptides Lie at the Ends of the Arms of the Y-shaped Molecule. These Regions are Responsible for the Antigen Recognition Function of the Antibody Molecules. The Actual Antigen-binding Site is Constituted from Hypervariable Residues within the V_L and V_H Regions. For Purposes of Illustration, Some Features are Shown on Only One or the Other L Chain or H Chain, but all Features are Common to Both Chains.

Like L chains, the amino acid sequence for the first 108 residues of H polypeptides is variable, ergo its designation as the V_H region, while residues 109 to 446 are constant in amino acid sequence. This "constant heavy" region consists of three quite equivalent domains of homology designated C_H1, C_H2, and C_H3. Each L chain has two intrachain disulfide bonds, one in the V_L region and the other in the C_L region.

The C-terminal amino acid in L chains is cysteine, and it forms an interchain disulfide bond to a neighbouring H chain. Each H chain has four intrachain disulfide bonds, one in each of the four regions. Figure presents a diagram of IgG organization.

Within the variable regions of the L and H chains, certain positions are hypervariable with regard to amino acid composition. These hypervariable residues occur at positions 24 to 34, 50 to 55, and 89 to 96 in the L chains and at positions 31 to 35, 50 to 65, 81 to 85, and 91 to 102 in the H chains. The hypervariable regions are also called complementarity-determining regions, or

CDRs, because it is these regions that form the structural site that is complementary to some part of an antigen's structure, providing the basis for antibody:antigen recognition.

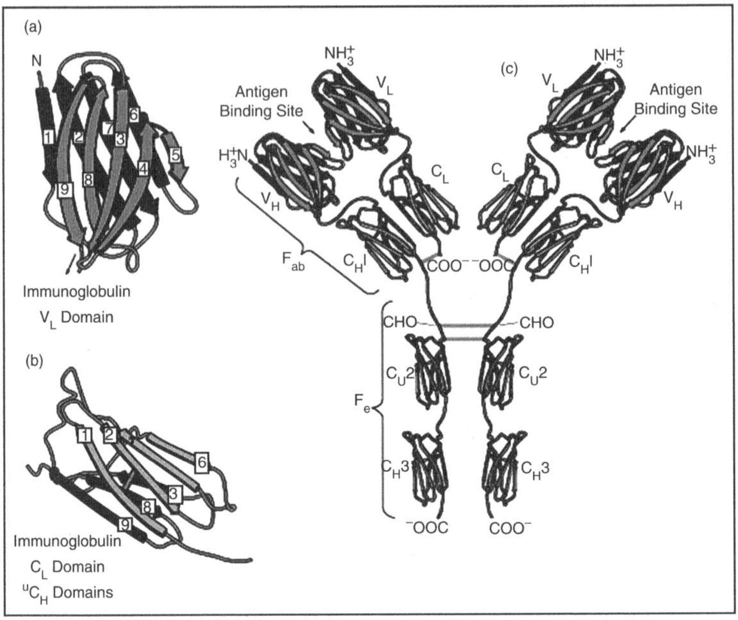

Fig. The Characteristic "Collapsed b-barrel Domain" known as The *Immunoglobulin Fold*. The b-barrel Structures for Both

a. Variable and
b. Constant regions are shown.
c. A schematic diagram of the 12 collapsed b-barrel domains that make up an IgG molecule.

CHO Indicates the Carbohydrate Addition Site; F_{ab} Denotes One of the Two Antigen-binding Fragments of IgG, and F_c, the Proteolytic Fragment Consisting of the Pairs of C_H2 and C_H3 Domains.

In the immunoglobulin genes, the arrangement of exons correlates with protein structure. In terms of its tertiary structure, the immunoglobulin G molecule is composed of 12 discrete*collapsed b-barrel domains,* each domain having a *Greek key*motif. The characteristic structure of this domain is referred to as the immunoglobulin fold. Each of IgG's two heavy chains contributes four of these domains and each of its light chains contributes two. The four*variable-region* domains (one on each chain) are encoded by multiple exons, but the eight constant-region domains are each the product of a single exon. All of these *constant-region* exons are derived from a single ancestral exon encoding an immunoglobulin fold. The major variable-region exon probably derives from this ancestral exon also. Contemporary immunoglobulin genes are a consequence of multiple duplications of the ancestral exon.

The discovery of variability in amino acid sequence in otherwise identical polypeptide chains was surprising and almost heretical to protein chemists. For geneticists, it presented a genuine enigma. They noted that mammals, which can make millions of different antibodies, don't have millions of different antibody genes. How can the mammalian genome encode the diversity seen in L and H chains?

MOLECULAR NATURE OF MUTATION

Genes are normally transmitted unchanged from generation to generation, owing to the great precision and fidelity with which genes are copied during chromosome duplication. However, on rare occasions, genetically heritable changes (mutations) occur that result in altered forms. Most mutated genes function less effectively than the unaltered, wild-type allele, but occasionally mutations arise that give the organism a selective advantage. When this occurs, they are propagated to many offspring. Together with recombination, mutation provides for genetic variability within species and, ultimately, the evolution of new species.

Mutations change the sequence of bases in DNA, either by the substitution of one base pair for another (so-called point mutations), or by the insertion or deletion of one or more base pairs (insertions and deletions).

POINT MUTATIONS

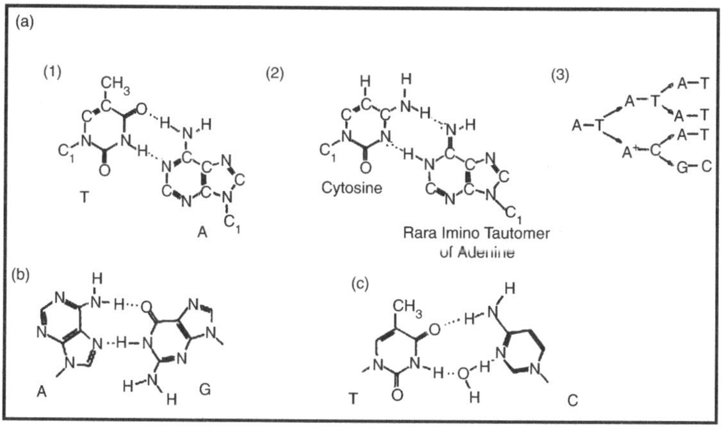

Fig. Point Mutations Due to Base Mispairings.

a. An example based on tautomeric properties. The rare imino tautomer of adenine base-pairs with cytosine rather than thymine.
    1. The normal A-T base pair.
    2. The A*-C base pair is possible for the adenine tautomer in which a proton has been transferred from the 6-NH_2 of adenine to N-1.

3. Pairing of C with the imino tautomer of A (A*) leads to a transition mutation (A-T to G-C) appearing in the next generation.

b. A in the syn conformation pairing with G (G is in the usual anti conformation).

c. T and C form a base pair by H-bonding interactions mediated by a water molecule.

Point mutations are the class of mutations in which one base pair is substituted for another. The two possible kinds of point mutations aretransitions, where one purine (or pyrimidine) is replaced by another, as in A $\rightarrow$ G (or T $\rightarrow$ C), and transversions, where a purine is substituted for a pyrimidine or *vice versa.*

Point mutations arise by the pairing of bases with inappropriate partners, by the introduction of base analogs into DNA, or by chemical mutagens. Bases may rarely mispair, either because of their tautomeric properties, or because of other influences (such as purines flipping from anti to syn conformations, or H_2O molecules serving as bridging H-bond donor/acceptors between two mispaired pyrimidines). Even in mispairing, the C_1'-C_1' distances between bases must still be close to that of a Watson-Crick base pair to maintain the mismatched base pair in the double helix. In tautomerization, for example, an amino group (-NH_2), usually an H-bond donor, can tautomerize to an imino form (=NH) and become an H-bond acceptor. Or a keto group (C=O), normally an H-bond acceptor, can tautomerize to an enol C-OH, an H-bond donor. Proofreading mechanisms operating during DNA replication catch most mispairings. The frequency of spontaneous mutation in both E. *coli* and fruit flies (*Drosophila melanogaster*) is about 10^{-10} per base pair per replication.

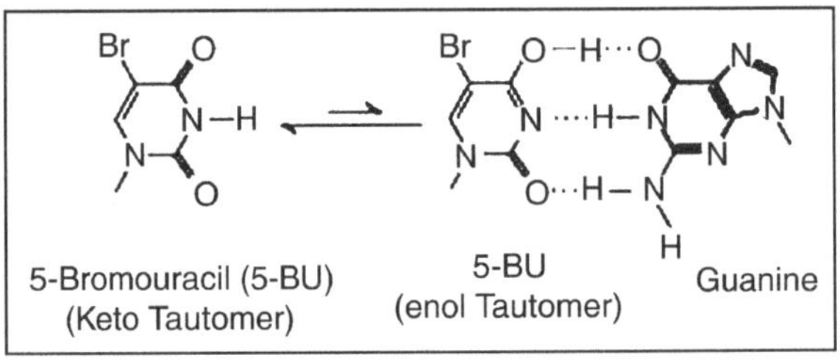

Fig. 5-Bromouracil Usually Favours the Keto Tautomer that Mimics the Basepairing Properties of Thymine, but it Frequently Shifts to the Enol form, Whereupon it can Base-pair with Guanine, Causing a T-A to C-G Transition.

MUTATIONS INDUCED BY BASE ANALOGS

Base analogs that become incorporated into DNA can induce mutations through changes in base-pairing possibilities. Two examples are5-bromouracil (5-BU) and 2-aminopurine (2-AP). 5-Bromouracil is a thymine analog and becomes inserted into DNA at sites normally occupied by T; its 5-Br group sterically resembles thymine's 5-methyl group. However, because 5-

BU frequently assumes the enol tautomeric form and pairs with G instead of A, a point mutation of the transition type may be induced. Less often, 5-BU is inserted into DNA at cytosine sites, not T sites.

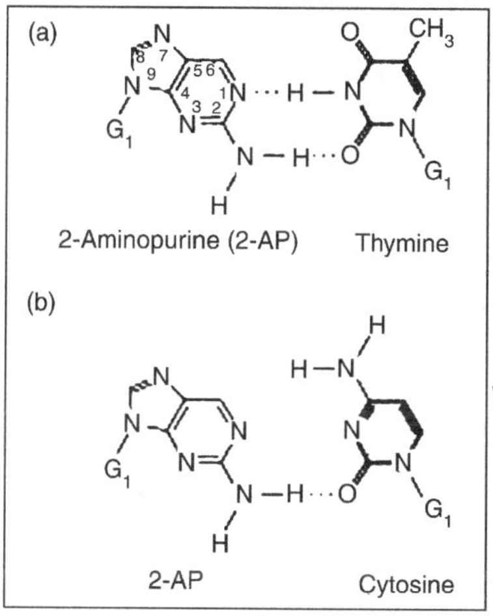

a. 2-Aminopurine normally base-pairs with T, but

b. May also pair with cytosine through a single hydrogen bond.

Then, if it base-pairs in its keto form, mimicking T, a C-G to T-A transition ensues. The adenine analog, 2-aminopurine (recall adenine is 6-aminopurine) normally behaves like A and base-pairs with T. However, 2-AP can form a single H bond of sufficient stability with cytosine that occasionally C replaces T in DNA replicating in the presence of 2-AP. Hypoxanthine is an adenine analog that arises *in situ* in DNA through oxidative deamination of A. Hypoxanthine base-pairs with cytosine, creating an A-T to G-C transition.

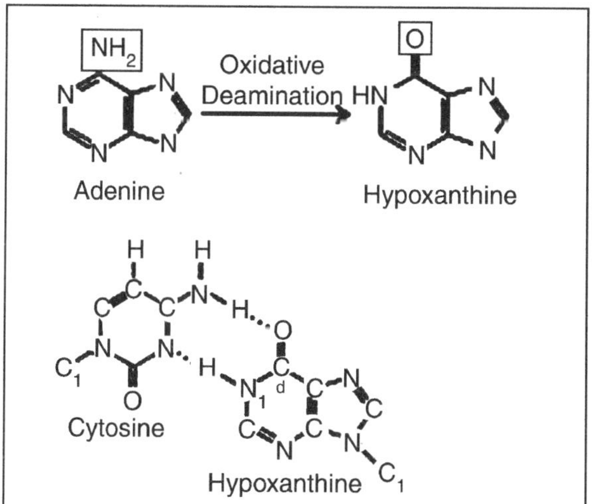

Hypoxanthine is in its Keto Tautomeric form Here

CHEMICAL MUTAGENS

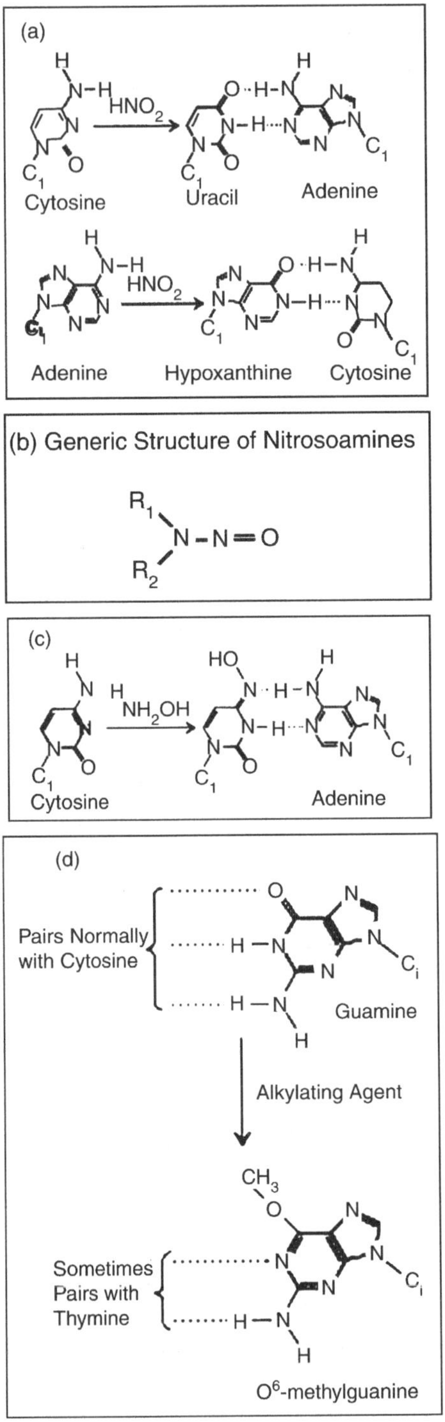

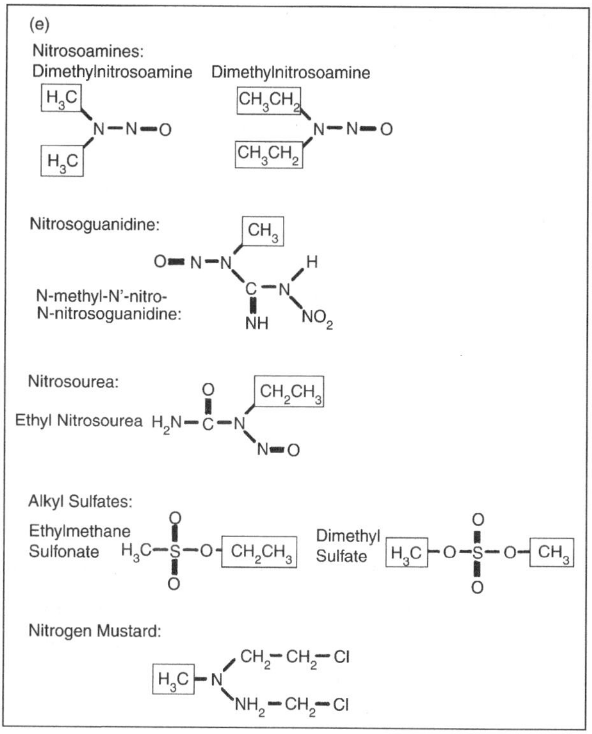

Fig. Chemical Mutagens.

a. HNO_2 (nitrous acid) converts cytosine to uracil and adenine to hypoxanthine.

b. Nitrosoamines, organic compounds that react to form nitrous acid, also lead to the oxidative deamination of A and C.

c. Hydroxylamine (NH_2OH) reacts with cytosine, converting it to a derivative that base-pairs with adenine instead of guanine. The result is a C-G to T-A transition.

d. Alkylation of G residues to give 0_6-methylguanine, which base-pairs with T.

e. Alkylating agents include *nitrosoamines, nitrosoguanidines, nitrosoureas, alkyl sulfates,* and *nitrogen mustards.*

Chemical mutagens are agents that chemically modify bases so that their base-pairing characteristics are altered. For instance,*nitrous acid* (HNO_2) causes the oxidative deamination of primary amine groups, found in adenine and cytosine. Oxidative deamination of cytosine yields uracil, which base-pairs the way T does and gives a C-G to T-A transition. *Hydroxylamine* specifically causes C-G to T-A transitions because it reacts specifically with cytosine, converting it to a derivative that base-pairs with adenine instead of guanine. Alkylating agents are also chemical mutagens. Alkylation of reactive sites on the bases with methyl or ethyl groups alters their H-bonding and hence base pairing. For example, methylation of O^6 on guanine (giving O^6-methylguanine) causes this G to mispair with thymine, resulting in a G-C to A-T transition. Alkylating agents can also induce point mutations of the transversion type. Alkylation of N^7 of guanine labilizes its N-glycosidic bond, which leads to elimination of the purine ring, creating a gap in the base sequence. An enzyme, apurinic acid endonuclease, then cleaves the sugar-phosphate backbone of the DNA on the 5'-side, and the gap can be repaired by enzymatic removal of the 5'-sugar phosphate and insertion of a new nucleotide. A transversion results if a pyrimidine nucleotide is inserted in place of the purine during enzymatic repair of this gap.

INSERTIONS AND DELETIONS

The addition or removal of one or more base pairs leads to *insertion* or *deletion* mutations, respectively. Either shifts the triplet reading frame of codons, causing frameshift mutations (misincorporation of all subsequent amino acids) in the protein encoded by the gene. Such mutations can arise when flat aromatic molecules such as *acridine orange* insert themselves between successive bases in one or both strands of the double helix. This insertion or, more aptly, intercalation, doubles the distance between the bases as measured along the helix axis. This distortion of the DNA results in bases being inappropriately inserted or deleted when the DNA is replicated. Disruptions that arise from the insertion of a transposon within a gene also fall in this category of mutation.

THE ORGANIZATION OF IMMUNOGLOBULIN GENES

The answer to the enigma of immunoglobulin sequence diversity is found in the organization of the immunoglobulin genes. The genetic information for an immunoglobulin polypeptide chain is scattered among multiple gene segments along a chromosome in germline cells (sperm and eggs). During vertebrate development and the formation of B lymphocytes, these segments are brought together and assembled by DNA rearrangement (that is, genetic recombination) into complete genes.

DNA rearrangement, or gene reorganization, provides a mechanism for generating a variety of protein isoforms from a limited number of genes. DNA rearrangement occurs in only a few genes, namely, those encoding the antigen-binding proteins of the immune response¾the immunoglobulins and the T cell receptors.

The gene segments encoding the amino-terminal portion of the immunoglobulin polypeptides are also unusually susceptible to mutation events. The result is a population of B cells whose antibody-encoding genes collectively show great sequence diversity even though a given cell can make only a limited set of immunoglobulin chains. Hence, at least one cell among the B cell population will likely be capable of producing an antibody that will specifically recognize a particular antigen.

DNA Rearrangements Assemble an L-Chain Gene by Combining Three Separate Genes

He organization of various immunoglobulin gene segments in the mouse genome is shown in Figure. L-chain variable-region genes are assembled from two kinds of germline genes, V_L andJ_L(J stands for *joining*). In mammals, there are two different families of L-chain genes, the κ, or kappa, gene family and the λ, or lambda, gene family; each family has V and J members. These families are on different chromosomes. In mice, 90 per cent of the L chains are κ chains; λ L chains are a minor component. Mice have four functional JV_k genes (and a fifth non-functional one); these J genes lie 2.5 to 4 kb upstream from the single C_k gene that encodes the L-chain constant region. There are at least 200 V_k genes, each with its own $L_k$segment for encoding the L-chain leader peptide that targets the L chain to the endoplasmic reticulum for IgG assembly and secretion. (This leader peptide is cleaved once the L chain reaches the ER lumen.) The λ family of L-chain genes is organized a little differently, with only two V_λ genes, each of which is followed downstream by a pair of J_l-C_l units. In different mature B lymphocyte cells, $V_k$and J_k genes have joined in different combinations, and along with the CV_k gene, form complete LV_k chains with a variety of V_k regions. However, any given B lymphocyte expresses only one V_k-J_k combination. Construction of the mature B lymphocyte L-chain gene has occurred by DNA rearrangements that combine three genes (L-$V_{\kappa,\lambda}$, $J_{\kappa,\lambda}$, $C_{\kappa,\lambda}$) to make one polypeptide!

DNA Rearrangements Assemble an H-Chain Gene by Combining Four Separate Genes

The first 98 amino acids of the 108-residue, H-chain variable region are encoded by a V_H gene. Each V_H gene has an accompanying $L_H$gene that encodes its essential leader peptide. It is estimated that there are from 200 to 1000 V_H genes and they can be subdivided into eight distinct families based on nucleotide sequence homology. The members of a particular V_H family are grouped together on the chromosome, separated from one another by 10 to 20 bp. In assembling a mature H-chain gene, a V_H gene is joined to a D gene (*D* for *diversity*), which encodes amino acids 99 to 113 of the H chain. These amino acids comprise the core of the third CDR in the variable region of H chains. The V_H-D gene assemblage is linked in turn to a J_H gene, which encodes the remaining part of the variable region of the H chain. The V_H, D, and J_H genes are grouped in three separate clusters on the same chromosome. The four J_H genes lie 7 kb upstream of the eight C genes, the closest of which is Cm. Any of four C genes may encode the constant region of IgG H chains: $C_{\gamma1}$ $C_{\gamma2a}$,

$C_{\gamma 2b}$, and $C_{\gamma 3}$. Each C gene is composed of multiple exons. Ten to twenty D genes are found 1 to 80 kb farther upstream. The V_H genes lie even farther upstream. In B lymphocytes, the variable region of a heavy-chain gene is composed of one each of the L_H-V_H genes, a D gene, and a J_H gene joined head to tail. Because the H-chain variable region is encoded in three genes and the joinings can occur in various combinations, the heavy chains have a greater potential for diversity than the light-chain variable regions that are assembled from just two genes (for example, L_k-V_k and J_k). In making heavy-chain genes, four genes have been brought together and reorganized by DNA rearrangement to produce a single polypeptide!

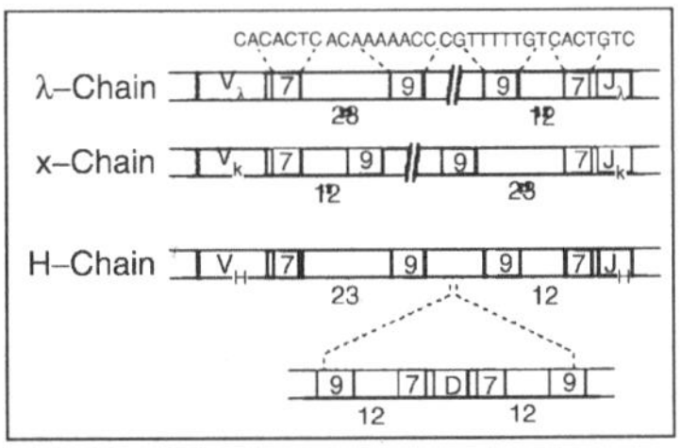

Fig. Consensus Elements are Located Above and Below Germline Variable-region Genes that Recombine to form Genes Encoding Immunoglobulin Chains. These Consensus Elements are Complementary and are Arranged in a Heptamer-nonamer, 12-bp to 23-bp Spacer Pattern.

The Mechanism of V-J and V-D-J Joining in Light- and Heavy-Chain Gene Assembly

Specific nucleotide sequences adjacent to the various variable-region genes suggest a mechanism in which these sequences act as joining signals. All germline V and D genes are followed by a consensus CACAGTG heptamer separated from a consensus ACAAAAACC non-amer by a short, non-conserved 23-bp spacer.

Likewise, all germline D and J genes are immediately preceded by a consensus GGTTTTTGT non-amer separated from a consensus CACTGTG heptamer by a short non-conserved 12-bp spacer. Note that the consensus elements downstream of a gene are complementary to those upstream from the gene with which it recombines. Indeed, it is these complementary consensus sequences that serve as recombination recognition signals (RSSs) and determine the site of recombination between variable-region genes. Functionally meaningful recombination happens only where one has a 12-bp spacer and the other has a 23-bp spacer. Lymphoid cell-specific recombination-activating gene proteins 1 and 2 (RAG1 and RAG2) recognize and bind at these RSSs, presumably through looping out of the 12- and 23-bp spacers and alignment of the homologous heptamer and non-amer regions. RAG1 and RAG2 together function as the V(D)J recombinase. The similarity between the organization of flanking repeats in immunoglobulin genes and the reaction catalyzed by RAG1/RAG2 proteins suggests that these genes and the RAG recombinase may have evolved from an ancestral transposon.

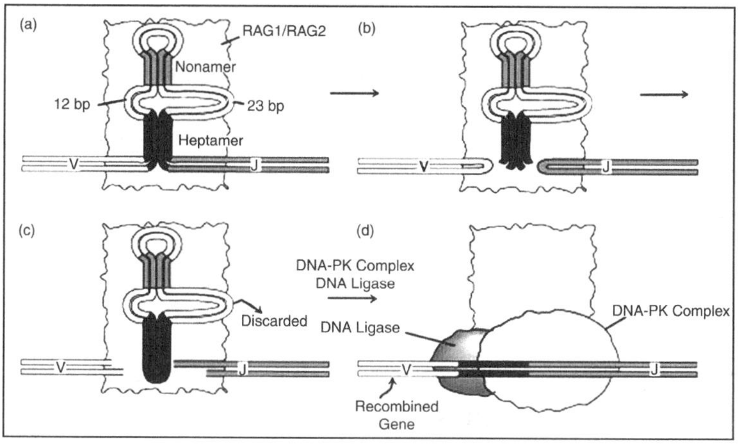

Fig. Model for V(D)J Recombination. A RAG1:RAG2 Complex is Assembled on DNA in the Region of Recombination Signal Sequences

a. This complex introduces double-stranded breaks in the DNA at the borders of protein-coding sequences and the recombination signal sequences

b. The products of RAG1:RAG2 DNA cleavage are novel: the DNA bearing the recombination signal sequences has blunt ends, whereas the coding DNA has hairpin ends.

IMPRECIS

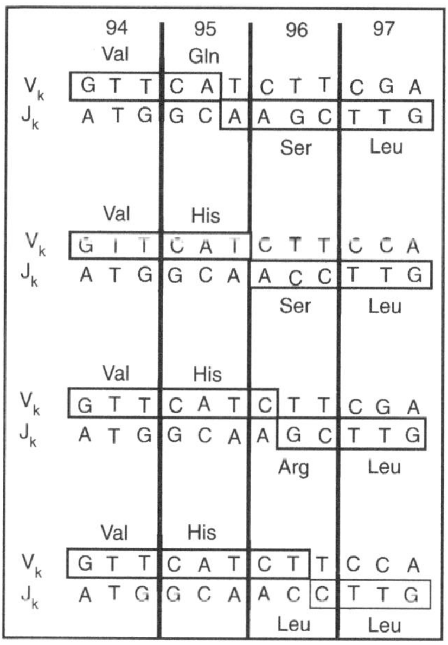

Fig. Recombination between the V_k and J_k Genes can Vary by Several Nucleotides, giving Rise to Variations in Amino Acid Sequence and Hence Diversity in Immunoglobulin L Chains.

e Joining

Joining of the ends of the immunoglobulin-coding regions during gene reorganization is somewhat imprecise. This imprecision actually leads to even greater antibody diversity because new coding arrangements result. Position 96 in k chains is typically encoded by the first triplet in the J_k element. Most k chains have one of four amino acids here, depending on which J_k gene was recruited in gene assembly. However, occasionally only the second and third bases or just the third base of the codon for position 96 is contributed by the J_k gene, with the other one or two nucleotides supplied by the V_k segment

So, the precise point where recombination occurs during gene reorganization can vary over several nucleotides, creating even more diversity perhaps 300 V_k genes, 4 J_k genes, 200 V_H genes, 12 D genes, and 4 J_H genes, the number of possible combinations is given by 300 × 4 × 200 × 12 × 4. Thus, greater than 10^7 different antibody molecules can be created from roughly 500 or so different mouse variable-region genes. Including the possibility for V_k-J_k joinings occurring within codons adds to this diversity, as does the high rate of somatic mutation associated with the variable-region genes. (Somatic mutations are mutations that arise in diploid cells and are transmitted to the progeny of these cells within the organism, but not to the offspring of the organism.) Clearly, gene rearrangement is a powerful mechanism for dramatically enhancing the protein-coding potential of genetic information.

6

Recombinant DNA Technology of Molecular Cytogenetics

INTRODUCTION

Recombinant DNA technology has provided methods for generating highly specific DNA and RNA molecular probes that recognize specific genetic sequences with high reliability. Similar advances in molecular technology have also provided methods for labeling these probes and identifying their binding sites when hybridized directly to chromosome preparations.

Combining molecular genetics with traditional cytogenetics has established a new era in clinical genetics, that of molecular cytogenetics. The advantages of this approach are based on the rapidity of diagnosis, the highly specific and reliable recognition of the presence or absence of a particular DNA sequence for which the probe has nucleotide sequence complementarity, the simplicity of the methodology, the stability of the signal, and the simple microscopy required for analysis.

Multiple gene sites can also be screened through use of sets of multiple probes. Contiguously related probe sets can evaluate larger DNA segments, such as a band, several bands or even entire chromosomes. This technology is likely to make significant changes in the role of cytogenetics in the practice of obstetrics and gynecology over the next decade. Molecular cytogenetic analysis involves simple DNA denaturation followed by probe hybridization and label recognition.

These studies can be performed on the chromosomes of peripheral lymphocytes, cancer cells, amniotic fluid cells and chorionic villi. It also is possible to perform this type of cytogenetic analysis on non-dividing, or interphase, cells, bypassing the need for tissue culturing and allowing analysis of biologic tissues heretofore extremely difficult because of their low mitotic indices (*e.g.* solid tumors).

This approach is in stark contrast to conventional cytogenetic techniques, which can involve extensive tissue culturing to generate sufficient high-quality metaphase spreads to obtain suitable karyotypes. Originally, detection of a specific nucleic acid sequence in cytologic preparations involved the use of DNA or RNA probes labeled with radioactive isotopes (32P or 35S), which were hybridized to metaphase chromosome spreads. A homologous site of the chromosomal DNA sequence was indicated by location of silver grains along the chromatids after autoradiography and staining. This approach, although highly specific and sensitive, had considerable drawbacks. The method was tedious, technically demanding, and time consuming, with exposures to a radioactive nucleic acid probe requiring weeks to months.

The use of radioactive probes required isolated laboratories and stringent safety precautions in their use and disposal. The next technical improvement was the introduction of non-radioactive fluorescence methods in combination with the procedure of *in situ* hybridization of a nucleic acid probe to target DNA. Fluorescent *in situ* hybridization (FISH) is a technique whereby a nucleic acid probe is typically labeled with nucleotides that are modified with biotin or digoxigenin. Hybridization of the probe to the specimen required several hours to overnight.

Visualization of hybridized probes for fluorescence microscopy is achieved using fluorochromes such as fluorescein or Texas red, which are conjugated to avidin for detection of biotin or anti-DIG for detection of digoxigenin. This method of indirect labeling of FISH probes now has been improved by directly labeling the FISH probes. Directly labeled FISH probes are reliable, simple to use, and permit a rapid completion of cytogenetic analysis within 1 to 2 hours. There are three general types of probes that address different applications of the FISH technology.

There are chromosome-specific repetitive sequence probes for centromeres, which are suited to detect numeric chromosome aberrations in interphase cells; whole chromosome probes which can screen for different chromosomes simultaneously; and probes for specific chromosome regions, with a theoretical limit of resolution of a single base pair. FISH has been applied to prenatal genetic testing for autosomal trisomies using interphase cells, for interpreting chromosomal rearrangements in normal and neo-plastic tissues, and for evaluating specimens not suitable for conventional cytogenetic analysis.

Specific applications include the identification of marker chromosomes, determination of chromosome mosaicism, rapid assessment of polyploidy, and exclusion of a chromosome disorder in autolyzed tissue from a fetus with structural anomalies. Analysis of cancer cells by FISH technology showed many specific deletions at different loci and referred to as *loss of heterozygosity.* Several syndromes have also been associated with deletions detected by molecular cytogenetic techniques, including Kallmann syndrome (*i.e.* hypogonadotropic hypogonadism and anosmia), congenital ichthyosis due to steroid sulfatase deficiency seen with a deletion involving chromosome band Xp22.3, Duchenne muscular dystrophy, chronic granulomatous disease, retinitis pigmentosa, and the McLeod erythrocyte abnormality, seen in some patients with a deletion involving Xp21.

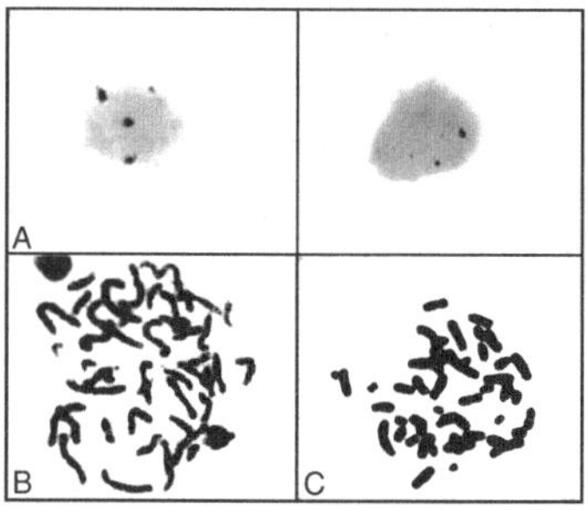

Fig. Application of Fluorescent *In Situ* Hybridization (FISH) to Cytogenetic Analysis.

A. Interphase FISH analysis of chorionic villi. *Right panel:* Three orange signals indicate three copies of chromosome 21 (trisomy 21); two green signals indicate two copies of chromosome 13. *Left panel:* Two aqua signals indicate two copies of chromosome 18; two green signals indicate two copies of the X chromosome (XX).

B. Chromosome paint DNA FISH probe applied in the case of a morphologically distinct, familial variant of chromosome 20 (*right*). A variant was observed in amniotic fluid cells and in the peripheral blood cells of the healthy father. A normal infant resulted (WCP 20, Vysis).

C. FISH probe applied in suspected case of DiGeorge syndrome. DiGeorge syndrome is a complex malformation pattern characterized by congenital heart defects, hypoplastic or absent thymus and parathyroid glands and dysmorphic facies. Metaphase cell demonstrating single signal (orange) indicative of a deletion of 22q11.2 that is common in DiGeorge syndrome. A gene control probe maps close to the end of chromosome 22 (22q13) shows two signals (LSI DiGeorge/VCFS Region Probe, Vysis).

The most direct method of physical gene mapping, the identification of the chromosome locus of a human gene, is accomplished by *in situ* hybridization. FISH has been used to detect single copy genes in interphase cells and unique sequences in G-banded and R-banded chromosomes. Commercial availability of FISH probes has greatly expanded the use of *in situ* hybridization in basic research on chromosome structure and function and in the clinical practice of reproductive and medical genetics.

DNA IS COMPOSED OF POLYNUCLEOTIDE CHAINS

The most important feature of DNA is that it is usually composed of two polynucleotide chains twisted around each other in the form of a double helix.

The upper part of the figure presents the structure of the double helix shown in a schematic form. Note that if inverted 180°, the double helix looks superficially the same, due to the complementary nature of the two DNA strands. The space-filling model of the double helix, in the lower part of the figure, shows the components of the DNA molecule and their relative positions in the helical structure. The backbone of each strand of the helix is composed of alternating sugar and phosphate residues; the bases project inward but are accessible through the major and minor grooves.

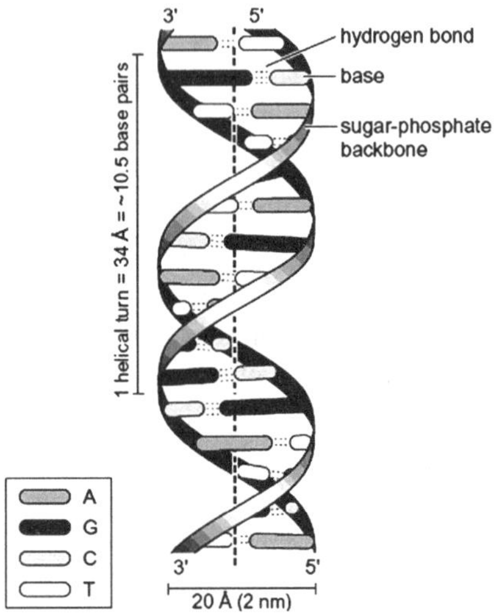

Fig. The Helical Structure of DNA

Let us begin by considering the nature of the nucleotide, the fundamental building block of DNA. The nucleotide consists of a phosphate joined to a sugar, known as 2'-deoxyribose, to which a base is attached. The sugar is called *2'*-deoxyribose because there is no hydroxyl at position 2' (just two hydrogens).

Note that the positions on the ribose are designated with primes to distinguish them from positions on the bases. We can think of how the base is joined to 2'-deoxyribose by imagining the removal of a molecule of water between the hydroxyl on the 1' carbon of the sugar and the base to form a glycosidic bond. The sugar and base alone are called a nucleoside. Likewise, we can imagine linking the phosphate to 2'-deoxyribose by removing a water molecule from between the phosphate and the hydroxyl on the 5' carbon to make a 5' phospho-monoester.

Adding a phosphate (or more than one phosphate) to a nucleo*s*ide creates a nucleo*t*ide. Thus, by making a glycosidic bond between the base and the sugar, and by making a phosphoester bond between the sugar and the phosphoric acid, we have created a nucleotide. Nucleotides are, in turn, joined to each other in polynucleotide chains through the 3' hydroxyl of 2'-deoxyribose of one nucleotide and the phosphate attached to the 5' hydroxyl of another nucleotide.

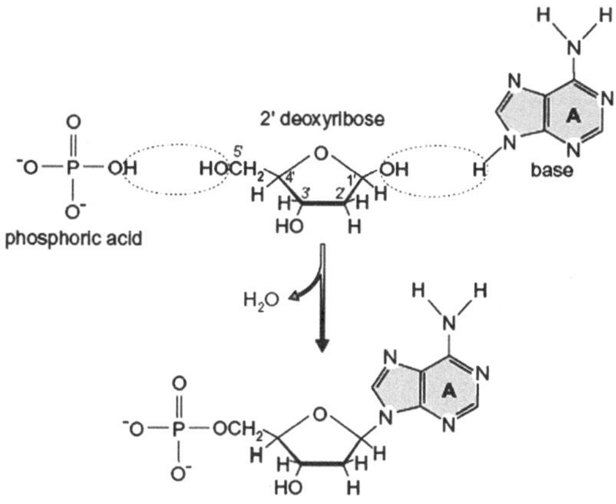

Fig. Formation of Nucleotide by Removal of Water

This is a phosphodiester linkage in which the phosphoryl group between the two nucleotides has one sugar esterified to it through a 3' hydroxyl and a second sugar esterified to it through a 5' hydroxyl. Phosphodiester linkages create the repeating, sugar-phosphate backbone of the polynucleotide chain, which is a regular feature of DNA. In contrast, the order of the bases along the polynucleotide chain is irregular. This irregularity as well as the long length is, as we shall see, the basis for the enormous information content of DNA. The phosphodiester linkages impart an inherent polarity to the DNA chain. This polarity is defined by the asymmetry of the nucleotides and the way they are joined. DNA chains have a free 5' phosphate or 5' hydroxyl at one end and a free 3' phosphate or 3' hydroxyl at the other end. The convention is to write DNA sequences from the 5' end (on the left) to the 3' end, generally with a 5' phosphate and a 3' hydroxyl.

	Base Adenine	Nucleoside Adenosine	Nucleotide Adenosine 5'-phosphate	Deoxynucleotide Deoxyadenosine 5' phosphate
Structure[a]	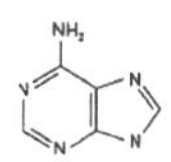	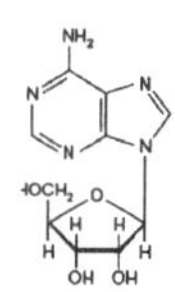	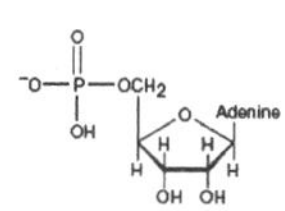	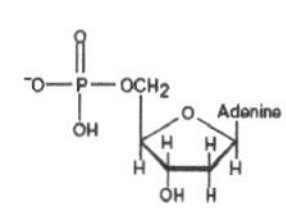
M.W.	135.1	267.2	347.2	331.2

MOLECULAR CYTOGENETICS

Major genome research programmes are underway to isolate expressed genes of corn and to DNA sequence at least the gene-rich regions. We are developing a highly efficient method for mapping DNA sequences to chromosome and sub-chromosomal regions based on lines derived from crosses between oats and corn. Partial hybrids have been obtained that contain an entire set of oat chromosomes along with one corn chromosome; lines are available that have each chromosome of corn individually added to the oat genome.

These lines are now being used to generate derivatives with only a subregion of the corn chromosome. This genetic material as well as B73xMo17 derived lines will be used to learn more about the subgenomic structure of corn and to test whether the degree of heteroallelism within and between the subgenomes contribute to heterosis. A pathogenic bacteria, called E. coli O157:H7, can contaminate hamburger and cause serious hemoraging. The contamination largely comes from the feces of cattle. Strains of E. coli have been selected that will kill the O157:H7 strain.

Two genes appear to be involved and this project will test their efficacy in corn. Wild rice is a crop under domestication. The seed falls (shatters) from the panicle upon maturation and is not available to be harvested. Storms can cause major losses of wild rice yield at harvest.

The shattering trait is considered a domestication trait that has been modified years ago in most major crops; the trait still exists in wild rice varieties. We have discovered major genes for shattering and have molecular genetic markers that should allow more efficient breeding for shattering resistant varieties. Methionine is an essential amino acid required in the diets of non-ruminant animals. Poultry have high demands for methionine accounting for why synthetic methionine is added to poultry rations in the U.S. A line of corn was selected that is high in methionine. Several generations of crossing, backcrossing, and selfing has resulted in relatively homozygous lines potentially high in methionine,

The specific objectives are:

1. Develop efficient mapping system for corn DNA sequences.
2. Learn more about the subgenomic structure of corn and relate the information to productivity.
3. Manipulate corn to produce compounds toxic to an enterohemoraghic bacterium (E. coli O157:H7).
4. Provide molecular genetic methods to the breeding programme to enhance the selection of varieties of wild rice with improved seed shattering resistance.
5. Finalize the selection and release of high methionine lines of corn.

We have produced lines of oats with a single corn chromosome added. These addition lines are available for all 10 corn chromosomes, individually present in the oat background. These lines are being used extensively for mapping DNA sequences to chromosome.

The oat-corn addition lines are being irradiated to produce radiation hybrid (RH) lines that have only part of the corn chromosome present. These lines allow much more refined mapping of corn DNA sequences. Future research will involve continued development and characterization of the RHs, mapping

ESTs, cytogenetic transmission studies of the various lines, and obtaining the transmission of the chromosome 10 addition and formation of RHs.

Molecular genetic analysis of corn has shown that its genome is composed of major duplicated regions suggesting an ancient allopolyploid origin.

The potential exists for many loci to not only be duplicated but to have different alleles at each locus. Maximum productivity could be the result of maximizing the number of alleles at these loci. We will screen several genetic markers in order to determine Mo17xB73 lines containing duplicate regions containing various combinations of the parental genotypes.

Yield tests will ultimately be performed on selected lines with maximum heteroallelism in regions known to possess major yield QTLs. Our molecular genetics research on wild rice has resulted in the first molecular marker map and the placement of QTLs for several traits on the map. A PCR based marker has been developed that tags the most major QTL with polymorphism between shattering and non-shattering types.

This QTL accounts for about 40 per cent of the variation in seed shattering in wild rice. A marker-assisted selection programme will be initiated using the molecular genetic marker for the major QTL and others as possible. Escherichia coli O157:H7 is a pathogenic strain of E. coli found in the digestive tract of beef cattle and causes food borne illness through contamination of beef during the slaughtering process. The goal of this project is to target the pathogen at the feed source or farm level to prevent contamination by developing a transgenic maize line that expresses a protein (termed "colicin") toxic to this strain of E. coli.

Colicin (produced by the cea gene) degrades DNA, and an immunity protein (produced by the cei gene) is required to prevent non-specific DNA degradation. Thus both genes are required to accomplish our goal. A screening procedure was developed in our lab that provides a screen for corn with elevated levels of methionine in the endosperm. A line was discovered, called BSSS5, that is resistant to lysine-threonine inhibition as a whole kernel but is inhibited when only the embryo is cultured.

The endosperm in this case possesses sufficient methionine to prevent the inhibition. BSSS 53 was found to produce a zein-2 protein with 23 per cent of the amino acid residues as methionine. We have attempted to backcross the high-methionine trait into three different genetic backgrounds (B73, Mo17, and A632) and have some lines almost ready to release.

Oat-corn addition lines have been developed for every corn chromosome. These strains allow - in a single experiment - the rapid mapping of corn DNA sequences to their respective chromosome.

These strains resulted from more than 100,000 oat x corn crosses and isolation of over 7300 immature embryos. We extended the available genetic backgrounds by recovering oat-corn addition lines with B73 or Mo17. Between the two inbreds, we have a complete addition line set; the B73 addition lines will be useful in assembling the corn genome.

The DOE Joint Genome Institute will sequence chromosome10 using our Mo17 chromosome 10 oat-maize addition line. We also have added to our collection of radiation hybrid lines for each chromosome. These lines allow the placement of corn genes to specific regions of the chromosome. Corn gene expression/silencing in the oat-corn addition lines is being tested using the DNA chip technology. Out of 17,000 tested sequences, 890 were expressed in the chromosome 5 addition.

About half of these sequences are of known position and are either placed on chromosome 5 or in a duplicated region of the genome. We further pursued the possibility of introducing C4 photosynthesis to oat with these materials.

Addition lines with both chromosomes 6 and 9 (carrying genes for key C4 enzymes) were tested for an effect on CO2 compensation point. A statistically significant change occurred in compensation point but not an impressive one. We will next investigate a third chromosome that has an effect on leaf cell morphology (known as Kranz Anatomy) involved in photosynthetic efficiency.

NIR analyses of self-pollinations of a new high oil accession indicate that the kernels contain 20 per cent oil. A QTL analysis is underway. There is considerable interest in this material as a source of biodiesel and ethanol.

Using the molecular genetic map we published for wild rice, a PCR-based marker system has been identified for use in marker assisted breeding for the seed shattering trait. Genetic tests indicate a close marker/trait linkage.

The recent outbreak of the pathogenic E. coli 0157:H7 on spinach emphasizes the importance of having control measures to minimize such occurrences. We are developing what might be considered a prototype approach of using corn feed to alter the microflora in the intestinal tract as a means to reduce the presence of this pathogenic bacterium at its source.

Whole corn plants were regenerated from appropriate calli and found to carry the colicin gene, express the RNA and protein, and have anti-microbial activity against E. coli O157:H7. An extensive bioinformatics analysis has been performed of 1100 DNA sequences involved in human cancers to determine if they are present in plants. P53 is a protein involved in more than 50 per cent of human cancers. We found a p53 binding sequence in the promoter of the MAD1 (Mitosis Arrest Deficiency) gene in human; MAD1 acts as a mitotic checkpoint to ensure chromosome stability. A MAD1 homolog was found in maize. Laboratory analyses of knockout mutants of MAD1 in Arabidopsis are underway.

Crop productivity is at least half due to genetic advances. Continued understanding of genetics and the application of molecular genetics will allow continued progress in food production. In addition, additional health benefits and identification of unique biofuel resources will be forthcoming.

ROUTINE OF CYTOGENETIC ANALYSIS

A routine cytogenetic analysis involves evaluating 15 to 20 cells to determine their modal chromosome number and assessing the structural integrity of each chromosome in the complement. The modal chromosome number can be determined by counting chromosomes through the microscope. Identification of an extra or missing chromosome is possible by microscopic examination, and the finding is confirmed through the preparation of a set of karyotypes. A karyotype is a photographic or computer-generated representation of the chromosomes in a cell arranged according to size, centromeric position, and banding pattern.

Assessment of the structural integrity of each chromosome in the complement can also be made through the microscope, and gross structural chromosome aberrations can be detected as well. However, a set of two or more karyotypes always must be prepared. Cells with well-prepared chromosomes that exhibit a total of 350 to 400 bands in the complement are considered an essential part of any routine cytogenetic analysis. A standard chromosome analysis, therefore, requires that the final interpretation be based on a chromosome count of at least 15 to 20 cells, the preparation and careful evaluation of two to three karyotypes consisting of well-banded chromosomes, and the exclusion of any structural chromosome aberrations. Subtle chromosome aberrations, such as the gain or loss of a single band, usually are detected as the result of a detailed analysis of the banding pattern of chromosomes magnified thousands of times by means of photographic or computer processing. In this way, homologous chromosomes can be examined meticulously and compared for small but clinically significant structural differences.

The number of cells analysed varies according to the reason for referral, the initial cytogenetic findings, the protocol of each laboratory, and the quality of the chromosome morphology. If an abnormal chromosomal complement is present in all of the cells of a specimen, analysis of a small number of cells may be sufficient for diagnostic purposes. The presence of a single cell with a chromosome aberration usually is attributed to an error in cell division during tissue culture or a technical artifact arising in the course of harvesting. In most cases, the parents are informed that such a finding is not associated with anomalous fetal development. Nevertheless, the presence of an abnormal cell may necessitate analysing 20 to 50 or more cells to distinguish between true mosaicism, which could be responsible for fetal maldevelopment, and pseudomosaicism (*false mosaicism*) caused by technical artifact or an *in vitro* error. True chromosome mosaicism occurs when there is a mixture of chromosomally normal and abnormal cells.

True mosaicism may be particularly difficult to detect if present in low frequency. Most cytogenetic analyses are statistically based on an evaluation of a minimum of 15 to 20 cells to reduce the chance to less than 5 per cent probability that true mosaicism will not be detected. A counseling dilemma arises when chromosome mosaicism is observed after chorionic villus sampling or amniocentesis: does the mosaicism detected in the villi or cultured amniotic fluid cells represent the chromosome complement of the fetus? Confirmation of chromosome mosaicism may require additional invasive procedures, such as percutaneous umbilical blood sampling (PUBS), which would place the patient at additional obstetric risk. Even attempts at confirmation of chromosome mosaicism may fail. For example, mosaicism for chromosome 5 present in cultured amniotic fluid cells, followed by normal chromosome analysis of fetal blood obtained by PUBS, still resulted in a child with developmental disabilities because the mosaicism was confined to the skin and was not present in blood cells.

NUCLEUS AND CYTOPLASM

In the introduction it was pointed out that in the cytoplasm there are constituents which can also be considered self-replicating and, in a sense, independent of the nuclear genes. This fact lies behind the terms genome (the sum of nuclear genes, Winkler, 1920) and plasmon (the sum of the hereditary factors of the cytoplasm, Wettstein, 1926) as well as the less clear term plastom (the hereditary factors of the plastics, Renner, 1934, Michaelis, 1957/58, and Gustafsson and Wettstein, 1957/58). Of general interest is the plasmatic basis of male sterility in hermaphroditic species, such as maize, sugar cane, sugar beet, onion and *Dactylis*, which makes this plasmatic factor an important tool in breeding for hybrid vigour or heterosis.

The situation is similar where there is differentiation into different sexes and cytoplasmic factors influence the formation of female or male sex organs (examples being *Aquilegia, Godetia, Bryonia Satureja, Cirsium* and *Streptocarpus*). In this connection, the classic studies of Correns (1908, 1916) are of great importance. The conifers are generally monoecious but in many species, such as *Pinus sylvestris*, individuals are found which are predominantly male or female.

A systematic selection for dioecy may be quite useful in breeding for heterosis or hybrid vigour and the use of male-sterile individuals in crossing work. By contrast, the change from monoecy or dioecy (for instance in *Salix* or *Populus*) to hermaphroditism may be advantageous for the production of inbred lines. Such changes have been effected in hemp, with changes from dioecy to monoecy or hermaphroditism. In the last few years there have been indications that plasmatic inheritance can be introduced into a species *via* induced mutations in nuclear genes.

With regard to some chlorophyll lethals, for instance, nuclear mutations induce irreversible changes in the plastics or plastogenes. If the nuclear genes are then removed, the plastic breakdown persists and is transmitted plasmatically through the mother plant to the offspring (Wettstein, 1961, and unpubl., Gustafsson 1960b). It would be highly desirable if plasmatically and genically conditioned male sterility or dioecy could also be induced or isolated in tree species. With regard to disease resistance, some data obtained by Langner (1952) indicate the occurrence of a cytoplasmic background of susceptibility to needle-cast in crossings between *Larix decidua* and *L. leptolepis*.

APPLICATIONS OF CYTOLOGY IN HETEROPLOID

Another aspect of cytology affecting Mendelian ratios must now be referred to. It sometimes happens that at mitotic or meiotic division the daughter nuclei do not separate cleanly in the usual manner. In these cases the chromosomes, either as a whole set or as part of a set or even individual members, lag behind. The laggards which should have gone to one daughter nucleus are included in the other. When this occurs, nuclei are formed with a number of chromosomes different from that normally found in the species.

A nucleus with such an aberrant number of chromosomes or a plant in whose nuclei they occur is said to be heteroploid. If a whole chromosome set is replicated, the heteroploid is said to be balanced or *euploid.* If only a part of the set is replicated, the nucleus is of the unbalanced or *aneuploid* class. Most attention is directed to those cases where whole sets are concerned, that is, to the euploid form.

Where one replication of the normal diploid has taken place, and the homologues are in threes, the plant is said to be *triploid.* Where the diploid has doubled and the homologues are in fours, the plant is said to be *tetraploid.* Plants are known where the body cells contain only half the normal diploid number; that is, the number normal to the gamete of the species. These plants are known as *haploids.*

There is thus an ascending series in the terminology: haploid, diploid, triploid, tetraploid, pentaploid, hexaploid, and upwards. These have been formed in nature as the result of some unknown stimulus, or artificially in cultivation by the use of such drugs as colchicine.

HETEROPLOID'S EFFECT

Giantism

Replication of chromosome sets often results in the production of a bigger (giant) form of plant. This in itself may be very useful as, when it occurs in a plant of a forage crop.

Sterility and Fertility

Heteroploidy may be very useful in regard to sterile hybrids. These hybrids often result from crosses between plants which are not very closely related, but close enough to form a new hybrid plant. The chromosomes derived from the two parents are sufficiently homologous to form a mixed nucleus in Fi, but not so as to behave normally at meiosis. The hybrid fails to form viable gametes.

Such hybrids or "mules"are sterile, and therefore, unless they can be propagated successfully and economically, are of little or no use in practice. Replication of the mixed sets of chromosomes, if it occurs in such a sterile hybrid, may render it fertile.One of the clearest examples of this is furnished by a non-agricultural plant.

Primula verticillatay with nine pairs of chromosomes in the somatic nuclei, can be crossed with Primula floribunda, which also has nine pairs of chromosomes.The gametes of each, therefore, contain nine chromosomes, and these form the hybrid. None of the nine floribunda chromosomes are homologous with the nine verticillata members, and therefore successful meiosis in the hybrid is barred.

Viable gametes are not possible, and the plant is sterile. The sterile hybrid formed a euploid and immediately became fertile. Schematically, the mating, the chromosomes being symbolized by the initial letter of the species from which they are derived.By replication, each V chromosome obtains a homologue and similarly each F chromosome. Normal meiosis is possible and gamete production follows. Many species not closely related can form hybrids which are sterile.

It is probable that many of these will be rendered fertile when duplication occurs. Wide crosses as between, a wild grass like couch and wheat, or between turnip and cabbage, are important because the new gene complex brought into being results in more than a mere addition of the characters of one parent to those of the other. Genes even in a normal diploid soma interact on each other to some extent, and when two complete gene complexes interact the effect is greatly increased, so that a new plant different from both parents is brought into being.

This is paralleled in chemistry when, the union of sodium, a white silvery metal, with chlorine, a green gas, produces a new substance, sodium chloride with characters quite different from the parent substances. The formation of heteroploids has occurred in nature. Our modern bread wheat as it was evolved is a complex hexaploid of diverse parentage with three sets of seven, that is, twenty-one, chromosomes in the gamete, and twenty-one pairs (forty-two individuals) in the somatic nucleus. One interesting result of heteroploidy occurs when the component sets of chromosomes each contribute to the new plant similar genes.

A case in point arises in connection with the production of the red colour in the grain of bread wheat. Here, if an appropriate variety with red grain is crossed with a variety having white grain, a 3: 1 ratio red. If a different red-grained variety had been used as a parent, a 15: 1 ratio red: white in Fs would have been obtained. If yet another variety had been employed, the ratio of red:

white in F2 would have been 64: 1. Results such as these indicate that three pairs of genes are involved in the production of grain colour.

Furthermore, each pair must be located on different chromosome pairs, because the genes segregate and recombine freely. In short, there must be six chromosomes (three pairs) in each hexaploid somatic nucleus, all of them carrying a grain-colour gene.

Cytological evidence shows the bread wheat to be hexaploid. Multiple genes (polymeric series) of this type become important when they affect quantitative characters, and each dose of, the dominant produces an increase in the character. Sugar content of sugar beet depends on a set of replicated genes, and every additional "dose"of gene produces an increase of sugar content.

APPLICATIONS OF GENETICS BEHAVIOUR

Any movement of any part of the body of an animal is, according to a broad definition of the term, *activity*. However, as we use the term, and as it is most commonly used in the literature, "activity" refers to movement of the whole organism. In order to distinguish whole organism movement from such Behaviours as orienting response (head turning responses, movement of the pinnae, or movement of the eyes), the term *locomotor activity* is often used. Fuller and Thompson recommend *ambulation* as a more unambiguous term. The problems with activity as a phenotype, however, derive not so much from nomenclature as from what this form of Behaviour represents, and how it is measured.

ACTIVITY AS A TRAIT

For several decades prior to the 1960s, activity *qua* activity was outside the mainstream of American psychology. During this era of the major Behaviouristic learning theories, the concept of "Behavioural homeostasis" was in vogue.

This concept held that the "normal" state of an organism was not behaving (*i.e.*, being inactive), and that activity occurred only as the result of a "drive," that resulted from an imbalance in a physiological system, such as a tissue deficit or increase in hormonal activity. According to the assumption of Behavioural homeostasis, then, one does not (indeed, cannot) account for activity as a general trait, because activity occurs only as a consequence of other internal and external events. In recent years, the waning influence of the major learning theories, together with a number of disparate empirical findings (for example, exploration of their surroundings by satiated animals; the neural and bodily activity seen in many species during sleep), has led to a decline in the acceptance of the Behavioural homeostasis assumption. There is an increasing degree of acceptance of its opposite: The normal state of an organism is behaving (*i.e.*, being active).

When comparing the amount of movement of different individuals within similar settings, in the absence of specific physiological states, external goals, or noxious stimuli, one is investigating activity as a trait. If individual organisms do indeed possess specific "general activity levels" that yield strong predictions of other sorts of Behaviour or that are highly consistent across environmental settings and resistant to alteration, the theoretical and practical implications would indeed be great. Knowledge of the extent and nature of the genetic determinants of activity as a trait would magnify the importance of these implications.

ACTIVITY AS AN INDEX

Activity has been of interest not only as a trait in its own right, but also as an index, or means of measurement, of more complex and difficult to define forms of Behaviour. The most frequent uses for activity measures in this regard are exploratory Behaviour, whiich we discuss at length and emotionality. As an index of emotionality, activity is often combined with other measures, especially urination and defecation. Stemming largely from the selection studies in rats by Hall in the early 1930s, emotionality is, in turn, an indication of the autonomic and Behavioural responsiveness of organisms to potentially dangerous or noxious situations. Organisms are assumed to be either more active or much less active (*e.g.*, "freezing") in these situations, depending on the species or the genotype within a species.

The advantages of using activity as an index are obvious:

- Activity is relatively easy to measure, with a number of measurement procedures yielding reliable, readily interpretable scores
- A considerable body of activity data exist for a variety of species of labouratory animals
- A number of successful selection and strain comparison studies have been completed, providing an abundance of subjects more or less "genetically standardized" on this phenotype.

The problems associated with using activity as an index are more subtle, but no less real. If activity serves as an index of a more complex characteristic, such as emotionality, there should be a strong correlation between activity and any other index measures. Such a correlation would indicate that the various phenotypes bear some causal relationship to one another; or that they are due to pleiotropic or linkage effects of the same genes; or, on the other hand, that the correlation between indices is a spurious, chance occurrence, perhaps furthered by measurement artifacts.

When the correlation between indices is high for some selected lines but not for others, or for some inbred strains but not others, one must give the greatest credence to the last of these explanations. We deal with this problem, a major factor in phenotype confusability. When activity is used as the sole index of a more complex phenotype, it is the same as operationally defining the more complex phenotype in terms of the measure of activity (for example, defining exploratory Behaviour *as* "squares crossed in an open field"). It then becomes the obligation of the "consumers" of such information to decide for themselves whether this sort of measure is appropriately definitive and adequately representative of the complex phenotype.

An example of the problems that can arise from the use of activity as an index can be illustrated by a study performed by Whimbey and Denenberg, and further discussed by Denenberg in 1969. In the portions of this study relevant to our discussion, they placed rats in an open field briefly for 5 days and obtained a variety of measures that were then intercorrelated and factor analysed. Correlated positively with defecation scores and could be considered an index of "emotionality." However, on Day 2 and thereafter this same measure of activity, taken under presumably the same conditions from the same subjects, produced a *negative* correlation with defecation and yielded different factor loadings. Then, the authors imply, activity becomes an index of exploratory Behaviour. Is activity a useful index for more complex Behavioural phenotypes?

SELF-FERTILIZATION AND CROSS-FERTILIZATION IN STERILITY

Self-fertilization and cross-fertilization have been referred to. These are often barred by sterility. Almost any cabbage plant will illustrate the point. If its own pollen only is allowed access to its stigmas a cabbage will set few or no seeds. If pollen from other cabbage plants gets access to the stigma, fertility is complete. This proves that its ovules are fully functional. Its pollen is fully effective in producing seed on most others of its fellows. The gametes of the plant accordingly are fully viable but self-incompatible.

It is now known that the tube of a pollen grain carrying any Sallelomorph can hardly pass through a style carrying this same allelomorph in the pair resident in the cells. A similar mechanism affects many varieties of fruit tree. If an orchard has been planted only with such a self-incompatible variety, every one of the trees produced by vegetative propagation will have exactly the same gene complex. Hence, every stigma will contain incompatible genes in common with the pollen supply and fertility will be barred. Such an orchard must be interplanted with a few trees of another variety which is known to be capable of supplying suipollen.

Many plants are known to have a cytological constitution which does not permit of normal meiosis. The plants are usually of hybrid origin. Many fruit trees are of this nature and do not form viable gametes.

VIGOUR LOSS

Where sterility is no bar, and continued self-fertilization is practised, another disability may show itself. The vigour of the progeny derived through continued self-pollination falls progressively generation by generation. In maize, after a few generations of self-fertilization the inbred lines become weak, and in extreme cases may fail completely and die out.

Hybrid Vigour

Curiously, when two inbred but unrelated lines are crossed the F1 is often extremely vigourous. Out-breeding brings back vigour. This vigour in a hybrid has been called *heterosis* or *hybrid-vigour.* It shows only in the F1 and does not appear in F2 when the F1 is self-fertilized.

Fig. Heterosis

When appropriate inbred lines are used as parents, the Fi is often very much more vigourous and yields much more than any commercial lot. Heterozic or Heterozygous maize seed is now sold for crop production. In fact, for this crop, practically all progressive growers sow only hybrid vigourous seed. All of the Fi crop goes for consumption as food, etc.; none is retained for sowing. Each year the Fi is made anew from inbred parental lines which are known to give the maximum effect when crossed.

The commercial production of heterozic maize seed is simple. The two parental lines are interplantcd in rows. At flowering time all the male inflorescences are removed from the plants designed as the seed parent. It is therefore pollinated only by its neighbour, and all the "seed"produced by it is of hybrid origin.

In any crop produced by vegetative propagation the vigourous hybrid need be produced only once. All clones arc genetically identical and just parts of the Fi itself.

They retain all its characteristics in full effect. This is well seen in the poplar tree used for wood pulp. Two parents arc crossed. If the progeny is highly vigourous, suckering is induced. The suckers are planted out. The trees from such suckers grow very fast and harvest is brought nearer by a number of years.

SELECTION OF PLANT

Cytological aberration and genetical hybridity provide plant populations variable in character. Variation in a crop, no matter how induced, provides the plant selector with his material. Two inbred strains of Maize on eitiier side of the vigourous hybrid got from their mating. A new combination of genes or gene complexes can be the starting point of a new variety waiting to be selected and multiplied. There are three methods of selection.

Pure Lines

The first consists of selecting single plants, self-fertilizing them, and maintaining the progeny of each as a separate "line." In effect this subjects each selection to the progeny test.

The best line which breeds true can be retained and the others scrapped. This method is known as single-plant selection. If the original parent was homozygous for the characters observed, a non-varying or *pure line* will have been formed.

A pure line consists of the descendants of one homozygous individual, self-fertilized and never cross-fertilized.

This amounts to *genotypic* selection. A line pure for *all* genes is probably not possible, but a line pure in all the genes that matter is obtainable in practice. It is to be noted that, so far as concerns the genes for which the line *is* pure, further selection can have no effect. A pure line, then, by definition is non-varying.

Apart from environmental fluctuation and the possibility of mutation, no alteration in appearance or behaviour is possible.The second method of selection is to choose all the plants in a crop which appear to conform to a predetermined standard. Such is mass or *phenotypic* selection. This is often useful with plants which are individually self sterile, or where the character is quantitative and depends on a comparatively large series of polymeric genes.

Mass selection gives an immediate effect with a large bulk in the first generation. As generation follows generation, if re-selection is not practised on each, the effect wears off, and the strain "runs back"to an average. Mass selection is often possible on large-scale lines. In many countries growing crops are inspected officially, and where a crop conforms to a definite standard the grower receives a certificate for it.

When such a certification scheme is widely patronized it amounts to a large-scale mass selection.

The third method of selection is not to select out the good plants, but to pick out from a crop the wrong ones or "rogues."

These are the plants which do not conform to standard. Roguing has a potent influence when the rogue is due to a dominant gene, and when the desired character is due to a recessive.

Roguing out a recessive is not really effective in practice, for it leaves in the crop the heterozygotes. In this latter case it usually pays better to develop a pure line following a progeny test.

CHROMOSOMAL MUTATION IN EVOLUTION

Variations are the individual differences in somatic or genetic level which due to the change in the genetic make up of the individual taking place during the recombination, segregation or during defective meiosis, or by induction through contamination. This changes in gene is due to which these variations appear is called mutation.

Thus mutation is defined as any sudden appearance of heritable variation in nature of any organism is known as mutation. According to derives, mutations are sudden and drastic heritable changes not traceable or ascribable to segregation or recombination.

Mutation may be distinguished into genemutation, chromosomal mutation (Mutation in chromosome level due to change in chromosome structure) Ns geomantic mutation (mutation) due to changes in the basic chromosome number).

CHROMOSOMAL MUTATION

Chromosomes are the structures with definite organisation of genes in linear order. Any alterations in the location or positions of gene or loss or addition of genes to the chromosome brings about alterations in gross chromosome structure called chromosomal mutation or chromosomal aberrations.

Chromosomal mutations take place at the depleting stage of meiosis at the time of crossing over due to breaking and reunion in abnormal manner. There are four types of chromosomal mutations or aberrations.

Deletions

Number and position of genes on a chromosome is usually fixed. If a chromosome is broken in to several pieces and then healing or reunion of segments takes place, it may happen that two ends of the fragments unite together leaving one or more acentric parts free. Thus rejoined chromosome will be deficient of some of the genes.

Depending on the length of chromosome part lost, it is called either deficiency or deletion. If deficiency is large then it is called deletion. This deficiency may be interstitial or terminal. It has both genetic and cytological effects *i.e.* lethal and pseudo dominance effects.

Duplications (Additions)

Sometimes in the nucleus of a cell extra chromosomal parts may be found in addition to normal chromosomal complements. Duplocatopms occur if a segment of the chromosome or a sequence of genes occur two or more times or repeatedly occur in a chromosome or chromosome pair is called duplicaton.

During meiotic pairing of the chromosome bearing the duplicated segments forms a loop to maximize the juxtaposition of homologous regions. In contrast to deficient loop the duplication loop is formed by the duplicated segments. Duplication cause more variability than the deficiency. Duplication may takes place due to an equal crossing over corresponding to deletion here new characters may be added but not lethal and has useful role in evolution of new species.

Translocation

This is a type of inter chromosomal aberration which involves the shifting of a part of one chromosome to another non homologous chromosomes. If two non- homologous chromosomes exchange their parts, which need not be of the same size there results a reciprocal translocation. The reciprocal translocation may be of following types.

Simple homozygous translocation

When a chromosome breaks in to two parts due to external and internal stress, open of the two segments of the broken chromosome may become attached to nearest chromosome which may not be its homolofue.

Heterozygotic translocation

In heteozygotic translocation a considerable degree of meiotic irregularity occur. During meiosis an individual which is heterozyous irregularity occur. During meiosis, an individual which is heterozyopus for a reciprocal translocation must form a cross shaped configurantion in order to affect pairing of all homologous segments. This cross shaped configuration often opens out into a ring as dhiasma teriminalise. The meiotic product differs. Reciprocal translocation is like that of crossing over except that involves exchange between the segments of two non-homologous chromosomes.

Inversion

In this type of chromosomal aberrations the piece of chromosome which removed or delinked reinserted in reverse order, so that the sequence of genes alters with changes in the order of their arrangement but there is no loss genes.

Inversions may be percentric or paracentric inversions.

Pericentic inversion: In this type the inversted segments include th kinetochores or centromeres. The kenetochore may move from middle to one end of chromosome, thus there is marked change in the appearance of chromosomes.

Paracentric inversion: Here the inverted segment does not include the centromere or kinetochere. Centromeres lies outside the inverted segment.

Inversion can be evidenced by suppression of crossing over and possibly the appearance of mutation owing to position effect inversion does not involve no loss or gain of genes.

ROLE OF CHROMOSOMAL MUTATION IN EVOLUTION

Hugo de vries (1884-1935) propounded the mutation theory in 1901 to explain the origin of new species of living organism. He studied on prim-rose (Oevothera lamerc kiana) and marked sudden appearance of new characters which are inherited to next generation.

i. Oenothera lamarkiana ha 14 chromosomes but derives found some of the off springs have 15, 16, 20, 22, 24, 24, 28, 29 and 30 chromosomes. The new individuals different in flower size shape arrangement of buds and size of seeds.
ii. New elementary species appear suddenly and attain full constancy at once.
iii. The same new species are produced in large, number of individuals, this increase the chance of selection by nature.
iv. Mutability is something fundamentally different from fluctuating variability.
v. Mutations takes place in nearly all directions and may involve any character.

NEUTRAL THEORY OF EVOLUTION

Molecular biology and the discovery of the genetic code had a profound effect on evolutionary biology. One shocking realization was that many sites for mutation—for example, the third position in a codon or a nucleotide in the middle of an intron (a term defined later), are expected to be invisible to natural selection. This led Motoo Kimura to propose the neutral theory of evolution. It was somewhat heretical when first proposed because it deemphasized the role of natural selection, but the theory states that the majority of sequence evolution is purely random, the product of mutation and drift.

Imagine what happens to a sequence as it accumulates random mutations over time. At first, the sequence is nearly identical to the original. If the rate of mutation is relatively consistent, you can count the number of mismatches to determine how much time has passed. This turns out to be very useful and forms the basis for determining the probability that a DNA sample matches a particular person, for example. Eventually, the number of mutations becomes so great that the sequence is no longer recognizably similar to the original. At this point, the sequence is *saturated* for mutation. Saturated sequences can't be used to measure time, but they are still very useful because they indicate which sequences aren't under selective pressure. By inference, those that remain similar over long periods of time are under selective pressure. As a practical example, when comparing the human and puffer fish genomes, you find that most of the conserved sequence is in genes.

One of the great debates of evolutionary biology is the relative importance of natural selection and neutral evolution in the formation of species. We don't need to be overly concerned with this argument because we're more interested in how sequences change over time, and for this we can observe actual sequence data.

MOLECULAR CLOCKS

If you compare the sequences from related organisms, it is clear that certain positions don't change much over time while others change very rapidly. For example, parts of the ribosomal RNA are identical in every organism sequenced to date, from bacteria to humans. These subsequences are so important that if they change, the organism dies. Clearly, these are under intense selective pressure. There are other sites, such as third codon positions, that are only mildly affected by selection and tend to drift. There are even sequences, such as viral coat proteins, in which selection acts to promote variation, and these change very rapidly. Regardless of the underlying mechanism, it is possible to use the rate of change as a molecular clock.

If you know the mutation rate for a particular sequence, you can use it to determine how long ago two sequences diverged. Suppose you have the same protein sequence from both cats and dogs, and there are 10 differences between them. From the fossil record, you estimate that cats and dogs had a common ancestor 50 million years ago. Now when you compare the cat sequence to the same sequence in humans, you find 12 differences. You can now estimate that carnivores and humans shared a common ancestor 60 million years ago. We're using a very simple model here that treats all positions identically and we're not using real data, but this is the general idea behind molecular clocks. The key to using molecular clocks is that the sequences must "tick" at the appropriate rate. The hypothetical protein in the last example is a poor choice for determining how long ago humans and chimps last shared a common ancestor because one difference here or there would lead to a large difference in the estimated time. Sequences that tick too fast are also not appropriate because they are prone to saturation.

HOMOLOGY, PHYLOGENY, AND TREES

When looking at the biological world around you, you see only what exists today. You can't get a clear picture of what the world looked like 100 million years ago. However, you can see relationships between organisms and make inferences. For example, you don't know what the last common ancestor of humans, chimpanzees, and gorillas looked like, but you can guess that it looked more like an ape than a bird. This is also the case at the sequence level; proteins from humans and chimps are much more similar to each other than either is to a bird. The study of relationships between organisms is called *phylogenetics*. By definition, two sequences are *homologous* if they share a common ancestor.

Two sequences are either homologous or they aren't. However, people often misuse the term and say something like "these two sequences are 80 per cent homologous." What they usually mean is that two sequences are 80 per cent identical and not that there is an 80 per cent chance that they have a common ancestor.

Determining if two sequences are indeed homologous requires making inferences. This isn't always a simple task; sometimes homology can be stated with near certainty, but not always. Sequences may appear to be related from chance similarity (or *convergent evolution*). Sequence homology is further refined by the terms orthologous and paralogous.

Sequences separated by speciation are called *orthologs,* while sequences separated by duplication are called *paralogs*. The genes for myoglobin in humans and mice are orthologs; they are the same gene in different species. If the myoglobin gene is duplicated in humans, the two myoglobins will be paralogs of each other.

It's somewhat confusing, but both human paralogs would be considered orthologous to the mouse myoglobin. It is generally the case that the most similar genes between species are orthologs, and this is often used as an operational definition.

An introduction to molecular evolution would be incomplete without an overview of life on Earth. You may have learned in an introductory biology class that there are five taxonomic kingdoms (animals, plants, fungi, monera, and protista). This is based largely on what can be seen with your eyes or a microscope. Molecular biology opened up a new way to classify organisms based on sequences rather than external features.

The various organisms based on ribosomal DNA sequence. There are three obvious domains that Carl Woese called the Bacteria, Archaea, and Eucarya. Note that the arrow in the figure points to the root of the plants, animals, and fungi.

From this perspective, the traditional five kingdoms are a bit nearsighted. gIn terms of genomes and overall cell structure, there are only two major divisions: the *prokaryotes* (bacteria and archaea) and *eukaryotes.* Except in rare cases, prokaryotes are microscopic organisms that are usually shaped like rods or spheres. Some of the more famous prokaryotes include *Escherichia coli* (a bacterium that lives in your gut and is a favourite model organism for microbiologists) and *Yersinia pestis* (the bacterium that causes bubonic plague). The major distinguishing feature of prokaryotes is that DNA replication, transcription, and translation all take place in the same compartment of the cell because there is only one compartment in the cell.

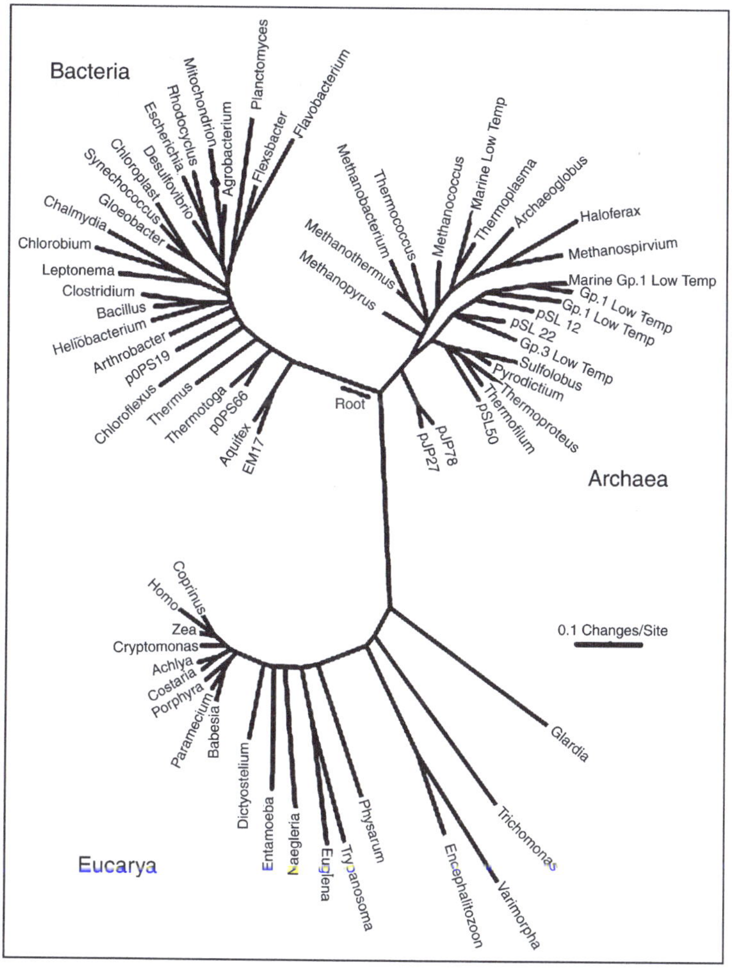

Fig. Tree of Life Based on rRNA Sequence

Eukaryotes come in many shapes and sizes, primarily because they can form multi-cellular organisms such as birds and trees. But some eukaryotes are simple, single-celled organisms such as *Saccharomyces cereviseae* (the yeast used for making beer).

All eukaryotes have a nucleus (*karya* is Greek for nucleus) in which DNA is stored, in addition to other membranous organelles. Interestingly, most eukaryotes contain *mitochondria*. These organelles have their own genome and are descended from bacteria that long ago entered a cooperative relationship with eukaryotes. This is also true of *chloroplasts*, which are responsible for photosynthesis in plants. It is thought that eukaryotes are a fusion of two bacteria, one a Eubacteria and one an Archaebacteria. So the next time you munch on a carrot, you might consider how many genomes are really in there. So far, this chapter has neglected viruses. Where do they fit in? By most definitions, viruses aren't even alive; they don't grow or have repair processes. Viruses seem to break every rule of biology. Some viruses infect prokaryotes and others that parasitize eukaryotes. Viruses come in many different shapes and have wildly different lifestyles. Some have genomes made from RNA instead of DNA, and others have single-stranded rather than double-stranded genomes.

7

DNA Structure in Mitochondrial Technology

INTRODUCTION

Mitochondria harbour a small but essential component of an eukaryote's genetic material. The symbiotic relationship that marks the origin of eukaryotes was established approximately 2 billion years ago and has been followed by a massive loss or transfer of genes to the host genome; only a tiny fraction (less than one per cent) of the endosymbiont's genome is retained on the mtDNA. Animal mtDNAs are extremely compact, being less than 20 kilobases in length, and encode fewer than 40 genes. All of the proteins encoded by the mtDNA are subunits of the mitochondrial respiratory chain. Despite their small number, the genes on the mtDNA necessitate a large expenditure of cellular resources because they are essential for strict aerobes. Hundreds of nuclear genes are needed to replicate, transcribe, and maintain the mitochondrial chromosome and to assemble the translation machinery needed to express its dozen or so proteins.

The *C. elegans* mtDNA is 13,794 nucleotides in length and encodes 36 genes: 2 ribosomal RNAs (12S rRNA and 16S rRNA), 22 transfer RNAs, and 12 MRC subunits. A typical animal mtDNA, the nematode mitochondrial genome is slightly smaller than its human counterpart. It lacks the ATP8 gene of the human genome, which encodes a subunit of the ATP synthase (complex V). Introns are absent and there are few or no non-coding nucleotides between genes; approximately 92 per cent of the mitochondrial genome has coding function.

Three genes end with incomplete termination codons (T or TA) that are apparently converted to TAA codons by polyadenylation of the transcript. Only one sizeable non-coding region called the displacement loop or D-loop is present. The replication and transcription of mtDNA has been mainly studied in mammals where the D-loop region contains one of the origins of replication and the promoters for mtDNA transcription. Few components needed for the maintenance or expression of the *C. elegans* mtDNA have been experimentally identified. The CLK-1 (biological clock abnormal) protein was reported to have

DNA binding activity specific for the light strand origin of replication (O_L) on the mtDNA, suggesting a role in regulating mtDNA replication or transcription in *C. elegans*.

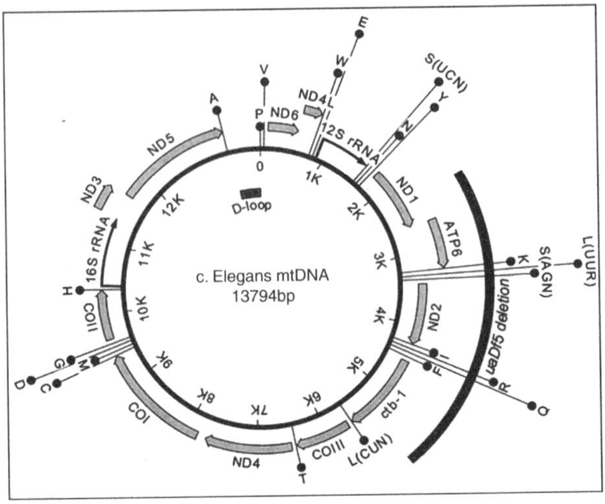

Fig. Gene Map of the *C. elegans* mtDNA.

The molecule contains the genes for twelve proteins (thick grey arrows), two rRNAs (black arrows), and 22 tRNAs (circles labeled with one-letter amino acid code). The serine and leucine tRNAs are also identified by the codon family recognized. The positions of the putative D-loop and of the *uaDf5* deletion mutation are indicated inside and outside the circle, respectively.

In contrast to the relatively uniform gene content of metazoan mtDNAs, the mitochondrial genetic codes are highly modified. In *C. elegans*, TGA specifies tryptophan rather than being a stop codon, ATA specifies methionine rather than iso-leucine, and AGA and AGG specify serine rather than arginine. The translation initiation codon ATG is not used; ATT, ATA, or TTG codons apparently serve as initiation codons.

These peculiarities of the genetic code may in part explain the retention of the twelve protein-coding genes on the mtDNA rather than their transfer to the nucleus.

MITOCHONDRIA: POWER PLANT OF CELL

For thousands of years, men puzzled over the question of where body heat comes from. This heat was recognized as nearly synonymous with life itself–but what produced it? The source of the energy of man or beast was equally mysterious. In recent decades it has become clear that both heat and energy can be traced to some sausage-shaped organelles in our cells, the mitochondria. Of course the ultimate source of energy for all plants and animals is sunlight. But the sun's energy can be harnessed by plants, through photosynthesis and stored in molecules of carbohydrates.

When animals eat these carbohydrates and break them down to carbon dioxide and water, with the help of oxygen and an arsenal of enzymes, large amounts of energy become available. Animals immediately convert this energy into molecules of high-energy ATP (adenosine triphosphate) -- the universal currency of energy in living things. Excluding only the very first stages in carbohydrate breakdown, which are called glycolysis, the entire, complicated process of energy transfer to ATP takes place within the mitochondria.

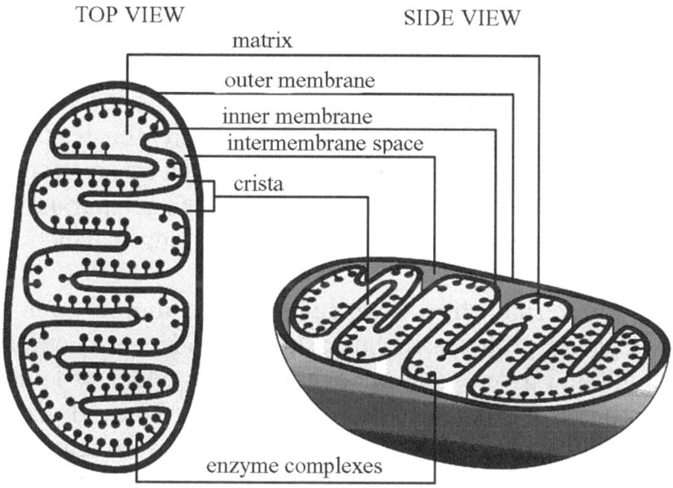

Fig. Mitochondrion.

Mitochondria are the largest organelles in the cell, after the nucleus, yet some cells have more than a thousand of them. Thin and long, they vary in diameter from 0.5 to 1 micrometer and in length up to 7 micrometers. Their thread-like outlines can be seen with a good light microscope.

Although they were first observed in 1850, it took a century to visualize their internal structure and understand their function. Actually the story goes back even further, to the 18th century, when Lavoisier showed that animals require oxygen for respiration. This led chemists to study cellular respiration, the process by which living cells use oxygen to release the chemical energy stored in foodstuffs.

Early in the 20th century biochemists discovered that these reactions fall into two major groups:

1 The carbon pathway, where a series of chemical reactions, each requiring a specific enzyme, breaks down the carbohydrates into carbon dioxide and hydrogen;

2 The hydrogen pathway, which transfers the hydrogen to oxyen in stages, forming water and releasing energy.

The whole process is organized much like an assembly line. In the hydrogen pathway, the hydrogen's electrons pass through an "electron transport chain" made up of enzymes that act as carriers and as the electrons move from enzyme to enzyme, they give up part of their energy, which is stored in molecules of ATP. The ironcontaining enzymes, cytochrome a, b and c, carry out the final stages of the process. All in all, three molecules of ATP are formed for every atom of oxygen that is used up in respiration. For a long time, biochemists studied these reactions in cell extracts without worrying about what parts of the cells might be involved.

One scientist, B.F. Kingsbury of Cornell University, did suggest that mitochondria might be the site of cellular respiration back in 1912, but this was ignored. In the 1940's, using an ultracentrifuge, Albert Claude and his associates isolated various cell particles including mitochondria. Shortly afterwards, in 1948, Albert Lehninger of Johns Hopkins and Eugene Kennedy of Harvard showed that the reactions leading to the synthesis of ATP occurred in the mitochondria. In the early 1950's, Palade and the Swedish scientist Fritiof Sjostrand reported that mitochondria are bounded by a membrane and that they have a system of parallel, regularly spaced inner ridges which were named cristae.

It is now clear that there are, in fact, two membranes around the mitochondria of plants, protozoa, molluscs, insects and man: an outer membrane, set off by a fluidfilled gap; and an inner membrane which is folded inward at various points to increase its surface, forming the cristae, so that the enzyme molecules of the electron transport chain, which are attached to this inner surface in specific sequences, can be packed more densely side by side in the mitochondria.

This general design for mitochondria seems to have existed unchanged, through the evolution of plants and animals, for more than 1.2 billion years. Though mitochondria are generally visualized as ovalshaped, they can change shape quite drastically. They swell and contract in response to various hormones and drugs. Even in very low concentrations, Lehninger found, the thyroid hormone thyroxin will make mitochondria swell and will simultaneously "uncouple" the electron transport chain from the synthesis of ATP, so that even though carbohydrates may be metabolized and oxygen consumed, no ATP will be manufactured.

ATP, on the other hand, makes mitochondria contract. This swelling and contraction appear related to the movement of water in, out and through cells, which is of particular importance in the kidney. The more scientists learn about mitochondria, however, the more their curiosity is aroused, for these are without a doubt the strangest of the cell's organelles.

To begin with, mitochondria contain their own hereditary material (DNA), which resembles that of bacteria far more than that of animal cell nuclei. Because of this similarity, many scientists believe that mitochondria are derived from bacteria that infected a primitive cell and evolved into a useful symbiotic relationship with it. As Lewis Thomas puts it, "A good case can be made for our non-existence as entities. We are shared, rented, occupied. At the interior of our cells, driving them, providing the oxidative energy that sends us out for the improvement of each shining day, are the mitochondria and in a strict sense they are not ours. They turn out to be little separate creatures, the colonial posterity of migrant procaryotes, probably primitive bacteria that swam into ancestral precursors of our eucaryotic cells and stayed there.

Ever since, they have maintained themselves and their ways, replicating in their own fashion, privately, with their own DNA and RNA quite different from ours. They are as much symbionts as the rhizobial bacteria in the roots of beans. Without them, we would not move a muscle, drum a finger, think a thought. My mitochondria comprise a very large proportion of me. Looked at in this way, I could be taken for very large, motile colony of respiring bacteria, operating a complex system of nuclei, microtubules and neurons for the pleasure and sustenance of their families and running, at the moment, a typewriter, continues Thomas somewhat facetiously.I am obliged to do a great deal of essential work for my mitochondria. My nuclei code out the outer membranes of each and a good many of the enzymes attached to the cristae must be synthesized by me. Each of them, by all accounts, makes only enough of its own materials to get along on and the rest must come from me. And I am the one who has to do the worrying.

During the past few years, scientists have begun to learn which proteins and enzymes of the mitochondria are synthesized according to directions in the mitochondrial DNA and which are controlled by the genes in the nucleus. This information would be of considerable importance for the design of antibiotics and drugs, as well as for the understanding of genetic diseases. For example, researchers have found that chloramphenicol–a powerful antibiotic which can cause a fatal anemia–interferes with the synthesis of mitochondrial DNA, while other antibiotics do not. Mitochondria are just as important to the life of each cell as the heart is to the human body.

At least 95 per cent of our cells' energy comes from mitochondria. (Only the mammalian red blood cells, which have no nuclei, get along without mitochondria; they obtain whatever energy they need from the partial breakdown of glucose in their cytoplasm). Growing cells continuously synthesize nucleic acids, proteins and other components, for which they need the energy that is stored in molecules of ATP.

Muscle cells which carry out mechanical work, plant cells which pump water against the force of gravity and many other cells also require a continuous source of energy. Besides supplying this energy, mitochondria apparently help to control the concentration of water, calcium and other ions in the cytoplasm.

They are also deeply involved in the breakdown and recycling of sugars, fatty acids and amino acids to recover their energy, as well as in the formation of urea as a waste product, to be excreted through the kidney. These strange organelles may thus hold clues to a variety of seemingly unrelated diseases. They have been linked to some liver diseases (*e.g.*, viral hepatitis, obstructive jaundice and cirrhosis of the liver); some muscle diseases (*e.g.*, familial myotrophic dystrophy); and some kidney diseases.

In all these conditions, pathologists have seen large amounts of materials pile up inside the mitochondria as "inclusions" of varying shapes and sizes, presumably waiting to be processed. What these inclusions consist of remains a mystery, but if scientists could find out, they might discover which specific components (for instance, enzymes) are involved in each disease. In addition, several calcium disorders may be related to mitochondrial defects. For example, abnormal concentrations of calcium (possibly caused by malfunctioning mitochondria) may lead to arthritis, bursitis, or arterial disease.

While all these leads are promising, the greatest spinoff from research on mitochondria may be in the area of energy–not only for the cell, but for general use by mankind. As the oil crisis has made us realise, we presently depend on a limited supply of fossil fuels–coal and oil which are of biological origin but cannot be replenished.

On the other hand, nature contains a far greater resource: a molecular know-how and technology gained over millennia, by which biological systems manufacture ATP from solar energy. This technology awaits us. It exists in very similar, if not identical forms in mitochondria and in their mirror image, the chloroplasts, which are found in green plants, where they do the work of photosynthesis, converting the sun's energy into carbohydrates.

Like mitochondria, chloroplasts have an outer and an inner membrane. They have some of their own DNA. They, too, may have originated as some kind of infection, probably with blue-green algae. And it is on chloroplasts that we ultimately depend for nearly all our energy–that of our bodies as well as that which is stored in deposits of oil and coal. If scientists ever unravel the details of the process by which chloroplasts store solar energy in carbohydrates and mitochondria release it, we may be able to produce unlimited quantities of usable energy.

THE CHEMICAL NATURE OF MITOCHONDRIA

In Bensley's work it was suspected that mitochondria contained a good deal of lipid material. One finding which suggested this was that mitochondria were stained by a method which is similar to that used for staining *Mycobacterium tuberculosis* and *Mycobacterium leprae*. These bacteria are stained by treating them for some minutes with a hot phenolic solution of basic fuchsine and once stained in this way they resist the de-staining effects of acid alcohol. This is the origin of their designation as acid-fast bacteria and their staining idiosyncrasies are believed to be due to a waxy or lipoidal coat. Mitochondria stained by hot acid fuchsine resist the decoloring action of picric acid for longer periods than most other cell constituents and this suggests that the mitochondria might also have a membrane which contains material of a lipoidal nature or at any rate which is of similar composition to that of the bacteria mentioned.

Mitochondria are difficult to demonstrate with conventional methods of staining if the fixative has contained acetic acid, alcohol, ether, chloroform, acetone, or other fat solvents. Also mitochondria have been found to stain with osmic acid and, incidentally, to resist de-staining by extraction with turpentine—a property of certain types of lipoprotein complexes. Baker has found that his "acid-haematin" test for lipids applied to a variety of tissues, nearly always gave a positive, reaction with the mitochondria. Thus there had accumulated for some years direct and indirect evidence that mitochondria contained appreciable amounts of lipid. Bensley's experiments, which, incidentally, were carried out some years before Baker's staining studies, showed that mitochondria which had been produced by homogenization and differential centrifugation contained appreciable amounts of lipid (30per cent).

METABOLIC SUBSTANCES FOUND IN MITOCHONDRIA

Perhaps we should give a brief summary of what was suspected about the metabolic significance of these organelles prior to the revolutionary studies which were made with isolated mitochondria beginning during the middle of the 1940's. One of the characteristic staining reactions of these organelles is that they give a green or green-blue stain with Janus green B which is diethylsafranine azodimethylaniline and it has been claimed by Cowdry that this reaction is primarily due to the diethylsafranine monocarboxylic acid component since this compound gives a very good and specific reaction with these structures.

It was observed many years ago by T. B. Roberston that if one drop of a saturated solution of safranine was added to a solution of trypsin, a coloured precipitate was formed and subsequently it was demonstrated that this coloured precipitate had proteolytic activity. Then Marston demonstrated that other azo dyestuffs would react in a similar way, particularly neutral red which is a dimethyl-diaminotoluazine hydrochloride.

Marston suggested that these results indicated that the reaction of mitochondria with Janus green might signify that they contained proteolytic enzymes. Now, the concept of the staining reaction with Janus green has been pretty well proved by Lazarow and Cooperstein to indicate that mitochondria play an important part in cellular oxidations and that the production of a pink colour from the Janus green by the mitochondria is due to the DPN specific dehydrogenases. The fact that Janus green B does not stain the mitochondria permanently green but that the green colour gradually becomes reduced to the pink and then to the colourless form has been known for many years. Joyet-

Lavergne demonstrated more than twenty years ago that mitochondria contain an oxidase system which oxidizes cobaltous to cobaltic salts and that these latter stain the mitochondria green.

It was noted by Gatenby that in the snail, *Lymnaea*, the mitochondria are coloured yellow in the natural state presumably by a carotenoid pigment; also extracted lipomitochondria frequently have a yellow appearance which too is probably due to carotene. It is of interest in this connection that it has been demonstrated by the present author and by Joyet-Lavergne that mitochondria give a blue reaction with antimony trichloride in chloroform solution, an acknowledged reaction for vitamin A. Carotene is also a provitamin A, so these histochemical results indicate the presence of vitamin A in mitochondria.

Criticism of Joyet-Lavergne's results was made by Gomori because the former author had used alcohol as a fixative. The present author has, however, always applied antimony trichloride in chloroform solution direct to fresh unfixed tissues and this suggests that this result is a true one, in any case biochemical tests have now confirmed that mitochondria (prepared by homogenization and differential centrifugation of the cells) contain 27-32per cent of their weight of lipids and that 100 mg. of this lipid contains approximately 249-910 U.S.P. units of vitamin A. JoyetLavergne has suggested that mitochondria contain a redox system in which vitamin A plays a part. A number of authors, Leblond, the present author and Giroud and his co-workers have demonstrated that mitochondria of some organs react with acetic acid-silver nitrate solution (which has been demonstrated as being a specific reagent for vitamin C) to give a positive reaction and this indicates that vitamin C may be present in them.

One has to accept the intracellular localization of vitamin C with a certain amount of discretion in view of the very destructive effect of this reagent on the cell cytoplasm. Electron-microscope studies in the present author's labouratory have demonstrated that this reagent has a most drastic effect on the ultrastructure of the cell and we cannot be sure that the localization of vitamin C in the mitochondria is, in fact, a real thing. It is of interest, however, that Chayen has found that the mitochondria of plant cells give this reaction very intensely and very specifically.

It may be that mitochondria in plant cells and some animal cells do, in fact, contain vitamin C, but we need further studies before this can be confirmed. Other vitamins appear to be present in mitochondria, members of the vitamin B complex have been found to be present, in some cases, in the form of coenzymes. The actual vitamins of the B complex recorded are vitamin B1 (thiamine), riboflavin, nicotinic acid (niacin), pantothenic acid and pyridoxine. However, although these vitamins are present in mitochondria, they are not present in any greater concentration than in the other parts of the cell so that they are not exclusively contained in these organelles. These studies were made on cell homogenates and, of course, it is possible that the presence of the vitamins in the other fractions of the homogenate may be due to the fact that they have been leached out of the mitochondria by the saline solution which is used in the homogenization process in these particular experiments and further work would have to be done before we could be certain about this. The present author, Joyet-Lavergne and Giroud have demonstrated that mitochondria contain glutithione or protein-bound SH and this is further evidence that mitochondria can play an important part in oxidation-reduction mechanisms in the cell.

It is of interest that mitochondria have been found in large quantities in the phloem cells of plants which are concerned with transport and may be concerned with this process. This suggestion is made in view of Conway's views that redox systems can play an important part in ion pumping. Another substance which was demonstrated histochemically in mitochondria by the present author, using the Schultz reaction, was cholesterol.

This was particularly evident in the mitochondria of cells of the adrenal cortex. Similar but less intense reaction was shown by the mitochondria of the liver and it has been demonstrated that isolated liver mitochondria contain about 2per cent cholesterol and it is possible they contain more in the adrenal cortex. Bensley has recorded the presence of a red pigment in the mitochondria and also in the submicroscopic particles of the liver cell. He believed that this pigment is derived from the oxidation of unsaturated fats and possibly the phospholipids of the mitochondria and this led him to suggest that in the liver cell the mitochondria in particular are, possibly, the seat of highly active oxidative processes which involve the metabolism of fats. The possible relation of mitochondria to oxidation-reduction processes was indicated by the publications of Ludford. He demonstrated that, if methylene blue was added to tissue cultures, the mitochondria of the cells stained a brilliant blue colour but this could be inhibited by potassium cyanide. If the cells were exposed to a bright light, the blue colour was rapidly bleached. It is of interest that the mitochondria, although they have been demonstrated to contain a good deal of fat and lipid, do not give a positive reaction with Sudan III.

They contain protein (quite a high proportion of it), but it is of interest that the earlier workers, using Millon's reagent, produced a negative reaction for protein. Subsequently Bensley and Gersh using a Millon's reagent of a different formula were able to obtain positive results from the mitochondria of many tissues and they found that they were particularly well stained in frozen, dried sections and particularly in those of *Amblystoma* liver.

The same authors demonstrated that, in undenatured, frozen, dried sections of *Amblystoma* liver, the mitochondria were destroyed if the sections were exposed to artificial gastric juice and artificial pancreatic juice. Despite all this chemical information most of which was in existence by the beginning of the 1940's, there was no certainty as to the function of mitochondria and it was not until 1946 and 1947 that the late George Hogeboom and his colleagues at

Bethesda completely revolutionized our ideas of their function by demonstrating that the major proportion of the succinic dehydrogenase and an appreciable proportion of the cytochrome oxidase activity of the liver cell were present in the mitochondria. This at once suggested that these organelles were the major sites of aerobic respiration in the cell. It is of interest that as long ago as 1915, Dr. Kingsbury had suggested that mitochondria were concerned with the respiration of the cell. His reasons for this were largely due to his observations that anesthetics such as ether and chloroform, which depressed cellular respiration and the respiration of the animal in general, also broke up mitochondria in the cell. The work of Hogeboom and his colleagues was carried out on homogenates of liver which had been produced by grinding up liver with saline. This gave poor preservation of the form of the mitochondria and subsequently 0.8 *M* sucrose was used—this preserved the nature and form of the mitochondria very well.

In the beginning there was some doubt as to whether the material being assayed was in fact mitochondria or not, but when sucrose was used this doubt gradually disappeared. Eventually electron-microscope studies of the structure of the granules isolated by homogenization and differential centrifugation demonstrated beyond doubt that they contained the same structures as the mitochondria of normal cells.

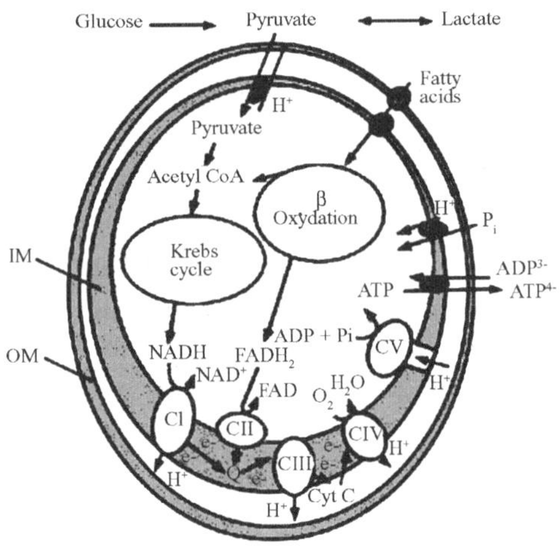

Fig. Respiratory Chain Enzymes

Another possible source of error, however, began to haunt the biochemists and was frequently verbalized by cytologists and this was that the mitochondria did not really contain the oxidative enzymes, mentioned in the foregoing, but that they were being absorbed or adsorbed by them from the homogenate. This possibility was negated at least in part by adding more enzyme to the homogenate and demonstrating that it could be recovered almost 100per cent from the supernatant and so was not taken up by the mitochondria. Thus the localization of the enzymes in the original cell was probably in the mitochondria. Subsequently, Kennedy and Lehninger demonstrated that in addition to containing cytochrome oxidase and succinic dehydrogenase, mitochondria also catalyze the condensation of pyruvic acid with oxaloacetic acid, the formation of succinic acid from α-ketoglutaric acid and the formation of malate from succinate.

These three reactions represent three steps of fundamental importance in the Krebs tricarboxylic acid cycle of which we will have more details shortly. Mitochondria also contain coenzyme I of which nicotinamide is an important constituent and cytochrome reductase which is a flavoprotein. These two enzymes are links between the Krebs tricarboxylic acid cycle and the cytochrome system and it, therefore, appears that mitochondria probably contain the whole enzymic equipment necessary for aerobic respiration of the cell.

In fact, it was originally demonstrated that a centrifugate of cells which included only nuclei and mitochondria were capable of carrying through the whole of the oxidation of glycogen to CO_2 and water; subsequently it was demonstrated that isolated mitochondria on their own could do this. Transaminase activity was subsequently demonstrated in mitochondria although in only the same concentration as in the rest of the cytoplasm. Transaminase is concerned with protein synthesis and it is of interest that pyruvic, oxaloacetic and α-ketoglutaric acids are the corresponding α-keto acids of the amino acids, alanine, aspartic and glutamic and need only to be transaminated to produce them.

Since the former compounds are themselves formed as intermediates in the course of the Krebs cycle, this indicates a mechanism whereby mitochondria might synthesize fresh protein material and so increase in size themselves or synthesize protein for other parts of the cell. This possibility is further extended by the fact that RNA is known to be present in varying amounts in mitochondria, although the exact amount is uncertain and there is a possibility of contamination of mitochondria with RNA from the rest of the homogenate when making such estimates. However, there is no doubt that mitochondria do have the equipment for synthesising protein although to what extent they do this *in vivo* is not known. All the fatty acid oxidase activity in the cell is also in the mitochondria and so is 80per cent of the octanoxidase activity.

Since these discoveries, an enormous amount of work has been done on the mitochondrial enzymes and this work was summarized by Hogeboom (unfortunately now deceased), Kuff and Schneider from the *National Institutes of Health* and has been published in Volume 6 of *The International Review of Cytology*, 1957. It may be of interest to record here other enzymes and compounds which they listed as being present in mitochondria. It is of interest too, that a considerable amount of ribonuclease and deoxyribonuclease are present in mitochondria—this seems unusual since only a very small amount of DNA and a relatively small amount of RNA are present in the mitochondria compared with other parts of the cell. However, it is possible that the bodies which contain RNAase and DNAase are not true mitochondria but are lysosomes; we will discuss these shortly. In the meantime, it is of interest to note that it was eventually demonstrated by fragmenting mitochondria and separating the membranes from the contents that cytochrome oxidase and succinic dehydrogenase activity were located primarily in the membranes. Subsequently it has been shown that the ability of the mitochondria to carry out the Krebs cycle and to combine this with oxidative phosphorylation is dependent on the presence of intact double membranes in the mitochondria. Now, it has been demonstrated and mentioned earlier that in the structure of mitochondria there is an outer membrane and an inner membrane to the mitochondrion, so that it is surrounded by a double membrane. Also, the inner membrane is folded inward to give elongated cristae and these cristae, because they consist of a fold of the inner membrane, are themselves double membranes.

There appears to be little difference biochemically between the double membrane structures of the cristae and the double membrane structure of the outer wall of the mitochondria. An important factor in the process of oxidative phosphorylation is the transport of electrons.

Dr. David E. Green has described the process of electron transport as carried out fundamentally with the aid of an electron transport particle—this is a submitochondria particle. The latter can be isolated in a form so that it can no longer carry out the full citric acid cycle, but it can still carry out the process of oxidative phosphorylation which it performs by the oxidation of succinate and DPNH (reduced diphospho-pyridine nucleotide, a codehydrogenaste) with freshly prepared by homogenization and differential centrifugation and are then exposed to alternate freezing and thawing, their form undergoes considerable change. First of all, the external membrane ruptures and some of the cristae break off and eventually the mitochondrion is completely disintegrated; however, fragments of double membrane structure are left which, according to D. E. Green, may be derived either from the cristae or from the outer envelope of the original mitochondrion.

These double membrane fragments represent collections of Dr. Green's electron transport particles. If other methods are used for preparing these particles and the particles so produced do not show a double membrane structure, then oxidation of succinate and DPNH coupled with phos-phorylation does not occur. Apparently the double membrane structure is essential for this process to take place. In the structures which do not have the double membrane structure, the capacity for electron transport is still present but oxidation phosphorylation cannot take place. About 35per cent of the dry weight of the electron transport particle of Green is composed of lipid and the lipid is concentrated in packets of lipoprotein. These are interspersed between two molecules of cytochrome or between a flavoprotein and cytochrome. In other words, all the oxidation reduction members of the election transfer chain are connected to one another by lipoproteins.

These lipoproteins, according to Green, act not only as structural devices, holding various enzymes together, but they also conatin a compound which can shuttle electrons back and forth so that then can pass from cytochrome through to flavoprotein and reverse. This compound, which Green and his colleagues have described, is a new coenzyme called coenzyme Q. It is a completely water insoluble benzoqtuinone derivative, which is connected in the lipoproteins of the electron transport particles and is related to vitamin K. Coenzyme Q can undergo oxidation and reduction in a similar fashion to cytochromes; for instance, the quinone part of the molecule can be reduced to become a hydroquinone and the hydroquinone can be reoxidized. It is pretty certain that this coenzyme plays an important part in the electron transfer chain of the particle.

The significance of the electron transport in these mitochondrial membranes may seem obscure but, as Green points out, one should think of this problem of electron transport in terms of its coupling—the fact that the process can be coupled to the esterification of inorganic phosphate. In this process a monophosphoric ester of the hydroquinone or, what Green described as a semiquinone form of coenzyme Q would be formed. This ester would then react with adenosine diphosphate by a process of transphosphorylation and ATP would be forme and simultaneously the semiquinone or hydroquinone would be oxidized to the quinone form in coenzyme Q by the ferric form of cytochrome c. The production of ATP is, in fact, the reason for the existence of such a system. This explanation of the way in which the oxidative processes of the mitochondria actually are used to produce ATP by oxidative phosphorylation is an extremely fundamental discovery and Green and his co-workers are to be complimented very much on the extraordinary patience and care which they have put in over a large number of years to work out this very complex problem in mitochondrial physiology. In view of the studies by Green and his colleagues which have just been described, it is of interest to consider some of the studies of Martias.

This author has for some time put forward the view that there may exist in mitochondria two routes for the transport of hydrogen ions or electrons, respectively, between the pyridine nucleotides (DPN and TPN) and cytochrome c. He thought that only one of these would be linked with enzymes which brought about phosphorylation. The route which is best known leads from DPN to cytochrome c reductase and then to cytochrome c. Martias believes that this particular route is not one which is concerned with the formation of highenergy phosphates. On the other hand, he proposed a second alternative route in which vitamin K (phylloquinone) is an important factor.

Until recently this concept of an electron transport system involving vitamin K in mitochondria has not proved to be particularly popular, Martias and Strouff have finally isolated a phylloquinone reductase that is a flavoprotein having a number of similarities to cytochrome c reductase. There is, however, a difference between them. Phylloquinone reductase and cytochrome c reductase can both react in the reduced state with phylloquinone, but with cytochrome c only cytochrome c reductase will have any effect. Vitamin K reductase has an extremely high activity even in relatively low concentrations. If we consider that there are two routes in mitochondria which can be used for electron transport, then the problem to be solved is what decides whether these electrons and hydrogen atoms are directed into either of these particular pathways.

Under normal circumstances they seem to be directed exclusively along the pathway which is coupled with oxidative phosphorylation and which leads via vitamin K reductase to vitamin K, cytochrome b and so through to cytochrome c. This pathway via vitamin K seems to be sensitive to any alteration of the internal structure of the mitochondria, possibly because some of its constituents, particularly the vitamin K, are associated with the double membranes of the cristae. If their combination is disturbed, the electrons take the emergency route which is via cytochrome c reductase to cytochrome c. It is of interest in connection with this theory that dicoumarol is capable of decoupling oxidative phosphorylation from respiration and, since dicoumarol is known to be a vitamin K antagonist, its action is possibly explicable by the fact that it blocks the pathway between vitamin K and its reductase. It is also of interest that thyroxine is a hormone which causes uncoupling of respiration from oxidative phosphorylation. Since it is known that, if thyroxine is added to mitochondria, it causes a swelling and change in the internal structure, this would be perhaps further evidence that the internal structure of the mitochondria is the one which is particularly concerned with the flow of electrons along the route leading to oxidative phosphorylation.

It may well be that hormones which affect respiration do so by affecting the structural nature of the mitochondria, apart from their effects on the endoplasmic reticulum which have already been mentioned. The precise relation between Green's coenzyme Q and Martias' vitamin K phylloquinone reductase is not absolutely clear, but it is very likely they are identical or closely related systems.

MTDNA-ENCODED PROTEINS

The *C. elegans* mtDNA encodes 12 MRC subunits out of a total of the approximately 80 subunits assembled into five MRC complexes. Four of these five complexes contain mtDNA-encoded subunits; complex II, the succinate-ubiquinone oxidoreductase complex, is the exception. The ND1, ND2, ND3,

ND4, ND4L, ND5, and ND6 subunits of complex I (the NADH-ubiquinone oxidoreductase) are all core subunits localized in the membrane arm of the enzyme. The ND1 subunit is implicated in ubiquinone binding. The cytochrome *b* of complex III (the ubiquinol-cytochrome c oxidoreductase) is encoded by the *ctb-1* gene in *C. elegans*; it is an essential component of the enzyme, harbouring 2 *b*-type hemes. Similarly, the COI, COII, and COIII subunits, which coordinate catalytic heme and copper cofactors, are essential components of complex IV (the cytochrome *c* oxidase). Finally, the ATP6 subunit of complex V plays a vital role in the efficiency of ATP synthesis. Not surprisingly, mutations in each of the human orthologs are causal in a variety of neurological, endocrinological, and muscular diseases.

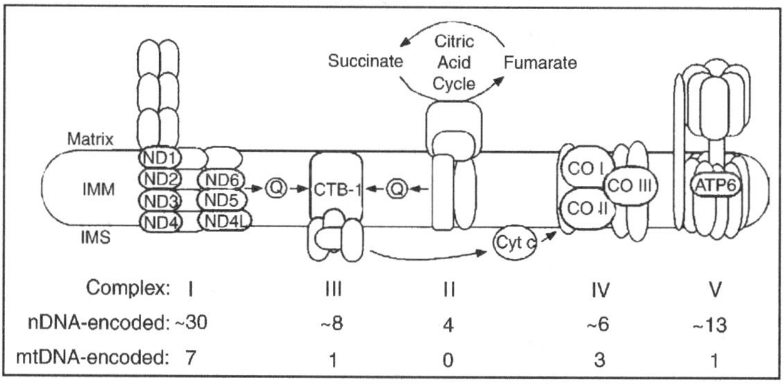

Fig. Schematic Representation of the Mitochondrial Respiratory Chain.

The figure indicates the estimated numbers of subunits that comprise the nematode complexes. IMM, inner mitochondrial membrane; IMS, intermembrane space; Q, ubiquinone; Cyt c, cytochrome c.

Very few mutations in the nematode mtDNA have been reported. In contrast, over 100 point mutations and 200 insertions, deletions, or rearrangements have been described for the human mtDNA.

It is puzzling why mtDNA mutations have not been reported in *C. elegans*, given the large number of genetic screens for defects in motility, reproduction, development, and other functions. Perhaps the nematode can better tolerate deleterious mtDNA mutations or has mechanisms to prevent their transmission. The genetics of the mitochondrial genome still offer much to be explored.

MITOCHONDRIAL DNA INHERITANCE

It is generally believed that animal mtDNA is inherited exclusively from the mother. However, it was recently shown in a patient with mitochondrial myopathy that the pathogenic mtDNA was of paternal origin. Evidence for the occasional paternal transmission of mtDNA has also been reported in sheep, mice, cattle, mussels, and insects. Although rare, the paternal inheritance of mtDNA may have a significant impact on evolution and on disease development.

IDENTIFIED OF DNA STRUCTURE

Deoxyribonucleic acid (DNA) was first identified in 1868 by Friedrich Miescher, a Swiss biologist, in the nuclei of pus cells obtained from discarded surgical bandages. The substance he found contained an acidic part, nucleic acid, and a basic (alkaline) part, which we now know to be histone proteins, which bind to the nucleic acid.

Which component was the genetic material? Many scientists were sure that it was protein. After all, protein had so many subunits (20 amino acids) that it seemed obvious that there existed within protein the possibility for much more diversity in expressing the genetic code than in DNA, which only has 4 subunits.

Each subunit is identical except for the base:

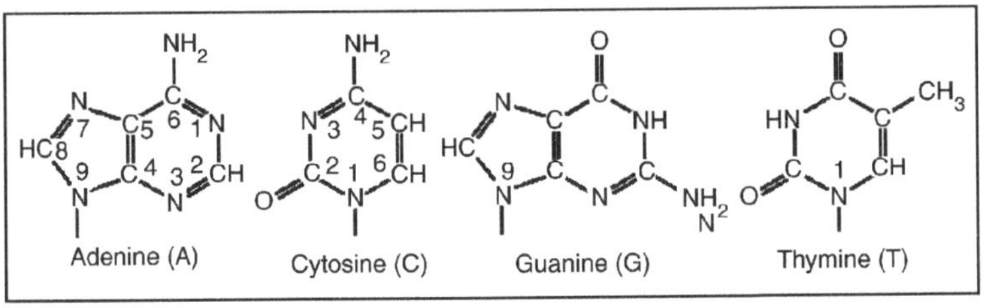

DNA-BINDING PROTEINS

In the long run, however, it turned out that the proteins attached to DNA have vital roles in chromosome structure and function. The roles of proteins include gene regulation (in bacteria and in eukaryotes) and chromosome imprinting.

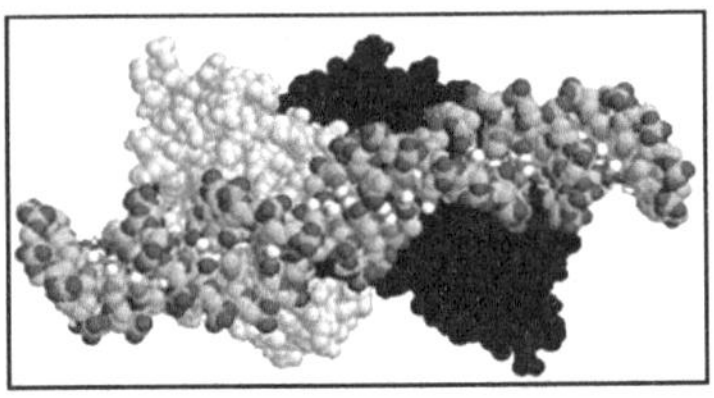

Fig. Catabolite Activator Protein Bound to Promoter of *Lac* Operon.

TRANSFORMING PRINCIPLE

In 1943, Oswald Avery, Colin Macleod, and Maclyn McCarty, at the Rockefeller Institute, discovered that different strains of the bacterium Strepotococcus pneumonae could have different effects on a mouse. One virulent strain could kill an injected mouse, and another avirulent strain had no effect.

When the virulent strain was heat-killed and injected into mice, there was no effect. But when a heat-killed virulent strain was coinjected with the avirulent strain, the mice died. What transforming principle was the dead virulent strain giving to the avirulent strain to make it lethal? This phenomenon of transformation, the uptake of DNA and incorporation into a genome, is now commonly performed in biotechnology.

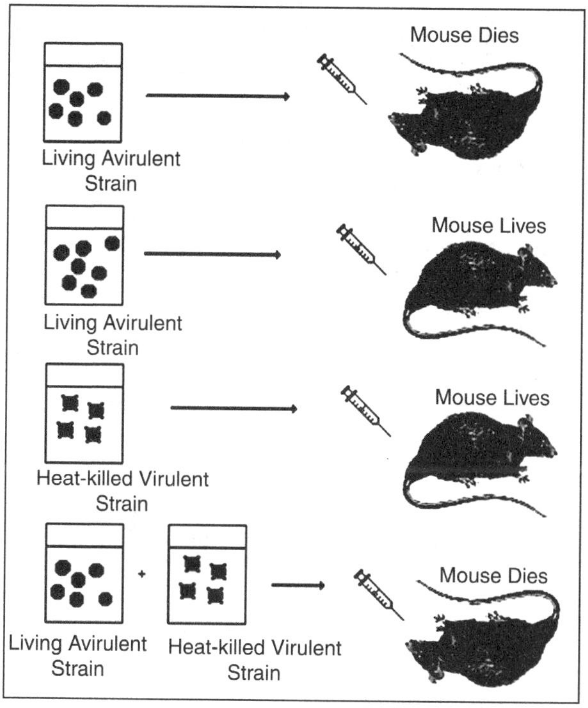

CHARGAFF - NUCLEOTIDE CONTENT IN DNA

In 1950, Erwin Chargaff at Columbia University discovered that no matter what tissue from an animal he looked at, the percentage content of each of the four nucleotides was the same, though the percentages could vary from species to species.

In all animals:

$$\% \text{ G} = \% \text{C}$$
$$\% \text{ A} = \% \text{ T}$$

The significance of these results was overlooked for three years, but they were crucial to elucidating the structure of DNA.

The Double Helix

In late 1953, James Watson and Francis Crick presented a model of the structure of DNA. It was already known from chemical studies that DNA was a polymer of nucleotide (sugar, base and phosphate) units. X-ray crytallographic data obtained by Rosalind Franklin, combined with the previous results from Chargaff and the chemists, were fitted together by Watson and Crick, who "borrowed" the data from Franklin's grant proposal. After several false starts, including the wrong tautomeric forms of the bases, they devised this model:

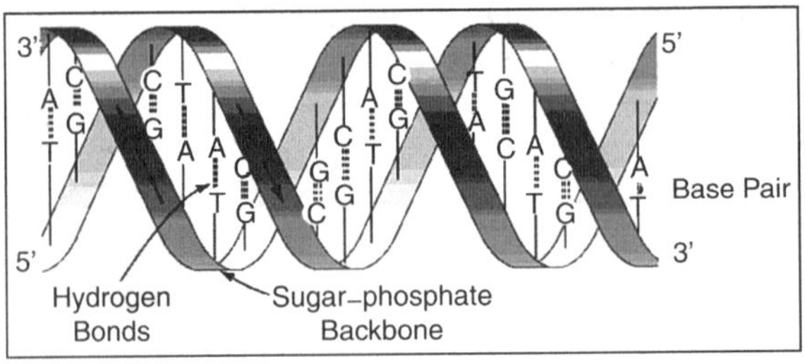

Under most cellular conditions, this double-stranded DNA molecule will coil naturally into a B-form helix, with one turn per 10.4 base pairs. However other structures are possible.

Each strand of the DNA is composed of nucleotides:

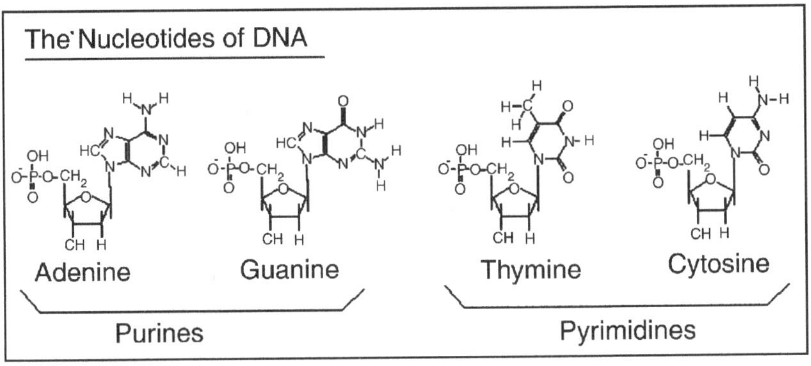

The nucleotides form base pairs:

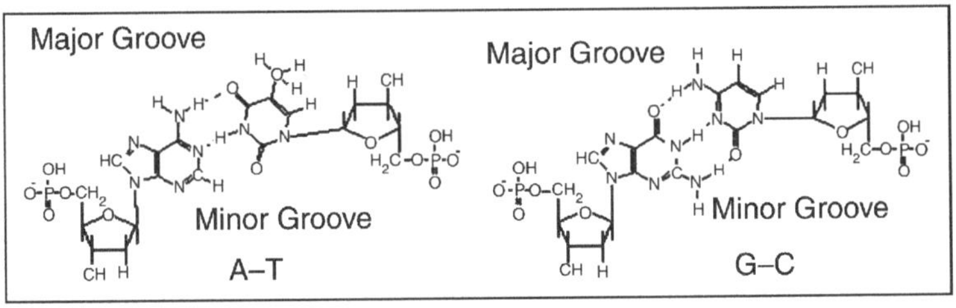

- Adenine pairs with Thymine because they make two hydrogen bonds.
- Guanine pairs with Cytosine because they make three hydrogen bonds.

The stacked base pairs form a major groove and a minor groove. Different regulatory proteins will bind to the major or minor groove.

Each base attaches to a phosphate at its 3' OH, and its 5' OH. The 2' carbon position has no OH; hence the "deoxy" part of DNA. The lack of 2' OH greatly stabilizes DNA, compared to RNA, because it prevents the intramolecular hydrolysis of phosphate linkages.

The base pairs "stack" together like rungs on a ladder, because of favourable interactions between the pi orbitals extending out of the heteroaromatic ring structure of each base.

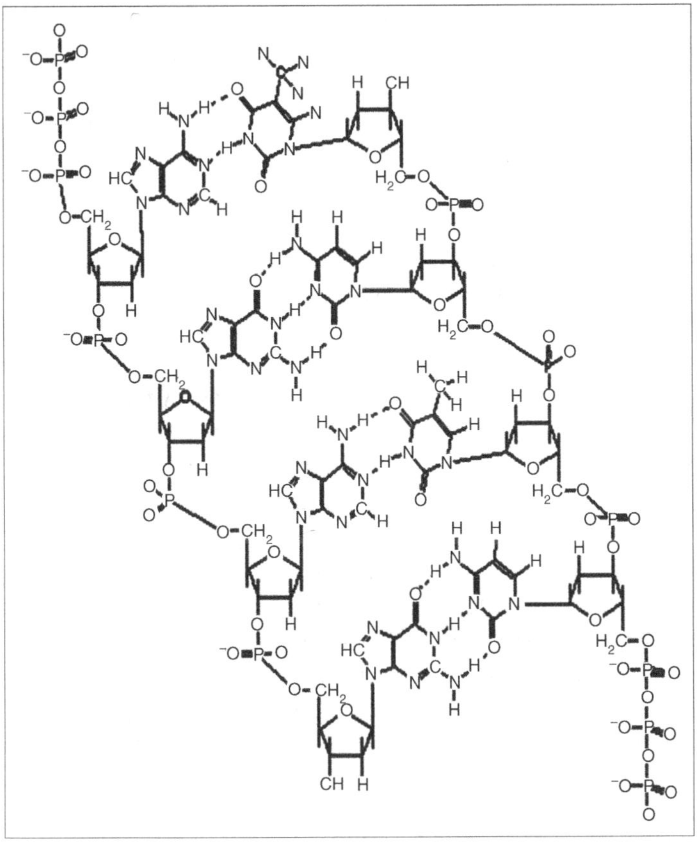

MITOCHONDRIAL DYNAMICS STRUCTURES

Mitochondria are dynamic structures and organelle plasticity is related to mtDNA inheritance. Multiple mtDNA molecules are organized into discrete protein-DNA complexes called nucleoids. Mitochondria form reticular or tubular networks and their interaction with cytoskeletal components provides clues to their distribution, movement, and inheritance. A mutation in the *anc-1* (anchorage of nuclei abnormal) gene of *C. elegans* resulted in hypodermal nuclei and mitochondria that float freely in the cytoplasm of syncytial cells. These mitochondria adopted an almost spherical shape and were often clustered. ANC-1 contains several coiled coil regions, a nuclear envelope localization domain, and an actin-binding domain, suggesting it mediates connections from nuclei and mitochondria to the actin network and to the nuclear envelope.

Mitochondrial division is required to regulate organelle numbers during cell division, differentiation, and in response to environmental conditions. The *C. elegans* DRP-1, a dynamin-related protein with a characteristic GTPase domain, plays a key role in mitochondrial distribution and division. RNAi of *drp-*

1 results in embryonic lethality and abnormal mitochondrial morphology and distribution, while overexpression leads to fragmentation of the organelle.DRP-1 is involved in the fission of the mitochondrial outer membrane, perhaps directly or perhaps through its ability to recruit additional fission machinery components.

MTDNA COPY NUMBER

The numbers of mitochondria and mitochondrial genomes in a cell are regulated and can vary substantially between tissues. The developmental regulation of mtDNA copy numbers in *C. elegans* has been investigated. An embryo contains ~25,000 copies of mtDNA and this number remains unchanged through the L1, L2, and L3 larval stages. A fivefold increase to 1.3 $\times 10^5$ copies in the L4 stage is followed by a further sixfold increase to 7.8 $\times 10^5$ copies in the adult hermaphrodite. The first copy number increase coincides with sexual maturation; spermatogenesis begins in the early L4 stage and oogenesis in the young adult.

mtDNA copy number is closely tied to reproduction. Mutations in the *glp-1* gene, which is involved in germ line proliferation, lead to an almost complete absence of germ line cells while somatic cell numbers are normal. The mtDNA content of an L4-stage *glp-1* hermaphrodite is ~70,000, half of wild type levels. There is no further increase in mtDNA copy number as the *glp-1* animal matures. Thus, increases in mtDNA content from the L3 to the adult stages can be subdivided into two components: a somatic component is *glp-1* independent and occurs during development from the L3 (~25,000 copies) to the L4 stage (~70,000 in the *glp-1* L4) and a germ line component accounts for the remainder of the increase to 7.8 $\times 10^5$ in the adult.

Oocyte production accounts for the majority of the germ line-related increases in mtDNA copy numbers. A loss of function mutation in the feminization gene *fem-1* blocks sperm production in the hermaphrodite but does not affect mtDNA contents in the L4 or in the adult. In contrast, a gain of function mutation in the *fem-3* gene, which blocks oocyte production, does not affect L4 mtDNA content but reduces the content in the adult to about one quarter (to 1.9 $\times 10^5$ copies) of wild type levels. Thus, the majority of organellar genomes are associated with oocyte development and the sperm-associated component is minor. When determined independently, 18,000 copies of mtDNA were measured per *fem-1* oocyte, a number similar to wild type embryos and early larvae.

The mtDNA content of individual *C. elegans* cells is considerably lower than the estimated values of 1,000-10,000 copies per cell in higher eukaryotes. The ~25,000 copies of mtDNA in the L1 larva, if evenly distributed around the 558 nuclei, suggest an average of 45 copies per cell. Similarly, the L4 larva has ~1,000 somatic nuclei and 70,000 copies of mtDNA in a *glp-1* mutant, corresponding to 70 copies per cell. It remains to be determined whether different somatic cells or tissues have distinct mitochondrial genome contents that might reflect their energy requirements. Estimated cellular contents of mtDNA for undifferentiated germ line cells and for sperm are 250 and 30-40, respectively. The increases in mtDNA copy numbers during development are suggestive of parallel increases in the numbers of organelles, although this has not been documented. Interestingly, there is an increase in organelle number and in the frequency of organelle division with a shift of temperature from 15 to 25°C. This increased content of mitochondria may reflect an adaptation to the higher metabolic demands of elevated temperatures. Consistent with this notion, respiration rates and ubiquinone levels are doubled with similar temperature shifts.

ROLE OF MTDNA IN DEVELOPMENT

A functional MRC is essential for viability and larval development past the L3 stage. Mutations in nuclear genes encoding MRC subunits such as *nuo-1* or *atp-2* (subunits of complexes I and V respectively) lead to a characteristic L3 stage arrest having an L2-staged gonad. Similarly, *clk-1* mutations, which impair ubiquinone biosynthesis, can lead to an L2 or L3-stage arrest. Inhibitors of mtDNA replication or transcription such as ethidium bromide or inhibitors of mitochondrial translation such as chloramphenicol or doxycycline can also quantitatively produce an L3 stage developmental arrest. Maturation to the L4 stage likely entails increased energy demands that are met with de novo synthesis of new mitochondrial energy production capacity. An energy sensor that responds to the concentrations of one or more metabolites, such as ATP or NADH, may communicate information about the status of mitochondrial energy generation and lead to altered patterns of MRC gene expression.

MTDNA MUTATIONS

Mitochondrial genetics differ from nuclear genetics in three aspects. *First,* mitochondrial genes are maternally inherited; they do not follow a Mendelian pattern of inheritance. *Second,* the mitochondrial genome is polyploid. Normally, a state of homoplasmy exists where only one form of mtDNA is present. Mutation can lead to a state of heteroplasmy where two or more forms of mtDNA coexist within a cell. *Third,* unlike a diploid nuclear gene that can normally only assume three states (homozygous wild type, heterozygous, or homozygous mutant), mtDNA heteroplasmy does not vary by discrete steps. The proportions of mtDNA species can vary with time or through mitotic segregation as cells divide.

The inheritance and maintenance of mtDNA species are most poorly understood in the heteroplasmic condition. The *uaDf5* mtDNA deletion, which removes 11 genes (four MRC subunit and seven tRNA genes), is maternally inherited and stably propagated without selection. While individual animals have *uaDf5* mtDNA contents ranging from #20 per cent to ~80 per cent, the population average is ~60 per cent and this value does not vary between larval stages. There is no phenotype associated with the deletion, even in animals with an ~80 per cent content of *uaDf5* mtDNA.

Most surprisingly, homoplasmic wild type or mutant animals have not been detected in over 100 generations, suggesting the two forms of mtDNA do not segregate from each other. Single, self-fertilized hermaphrodites with intermediate levels of *uaDf5* mtDNA (~50-60 per cent) have offspring containing from 20-80 per cent mutant mtDNA, indicating that the level of heteroplasmy can change significantly in one generation. However, hermaphrodites with extreme levels of heteroplasmy (~20 per cent or ~80 per cent) only produce offspring with more moderate levels, suggesting two opposing forces are in operation. One force increases *uaDf5* contents when they are low, while the second decreases them when they are high. One possible explanation for this phenomenon is that wild type mtDNA contains an undetected mutation that is complemented by the *uaDf5* mtDNA. Thus, both genomes are needed to produce a fully functional MRC and homoplasmy for either mutation would be severely compromising or lethal. Alternatively, the stable heteroplasmy may result from intra-mitochondrial heteroplasmy, a condition where the mtDNA species are intermixed within the nucleoid of one organelle.

Although the heteroplasmic *uaDf5* animals do not have a phenotype, they have adjusted their mtDNA contents. Both hermaphrodites and males have approximately twice the number of mitochondrial genomes as their wild type counterparts (Tsang and Lemire, 2002). The upregulation of mtDNA copy number may in part compensate for the effects of the mutation by enhancing the production of functional mitochondrial transcripts and/or proteins. Understanding the signals and mechanisms employed to regulate mtDNA copy number remains an important challenge. The *C. elegans* mtDNA mutation rate has been estimated using a series of mutation accumulation lines maintained by single-progeny descent for over 200 generations. In over 770,000 base pairs of sequenced DNA, 16 base substitution and 10 insertion/deletion mutations were detected, corresponding to a measured rate approximately 100-fold higher than previous indirect estimates. Four of the insertion/deletion mutations introduced drastic changes to coding sequences, were homoplasmic (or nearly so), but had only modest effects on fitness. Most spontaneous mtDNA mutations in *C. elegans* appear to be mildly deleterious although selection may occur in a natural context.

The mtDNA mutation rate can be increased by environmental agents or by mutation of nuclear genes involved in mtDNA maintenance. For example, nucleoside analogs used in the clinical treatment of viral infections, such as human immunodeficiency virus, inhibit the mtDNA polymerase, which is essential for mtDNA synthesis and repair. A mutation that impairs the fidelity of the mtDNA polymerase produced a mtDNA mutator phenotype in a mouse model; this was associated with premature aging and reduced lifespan.

mtDNA and Aging

Mitochondrial functions influence and possibly control the rate of aging and conversely, aging affects the integrity of the mitochondrial genome. An increased production of reactive oxygen species, often associated with MRC dysfunction, can precipitate premature aging, cause mutations in the mitochondrial and nuclear genomes, and shorten lifespan. The first *C. elegans* mtDNA deletions detected were spontaneous events in an aging, wild type population. *age-1* (aging abnormal) mutants, which are more resistant to oxidative stress and live twice as long as wild type, have a lower frequency of mtDNA deletions. It has not been determined whether these aging-associated mtDNA deletions reduce fitness or whether they are heritable.

To date, there is no direct evidence linking genes encoded by mtDNA with aging, but mutations in nuclear genes provide ample reason to believe that they are involved. The *isp-1(qm150)* mutation, which affects the iron sulfur protein of complex III, was the first mutation of an MRC subunit reported to increase lifespan.

The mtDNA-localized, maternally-inherited *ctb-1(qm189)* mutation affecting the cytochrome β subunit of complex III, does not alter the long life span of *qm150* animals but does partially suppress other effects of the *qm150* mutation. The *ctb-1(qm189)* mutation has no independent phenotype. The genetic interaction of the *isp-1* and *ctb-1* mutations clearly illustrates the interplay of the mitochondrial and nuclear genomes in cellular metabolism.

The *mev-1(kn1)* mutation is a missense allele of the *cyt-1* gene encoding the cytochrome β subunit of complex II. Mutants are hypersensitive to methyl viologen and oxidative stress, have elevated levels of superoxide anion production, display signs of premature aging, have a reduced lifespan, and are hypermutable. A probable null mutation in the gene for the mitochondrial leucyl-tRNA synthetase *(lrs-2(mg312))* markedly extends life span. The synthetase charges mitochondrial tRNAs needed for the translation of the 12 mtDNA-encoded MRC subunits. Mutation of the gene encoding the isopentenyl-pyrophosphate:tRNA transferase *(gro-1(e2400)*, abnormal growth rate), produces a Clk (biological clock abnormal) phenotype characterized by slow development and increased lifespan. The apparent use of alternative translation initiation codons produces differentially targeted forms of GRO-1; it is the loss of the mitochondria-targeted form, which presumably modifies a subset of tRNAs and affects mitochondrial translation, that leads to the Clk phenotype.

Non-genetic interventions that affect mitochondrial function can also affect aging. Systematic RNAi screens for increased lifespan have identified numerous mitochondrial components, including MRC subunits, mitochondrial carrier proteins, a subunit of the mitochondrial ribosome, and a protein involved in MRC biogenesis. Mutations in the *clk-1* gene disrupt ubiquinone biosynthesis and significantly increase lifespan. Finally, exposure to antimycin A, a tight-binding inhibitor of complex III, reduces rates of respiration and promotes longevity. Given the presence of 12 key MRC genes on the mtDNA and the intimate connections between MRC function and lifespan, it seems inevitable that mutations that slow aging will be identified in one or more of those 12 genes.

PERSPECTIVES

The mitochondrial genome is often ignored although it is a vital component of animal metabolism, physiology, and development. Few studies have specifically addressed the *C. elegans* mtDNA and much of our knowledge is inferred from investigations of vertebrate mtDNAs and from bioinformatic approaches.

The *C. elegans* cellular machinery required for mtDNA maintenance and expression is almost completely unexplored; undoubtedly biological novelties remain to be exposed. Deciphering how mitochondria communicate with the remainder of the cell will have far-reaching implications for understanding cellular metabolism in normal and in diseased states. Few model systems offer as many advantages as the nematode for grasping the full importance and scope of mitochondrial genetics in animal life.

DOUBLE-STRANDED STRUCTURE OF DNA

DNA usually exists in a double-stranded structure, with both strands coiled together to form the characteristic double-helix. Each single strand of DNA is a chain of four types of nucleotide: adenine, cytosine, guanine, and thymine. A nucleotide consists of a phosphate and a deoxyribose sugar forming the backbone of the DNA double helix plus a base that points inwards. Nucleotides are matched between strands through hydrogen bonds to form base pairs. Adenine pairs with thymine and cytosine pairs with guanine. The physical pairing of bases in DNA means that the information contained within each strand is redundant. The nucleotides on a single strand can be used to reconstruct nucleotides on a newly synthesized partner strand.

DNA strands have a directionality, and the different ends of a single strand are called the "3' end" and the "5' end" (these refer to the carbon atom in ribose that the next phosphate in the chain attaches to). In addition to being complementary, the two strands of DNA are antiparallel: they are orientated in opposite directions. This directionality has consequences in DNA synthesis, because DNA polymerase can only synthesize DNA in one direction by adding nucleotides to the 3' end of a DNA strand.

DNA POLYMERASE

DNA polymerases are a family of enzymes critical for all forms of DNA replication. A DNA polymerase synthesizes a new strand of DNA by extending the 3' end of an existing nucleotide chain, adding new nucleotides matched to the template strand one at a time. Some DNA polymerases may also have some proofreading ability, removing nucleotides from the end of a strand in order to remove any mismatched bases. DNA polymerases are generally extremely accurate, making less than one error for every million nucleotides added.

The energy for the process of DNA polymerization comes from the two additional phosphates attached to each of the unincorporated nucleotides. These free nucleotides, also known as nucleoside triphosphates, contain a total of three phosphates. When a nucleotide is being added to a growing DNA strand, two of the phosphates are removed and the energy produced is used to attach the remaining phosphate to the growing chain. The energetics of this process may also explain the directionality of synthesis - if DNA were synthesized in the 3' to 5' direction, the energy for the process would come from the 5' end of the growing strand rather than from free nucleotides. During proofreading, if the 5' nucleotide needed to be removed this triphosphate end would be lost, losing the energy source required to add a new nucleotide to the end.

DNA polymerase can only extend an existing DNA strand paired with a template strand, it cannot begin the synthesis of a new strand. To do this a short fragment of DNA or RNA, called a primer, must be created and paired with the template strand before DNA polymerase can synthesize new DNA.

DNA REPLICATION WITHIN THE CELL

For a cell to divide, it must first replicate its DNA. This process is initiated at particular points within the DNA, known as "origins", which are targeted by proteins that separate the two strands and initiate DNA synthesis. Origins contain DNA sequences recognized by replication initiator proteins (eg. dnaA in *E coli'* and the Origin Recognition Complex in yeast). These initiator proteins recruit other proteins to separate the two strands and initiate replication forks. Initiator proteins recruit other proteins to separate the DNA strands at the origin, forming a bubble.

Origins tend to be "AT-rich" (rich in adenine and thymine bases) to assist this process because A-T base pairs have two hydrogen bonds (rather than the three formed in a C-G pair)—strands rich in these nucleotides are generally easier to separate. Once strands are separated, RNA primers are created on the template strands and DNA polymerase extends these to create newly synthesized DNA. As DNA synthesis continues, the original DNA strands continue to unwind on each side of the bubble, forming replication forks. In bacteria, which have a single origin of replication on their circular chromosome, this process eventually creates a "theta structure" (resembling the Greek letter theta: è). In contrast, eukaryotes have longer linear chromosomes and initiate replication at multiple origins within these.

PROKARYOTIC DNA REPLICATION

Duplication of DNA is one of the fundamental properties of "molecule of life". Duplication or replication takes place once in a cell cycle. The duration and initiation point differs from one system to another. When conditions are favorable cell cytoplasmic mass increases and when cytoplasmic mass reaches a ratio to that of the cell size to 2L, bacterial cell division is triggered.

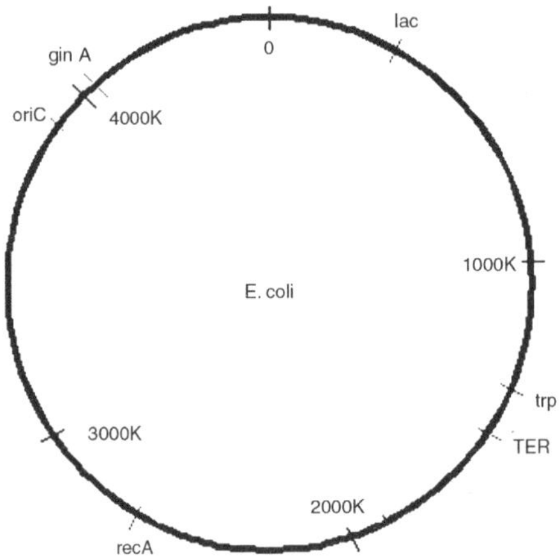

Fig. The Basic Circular Model of the Chromosomes with Map Positions such as Ori C and TER Regions

To complete its cell cycle it requires hardly 30 -40 minutes. Several genes involved in different steps of cell division have been characterized. Par excellent system to understand molecular process of prokaryotic cell division is E.coli. The size of the E.coli chromosome is 4.6×10^6 base pairs, but it is compacted by the binding of histone like proteins.

First the nucleoid i.e the chromosome frees from the membrane and unwinds. Such DNA initiates replication and completes in a matter of 15-20 minutes. Then the cytoplasm starts dividing almost in the centre of the cell by central septal ring formation.

It is at this time the daughter DNA molecules are partitioned and segregated to the two compartments. The central periseptal ring that started 1-2 minutes after DNA replication now furrows inwards and divides the cytoplasm in the middle region. Then mid-wall splits in the middle and daughter cell are set free, hence the Schizomycetes to bacteria (E.coli).

Bacterial cell division requires the participation of the following genes and gene products. Eukaryotic cell division is more complex and highly regulated. The paradigm for it is yeast, for it is a unicellular system, grows fast and requires hardly 60 minutes for one cell cycle. One can obtain different kinds of mutation like conditional and null or auxotrophic kinds. In culture cells, the duration of one cell cycle is about 12 hrs.

Onion root tip cells divide once in 12 hrs. Whereas embryonic cells in its early stages require just 30 minutes for one cell cycle and they go through 6 to 8 such cell divisions very fast.

Replication of DNA in cell division is an important phase. The replication is very accurate and it is nucleotide by nucleotide and there is no provision for making any mistakes, even if mistakes are made they are immediately corrected or repaired, otherwise consequences are serious and deleterious. In general DNA replication is precise, exact and regulated and involves initiation, elongation and termination steps, but mechanisms and rate of replication vary from one system to the other.

REPLICATION IS SEMI CONSERVATIVE

Meselson and Stahl showed that DNA replicates in semi conservative mode. There were some doubts about the mode whether it is conservative, dispersive or semi conservative. But Meselson and Stahl used non-radioactive heavy Nitrogen isotope (15N), as the source of Nitrogen.

When cells were grown on such source, 15 ^ N incorporated nitrogen bases get incorporated into DNA, thus it becomes heavier than the DNA that is grown in normal (14N) nitrogen media. When two such DNAs, one grown in the presence of 15N source and the other in 14N (normal), are extracted separately and stained with ethidium Bromide, then subjected to equilibrium density centrifugation at ultra speed of 42000rpm for about 24 –36 hrs, the DNA, separates as two distinct bands, the one that is heavier (15N) is the lower band than the normal (14N) upper DNA.

If cells grown in heavy isotopes are shifted to normal Nitrogen source and grown for one or two cell generation and if DNA is isolated from it and if one combine this with the first two DNA sources and subject them to equilibrium density ultra centrifugation (using cesium chloride or cesium sulphate), one can observe three bands, the top DNA band is from cells grown in 14N without isotope, the second band is the hybrid band with one strand 15 ^ N and the other with normal 14N-Nitrogen and the third band is with heavy nitrogen where both the strands are labeled with 15N.

This experiment virtually and unambiguously proved that DNA replicates in semi conservative mode. Taylor using root tips, labeled with 32 ^ P radioisotopes, of Vicia faba demonstrated semi conservative mode of replication at chromosomal level.

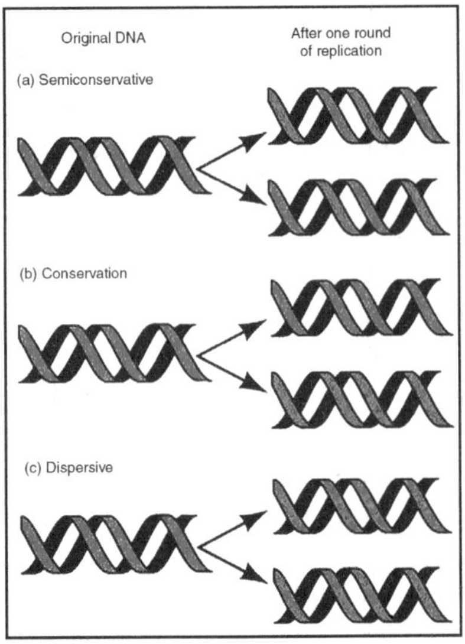

Semi conservative means that the two daughter molecules produced at the end of replication, each of the molecules retains one parental strand and the other one is the newly made one. So it is semi conservative. The mode of replication is absolutely semi conservative irrespective the kind of DNA, whether it is ds DNA or ssDNA, whether it is linear or circular, and the basic mechanism is more or less similar; each of them uses their own specialized mechanisms. Experimentally semicon-servative mode of replication has been shown by density gradient centrifugation of non-radioactive isotopes such as 15N and 14N sources.

HELICAL FORMS OF DNA

The structure of B-form DNA helix was first determined by X-ray analysis of crystalized molecules. However, other forms of helix can be stable under certain conditions of salt, pH, and temperature. In fact, triple-helical forms (triplex DNA, H form) have been found. Certain regulatory sites within cells appear to have DNA sequence that takes a non-standard form, sometimes assisted by a protein. Moreover, DNA technologists are making use of unusual DNA properties to construct genetic medicines. Genetic medicines are pieces of artificial DNA that can hybridize to a region of the genome and turn off transcription of genes, such as a cancer gene.

STABILITY OF DNA

DNA is a stable molecule; short pieces of DNA can remain intact in fossils and mummies for thousands of years. However, in water solution certain chemical conditions can destabilize DNA. *For example*:

- Acid (low pH) causes detachment of purines from the backbone.

- Alkali (high pH) prevents hydrogen bonding, so the two strands come apart.

This is one reason that all living things have to regulate their own pH, as studied by Kenyon students in the NSF-funded Bacterial pH Research Lab.

REPLICATION IS SEMICONSERVATIVE

Actually, three basic models for DNA replication are consistent with the requirement that the nucleotide sequence in one strand dictate, through Watson and Crick's base-pairing rules, the sequence of nucleotides in the other. These three models—conservative, semiconservative, and dispersive—are diagrammed in Figure. In 1958, Matthew Meselson and Franklin Stahl provided the experimental proof for the semiconservative model of DNA replication. *Escherichia coli* cells were grown for many generations in media containing $^{15}NH_4Cl$ as the sole nitrogen source. Thus, the nitrogen atoms in the purine and pyrimidine bases of the DNA in these cells were mostly ^{15}N, the stable heavy isotope of nitrogen. DNA isolated from cells grown on $^{15}NH_4^{\ 1}$ (the "0" generation cells) banded in the ultracentrifuge at a density corresponding to 1.724 g/mL, whereas DNA from cells grown for 4.1 generations on ^{14}N had a density of 1.710 g/mL. These bands represent "heavy" and "light" DNA, respectively. Significantly, DNA isolated from cells grown for just one generation on ^{14}N yielded just a single band corresponding to a density of 1.717 g/mL, halfway between heavy and light DNA, indicating that each DNA duplex molecule contained equal amounts of ^{15}N and ^{14}N. This result is consistent with the semiconservative model for DNA replication, at the same time ruling out the conservative model.

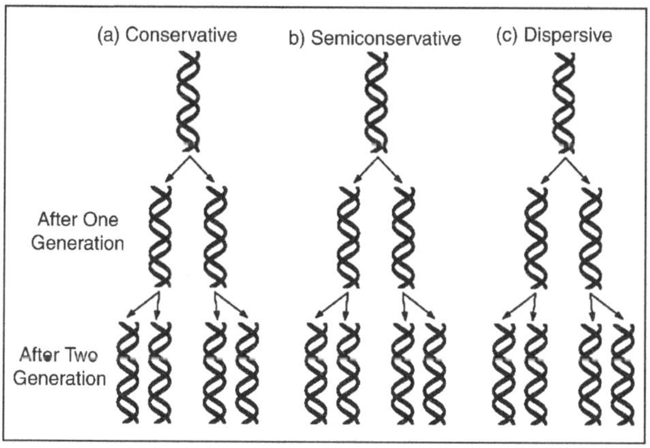

Fig. Three Models of DNA Replication Prompted by Watson and Crick's Double Helix Structure for DNA.

a. *Conservative:* Each strand of the DNA duplex is replicated and the two newly synthesized strands join to form one DNA double helix while the two parental strands remain associated with each other.

The products are one completely new DNA duplex and the original DNA duplex.

b. *Semi-conservative:* The two strands separate, and each strand is copied to generate a complementary strand. Each parental strand remains associated with its newly synthesized complement, so that each DNA duplex contains one parental strand and one new strand.

c. *Dispersive:* This model predicts that each of the four strands in the two daughter DNA duplexes contains both newly synthesized segments and segments derived from the parental strands.

After approximately two generations on ^{14}N, cells yielded DNA that gave two essentially equal bands upon ultracentrifugation, one at the intermediate density of 1.717 g/mL and the other at the light position, 1.710 g/mL, also in accord with semiconservative replication.

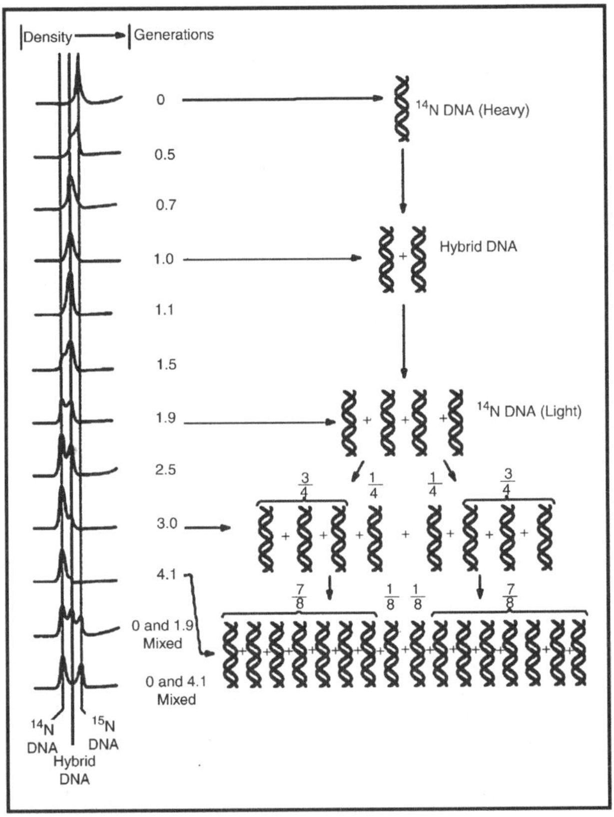

Fig. The Meselson and Stahl Experiment Demonstrating that DNA Replication is Semiconservative. On the Left are Shown Densitometric Traces Made of UV Absorption Photographs Taken of the Ultracentrifugation Cells Containing DNA Isolated from *E. coli* Grown for Various Generation Times After ^{15}N-labeling. The Photographs were Taken Once the Migration of the DNA in the Density Gradient had Reached Equilibrium.

It remained conceivable that DNA replication might follow a dispersive model, so that both strands of DNA in the chromosomes of cells after one generation on ^{14}N might be intermediate in density. Meselson and Stahl eliminated this possibility by the following experiment: DNA isolated from ^{15}N-labeled cells kept one generation on ^{14}N was heated at 100°C so that the DNA duplexes were denatured into their component single strands. When this heat-denatured DNA was analysed by CsCl density gradient ultracentrifugation, two distinct bands were observed, showing that one strand was ^{15}N-labeled and the other strand was ^{14}N-labeled.

A dispersive mode of replication would have yielded a single band of DNA of intermediate density. An interesting feature of density gradient ultracentrifugation is that the width of the band occupied by a macromolecular species in the gradient is inversely proportional to the square root of the molecular weight of the macromolecule. The band widths of the two bands seen with heat-denatured DNA both correspond to molecular masses one-half that of native DNA. This experiment established that DNA replication proceeded by a semiconservative mechanism and also verified that DNA was indeed composed of two polynucleotide strands of equal size.

8

Applications of DNA and RNA Technology

INTRODUCTION

The nucleic acids are the building blocks of living organisms. You may have heard of DNA described the same way.

Guess what? DNA is just one type of nucleic acid. Some other types are RNA, mRNA, and tRNA. All of these "Na's" work together to help cells replicate and build proteins.

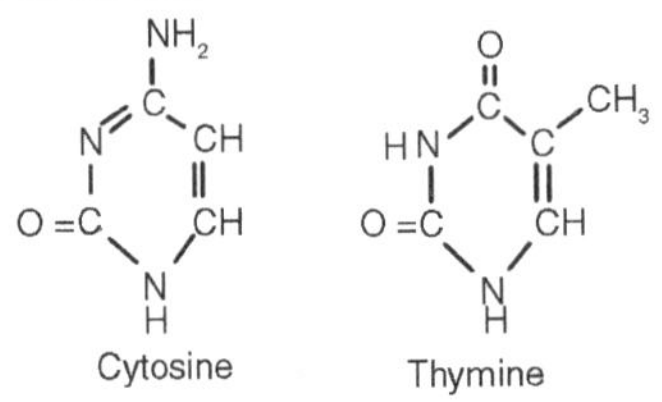

Pyrimidine Bases

BASICS

We already told you about the biggie nucleic acids (DNA, RNA, mRNA, tRNA). They are actually made up chains of base pairs stretching from only a few to millions. When those pairs combine in super long chains (DNA), they make a shape called a double helix.

The double helix shape is like a twisty ladder. The base pairs are the rungs. We're very close to talking about the biology of cells here. Back to the chemistry..

FIVE EASY PIECES

There are five easy parts of nucleic acids. All nucleic acids are made up of the same building blocks (monomers). Chemists call the monomers nucleotides. The five pieces are Uracil,Cytosine, Thymine, Adenine, and Guanine. Just as there are twenty (20) amino acids needed by humans to survive, there are five (5) nucleotides.

These nucleotides are made of three parts:

1. A five carbon sugar
2. A base that has a nitrogen (N) atom
3. An ion of phosphoric acid

The first isolation of what we now refer to as DNA was accomplished by Johann Friedrich Miescher *circa* 1870. He reported finding a weakly acidic substance of unknown function in the nuclei of human white blood cells, and named this material "nuclein".

A few years later, Miescher separated nuclein into protein and nucleic acid components. In the 1920's nucleic acids were found to be major components of chromosomes, small gene-carrying bodies in the nuclei of complex cells. Elemental analysis of nucleic acids showed the presence of phosphorus, in addition to the usual C, H, N & O. Unlike proteins, nucleic acids contained no sulfur.

Complete hydrolysis of chromosomal nucleic acids gave inorganic phosphate, 2-deoxyribose (a previously unknown sugar) and four different heterocyclic bases (shown in the following diagram). To reflect the unusual sugar component, chromosomal nucleic acids are called deoxyribonucleic acids, abbreviated DNA. Analogous nucleic acids in which the sugar component is ribose are termed ribonucleic acids, abbreviated RNA. The acidic character of the nucleic acids was attributed to the phosphoric acid moiety.

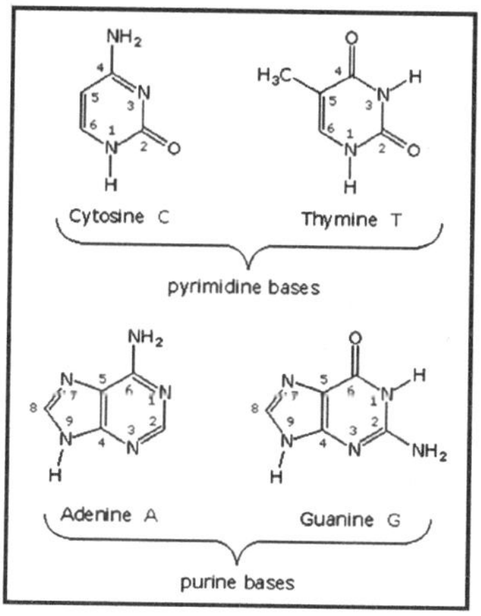

The two monocyclic bases shown here are classified as pyrimidines, and the two bicyclic bases are purines. Each has at least one N-H site at which an organic substituent may be attached. They are all polyfunctional bases, and may exist in tautomeric forms.

Base-catalysed hydrolysis of DNA gave four nucleoside products, which proved to be N-glycosides of 2'-deoxyribose combined with the heterocyclic amines. Structures and names for these nucleosides will be displayed above by clicking on the heterocyclic base diagram. The base components are coloured green, and the sugar is black. As noted in the 2'-deoxycytidine structure on the left, the numbering of the sugar carbons makes use of primed numbers to distinguish them from the heterocyclic base sites. The corresponding N-glycosides of the common sugar ribose are the building blocks of RNA, and are named adenosine, cytidine, guanosine and uridine (a thymidine analog missing the methyl group).

From this evidence, nucleic acids may be formulated as alternating copolymers of phosphoric acid (P) and nucleosides (N), as shown:

$$\sim P - N - P - N' - P - N'' - P - N''' - P - N \sim$$

At first the four nucleosides, distinguished by prime marks in this crude formula, were assumed to be present in equal amounts, resulting in a uniform structure, such as that of starch.

However, a compound of this kind, presumably common to all organisms, was considered too simple to hold the hereditary information known to reside in the chromosomes. This view was challenged in 1944, when Oswald Avery and colleagues demonstrated that bacterial DNA was likely the genetic agent that carried information from one organism to another in a process called "transformation".

He concluded that"*nucleic acids must be regarded as possessing biological specificity, the chemical basis of which is as yet undetermined.*" Despite this finding, many scientists continued to believe that chromosomal proteins, which differ across species, between individuals, and even within a given organism, were the locus of an organism's genetic information.

It should be noted that single celled organisms like bacteria do not have a well-defined nucleus. Instead, their single chromosome is associated with specific proteins in a region called a "nucleoid". Nevertheless, the DNA from bacteria has the same composition and general structure as that from multicellular organisms, including human beings.

Views about the role of DNA in inheritance changed in the late 1940's and early 1950's. By conducting a careful analysis of DNA from many sources, Erwin Chargaff found its composition to be species specific. In addition, he found that the amount of adenine (A) always equaled the amount of thymine (T), and the amount of guanine (G) always equaled the amount of cytosine (C), regardless of the DNA source. As set forth in the following table, the ratio of (A+T) to (C+G) varied from 2.70 to 0.35. The last two organisms are bacteria.

NUCLEOSIDE BASE DISTRIBUTION IN DNA

In a second critical study, Alfred Hershey and Martha Chase showed that when a bacterium is infected and genetically transformed by a virus, at least 80 per cent of the viral DNA enters the bacterial cell and at least 80 per cent of the viral protein remains outside. Together with the Chargaff findings this work established DNA as the repository of the unique genetic characteristics of an organism.

Organism	Base Composition (mole %)				Base Ratios		Ratio (A+T)/(G+C)
	A	G	T	C	A/T	G/C	
Human	30.9	19.9	29.4	19.8	1.05	1.00	1.52
Chicken	28.8	20.5	29.2	21.5	1.02	0.95	1.38
Yeast	31.3	18.7	32.9	17.1	0.95	1.09	1.79
Clostridium perfringens	36.9	14.0	36.3	12.8	1.01	1.09	2.70
Sarcina lutea	13.4	37.1	12.4	37.1	1.08	1.00	0.35

THE CHEMICAL NATURE OF DNA

The polymeric structure of DNA may be described in terms of monomeric units of increasing complexity. In the top shaded box of the following illustration, the three relatively simple components mentioned earlier are shown. Below that on the left, formulas for phosphoric acid and a nucleoside are drawn.Condensation polymerisation of these leads to the DNA formulation outlined above. Finally, a 5'- monophosphate ester, called a nucleo-tide may be drawn as a single monomer unit, shown in the shaded box to the right. Since a monophosphate ester of this kind is a strong acid (pK_a of 1.0), it will be fully ionised at the usual physiological pH (ca.7.4).

Names for these DNA components are given in the table to the right of the diagram. Isomeric 3'-monophospate nucleotides are also known, and both isomers are found in cells.

They may be obtained by selective hydrolysis of DNA through the action of nuclease enzymes. Anhydride-like di- and tri-phosphate nucleotides have been identified as important energy carriers in biochemical reactions, the most common being ATP (adenosine 5'-triphosphate).

Names of DNA Base Derivatives

Base	Nucleoside	5'-Nucleotide
Adenine	2'-Deoxyadenosine	2'-Deoxyadenosine-5'-monophosphate
Cytosine	2'-Deoxycytidine	2'-Deoxycytidine-5'-monophosphate
Guanine	2'-Deoxyguanosine	2'-Deoxyguanosine-5'-monophosphate
Thymine	2'-Deoxythymidine	2'-Deoxythymidine-5'-monophosphate

A complete structural representation of a segment of the DNA polymer formed from 5'-nucleotides. Several important characteristics of this formula should be noted.

- First, the remaining P-OH function is quite acidic and is completely ionised in biological systems.
- Second, the polymer chain is structurally directed. One end is different from the other.
- Third, although this appears to be a relatively simple polymer, the

possible permutations of the four nucleosides in the chain become very large as the chain lengthens.

- Fourth, the DNA polymer is much larger than originally believed. Molecular weights for the DNA from multicellular organisms are commonly 10^9 or greater.

Information is stored or encoded in the DNA polymer by the pattern in which the four nucleotides are arranged. To access this information the pattern must be "read" in a linear fashion, just as a bar code is read at a supermarket checkout. Because living organisms are extremely complex, a correspondingly large amount of information related to this complexity must be stored in the DNA.

Consequently, the DNA itself must be very large, as noted above. Even the single DNA molecule from an *E. coli* bacterium is found to have roughly a million nucleotide units in a polymer strand, and would reach a millimeter in length if stretched out. The nuclei of multicellular organisms incorporate chromosomes, which are composed of DNA combined with nuclear proteins called histones.

The fruit fly has 8 chromosomes, humans have 46 and dogs 78 (note that the amount of DNA in a cell's nucleus does not correlate with the number of chromosomes). The DNA from the smallest human chromosome is over ten times larger than *E. coli* DNA, and it has been estimated that the total DNA in a human cell would extend to 2 meters in length if unraveled. Since the nucleus is only about 5ìm in diameter, the chromosomal DNA must be packed tightly to fit in that small volume.

In addition to its role as a stable informational library, chromosomal DNA must be structured or organised in such a way that the chemical machinery of the cell will have easy access to that information, in order to make important molecules such as polypeptides.

Furthermore, accurate copies of the DNA code must be created as cells divide, with the replicated DNA molecules passed on to subsequent cell generations, as well as to progeny of the organism.

RNA, A DIFFERENT NUCLEIC ACID

The high molecular weight nucleic acid, DNA, is found chiefly in the nuclei of complex cells, known as eucaryotic cells, or in the nucleoid regions ofprocaryotic cells, such as bacteria. It is often associated with proteins that help to pack it in a usable fashion. In contrast, a lower molecular weight, but much more abundant nucleic acid, RNA, is distributed throughout the cell, most commonly in small numerous organelles called ribosomes.

Three kinds of RNA are identified, the largest subgroup (85 to 90 per cent) being ribosomal RNA, rRNA, the major component of ribosomes, together with proteins. The size of rRNA molecules varies, but is generally less than a thousandth the size of DNA. The other forms of RNA are messenger RNA, mRNA, and transfer RNA, tRNA. Both have a more transient existence and are smaller than rRNA.

All these RNA's have similar constitutions, and differ from DNA in two important respects. As shown in the following diagram, the sugar component of RNA is ribose, and the pyrimidine base uracil replaces the thymine base of DNA. The RNA's play a vital role in the transfer of information (transcription) from the DNA library to the protein factories called ribosomes, and in the interpretation of that information (translation) for the synthesis of specific polypeptides. These functions will be described later.

A complete structural representation of a segment of the RNA polymer formed from 5'-nucleotides.

THE SECONDARY STRUCTURE OF DNA

In the early 1950's the primary structure of DNA was well established, but a firm understanding of its secondary structure was lacking. Indeed, the situation was similar to that occupied by the proteins a decade earlier, before the alpha helix and pleated sheet structures were proposed by Linus Pauling. Many researchers grappled with this problem, and it was generally conceded that the molar equivalences of base pairs (A & T and C & G) discovered by Chargaff would be an important factor. A scientist, working at King's College, obtained X-ray diffraction evidence that suggested a long helical structure of uniform thickness.

THE DOUBLE HELIX

After many trials and modifications, an ingenious double helix model for the secondary structure of DNA. Two strands of DNA were aligned anti-parallel to each other, *i.e.* with opposite 3' and 5' ends, as shown in part a of the following diagram.

Complementary primary nucleotide structures for each strand allowed intra-strand hydrogen bonding between each pair of bases. These complementary strands are coloured red and green in the diagram. Coiling these coupled strands then leads to a double helix structure, shown as cross-linked ribbons in part b of the diagram. A space-filling molecular model of a short segment is displayed in part c on the right.

The helix shown here has ten base pairs per turn, and rises 3.4 Å in each turn. This right-handed helix is the favoured conformation in aqueous systems, and has been termed the B-helix. Two alternating grooves are evident, a wide and dccp major groovc, and a shallow and narrow minor groove. Other molecules, including polypeptides, may insert into these grooves, and in so doing perturb the chemistry of DNA. Other helical structures of DNA have also been observed, and are designated by letters (*e.g.* A and Z).

The Double Helix Structure for DNA

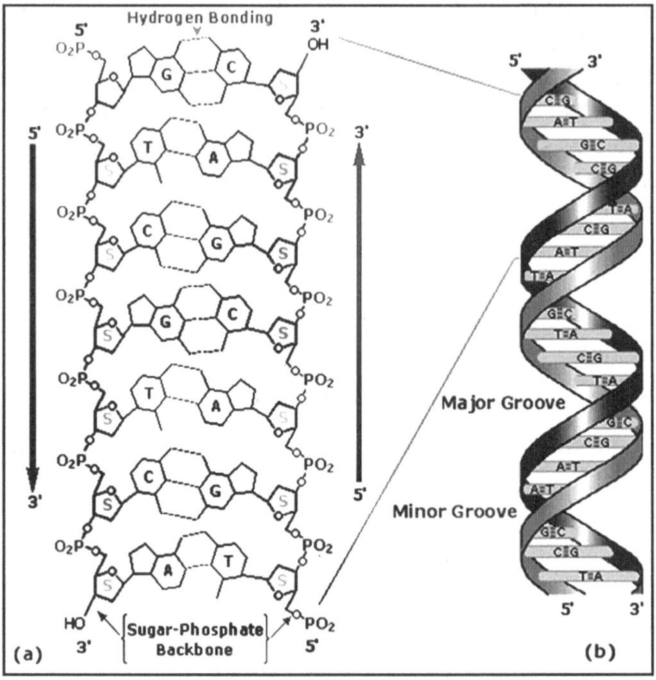

Fig. Space-Filling Molecular Model.

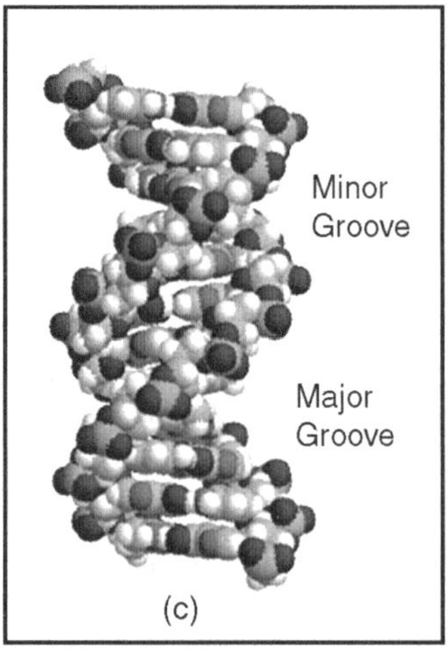

Fig. Model of a Short DNA Segment

DNA Replication

In their announcement of a double helix structure for DNA, It is stated, *"It has not escaped our notice that the specific pairing we have postulated immediately suggests a possible copying mechanism for the genetic material."*. The essence of this suggestion is that, if separated, each strand of the molecule might act as a template on which a new complementary strand might be assembled, leading finally to two identical DNA molecules. Indeed, replication does take place in this fashion when cells divide, but the events leading up to the actual synthesis of complementary DNA strands are sufficiently complex that they will not be described in any detail.

As depicted in the following drawing, the DNA of a cell is tightly packed into chromosomes. First, the DNA is wrapped around small proteins called histones (coloured pink below). These bead-like structures are then further organised and folded into chromatin aggregates that make up the chromosomes. An overall packing efficiency of 7,000 or more is thus achieved. Clearly a sequence of unfolding events must take place before the information encoded in the DNA can be used or replicated.

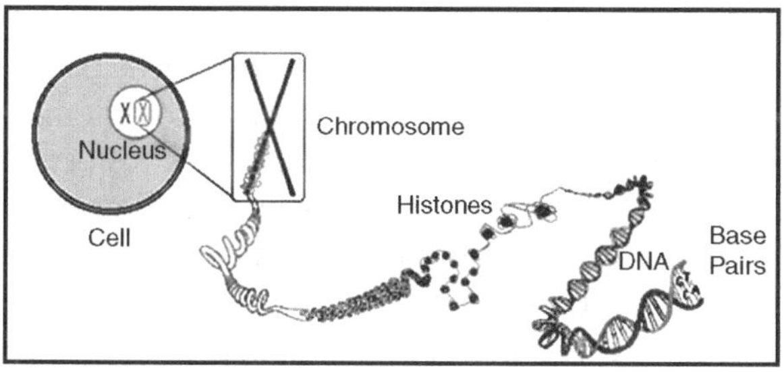

Once the double stranded DNA is exposed, a group of enzymes act to accomplish its replication.

These are described briefly here:

- *Topoisomerase:* This enzyme initiates unwinding of the double helix by cutting one of the strands.
- *Helicase:* This enzyme assists the unwinding. Note that many hydrogen bonds must be broken if the strands are to be separated..
- *SSB:* A single-strand binding-protein stabilises the separated strands, and prevents them from recombining, so that the polymerisation chemistry can function on the individual strands.
- *DNA Polymerase:* This family of enzymes link together nucleotide triphosphate monomers as they hydrogen bond to complementary bases. These enzymes also check for errors (roughly ten per billion), and make corrections.
- *Ligase:* Small unattached DNA segments on a strand are united by this enzyme.

Polymerisation of nucleotides takes place by the phosphorylation reaction described by the following equation.

Di- and triphosphate esters have anhydride-like structures and are consequently reactive phosphorylating reagents, just as carboxylic anhydrides are acylating reagents. Since the pyrophosphate anion is a better leaving group than phosphate, triphosphates are more powerful phosphorylating agents than are diphosphates.

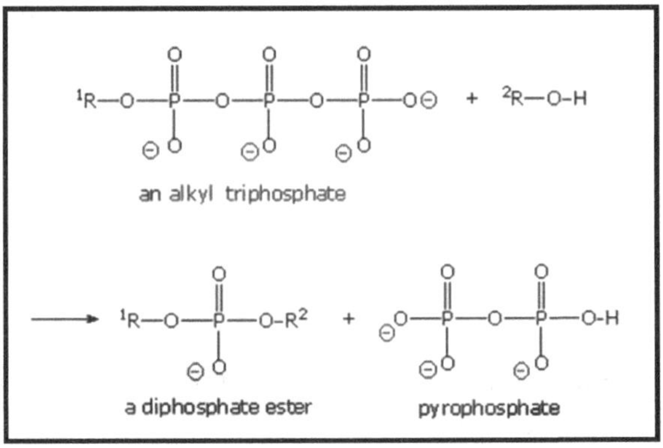

Formulas for the corresponding 5'-derivatives of adenosine will be displayed, and similar derivatives exist for the other three common nucleosides. The DNA polymerisation process that builds the complementary strands in replication, could in principle take place in two ways. Referring to the general equation above, R^1 could represent the next nucleotide unit to be attached to the growing DNA strand, with R^2 being this strand. Alternatively, these assignments could be reversed.

In practice, the former proves to be the best arrangement. Since triphosphates are very reactive, the lifetime of such derivatives in an aqueous environment is relatively short. However, such derivatives of the individual nucleosides are repeatedly synthesised by the cell for a variety of purposes, providing a steady supply of these reagents.

In contrast, the growing DNA segment must maintain its functionality over the entire replication process, and can not afford to be changed by a spontaneous hydrolysis event. As a result, these chemical properties are best accomodated by a polymerisation process that proceeds at the 3'-end of the growing strand by 5'-phosphorylation involving a nucleotide triphosphate.

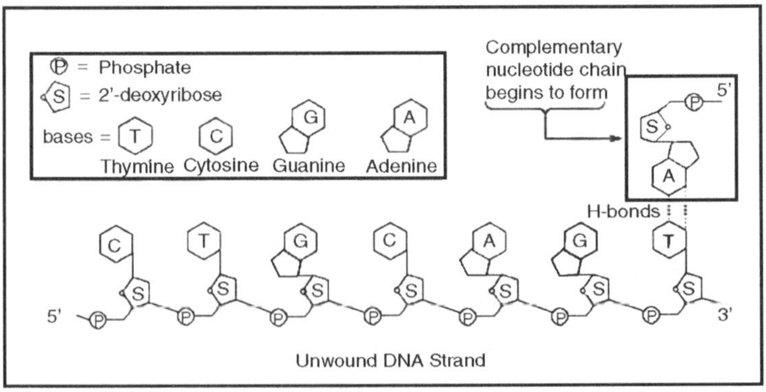

The polymerisation mechanism described here is constant. It always extends the developing DNA segment towards the 3'-end (*i.e.* when a nucleotide triphosphate attaches to the free 3'-hydroxyl group of the strand, a new 3'-hydroxyl is generated). There is sometimes confusion on this point, because the original DNA strand that serves as a template is read from the 3'-end towards the 5'-end, and authors may not be completely clear as to which terminology is used.

Because of the directional demand of the polymerisation, one of the DNA strands is easily replicated in a continuous fashion, whereas the other strand can only be replicated in short segmental pieces. This is illustrated in the following diagram. Separation of a portion of the double helix takes place at a site called the replication fork. As replication of the separate strands occurs, the replication fork moves away (to the left in the diagram), unwinding additional lengths of DNA.

Since the fork in the diagram is moving towards the 5'-end of the red-coloured strand, replication of this strand may take place in a continuous fashion (building the new green strand in a 5' to 3' direction). This continuously formed new strand is called the leading strand. In contrast, the replication fork moves towards the 3'-end of the original green strand, preventing continuous polymerisation of a complementary new red strand. Short segments of complementary DNA, called Okazaki fragments, are produced, and these are linked together later by the enzyme ligase. This new DNA strand is called the lagging strand.

When you consider that a human cell has roughly 10^9 base pairs in its DNA, and may divide into identical daughter cells in 14 to 24 hours, the efficiency of DNA replication must be extraordinary. The procedure described above will replicate about 50 nucleotides per second, so there must be many thousand such replication sites in action during cell division. A given length of double stranded DNA may undergo strand unwinding at numerous sites in response to promoter actions. The unraveled "bubble" of single stranded DNA has two replication forks, so assembly of new complementary strands may proceed in two directions. The polymerisations associated with several such bubbles fuse together to achieve full replication of the entire DNA double helix. Note that the events shown proceed from top to bottom in the diagram.

MECHANISM OF DNA REPLICATION

Replication of the E. *coli* chromosome begins at a single replication origin and proceeds bidirectionally until the two replication forks meet. At each replication fork, both leading and lagging strand syntheses are catalyzed by a single multiprotein replication machine, the so-calledreplisome, which consists of DNA-unwinding proteins; the priming apparatus, or primosome, which is needed to initiate, or "prime," DNA replication; and DNA polymerase III holoenzyme with two equivalents of "core" polymerase, one for the leading strand and one for the lagging strand. As this replisome follows the replication fork, the template for lagging strand synthesis (the strand running 5' → 3' in the direction of fork movement) must be looped around so that it can be read in the 3' → 5' direction.

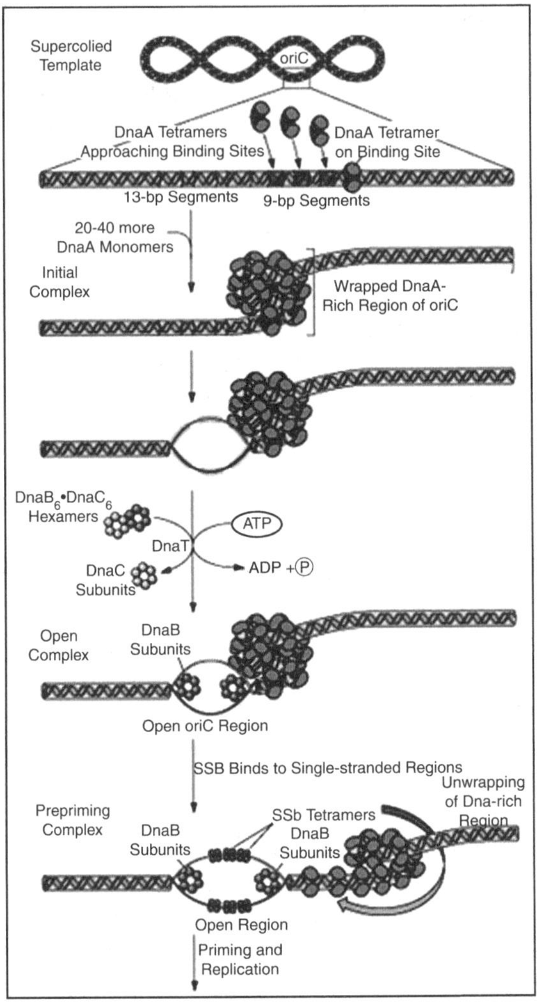

Fig. The Probable Course of Events During Initiation of E. *coli* Replication at *oriC*. Tetramers of DnaA Protein Bind at each of the Four 9-bp Repeats in *oriC*; then Additional DnaA Protein Binds to give a Nucleosomelike Structure with DNA on the Outside. The Three 13-bp A:T-rich Regions are then "Melted" to Yield an Open Complex, and the DnaB Protein Delivered from the DnaB:DnaC Protein Complex Enters. The Helicase Activity of DnaB Unwinds the Duplex, Displacing DnaA Protein. SSB Protein Binds to Single-stranded DNA as it is Generated, Preventing its Re-annealing.

INITIATION

Replication of the E. *coli* chromosome is initiated at a unique site, *oriC*. The 245-bp sequence of *oriC* contains elements that are highly conserved among Gram-negative bacteria. Within *oriC* are four 9-bp repeats. DnaA protein, a 52-kD polypeptide, is the*initiation factor,* which recognizes and binds to these repeats DnaA protein binding is cooperative; once the four 9-bp repeats are occupied, 20 to 40 additional DnaA monomers bind so that the entire *oriC* region is complexed with DnaA protein.

HU, a histone-like protein, prevents non-specific initiation at sites other than *oriC*. The resulting complex resembles a nucleosome, with negatively supercoiled *oriC* DNA wrapped around a DnaA core.

The DnaA protein then mediates the separation of the strands of the DNA duplex by acting on three AT-rich tandem repeats (consensus sequence = 5'-GATCTNTNTTNTT, where N = any nucleotide) located at the 5'-end of the sequence defining *oriC*. Formation of this 45-bp "open complex" by DnaA protein is ATP-dependent.

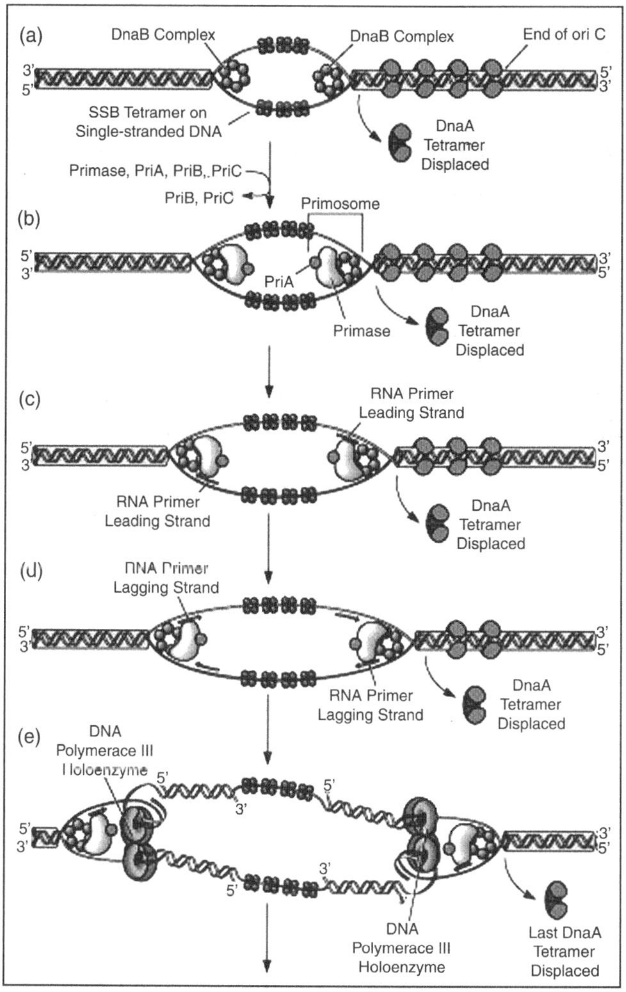

Fig. A Representation of Replisome Assembly at each Replication Fork.

Next, DnaB protein (a hexamer of 50-kD subunits) binds to the "open"*oriC*. DnaB protein is delivered to *oriC* by DnaC protein (29 kD) in the form of a hexameric (DnaB: DnaC:ATP)$_6$ complex, but DnaC protein does not enter the protein assemblage at *oriC*. Delivery of DnaB protein by DnaC protein is assisted by DnaT protein. The addition of DnaB protein completes assembly of the pre-priming complex.ATP hydrolysis drives the formation of this complex. DnaB protein has helicase activity and it further unwinds the DNA in the pre-priming complex in both directions, assisted by DNA gyrase. SSB tetramers coat single-stranded regions as they arise. Unwinding exposes the base sequence of the strands so that RNA primers can be synthesized by primase, and the strands can be read as templates by the replicative polymerase.

The addition of primase to the pre-priming complex completes the formation of the primosome, a protein machine of about 700 kD that synthesizes the RNA primer essential to DNA synthesis. Interestingly, the AT-rich tandem repeats of *oriC*resemble sequences with which RNA polymerase interacts. Note that *two*primosomes form at *oriC*, one for each replication fork. Once the primosome has primed leading strand synthesis, it remains associated with the lagging strand and periodically primes Okazaki fragment synthesis, as required.

ELONGATION

The double helix must be unwound ahead of the advancing replication fork. Unwinding is driven by ATP hydrolysis; DnaB protein has the helicase activity associated with the primosome. DnaB moves in the 5' → 3' direction along the leading strand template, hydrolyzing 2 ATPs for each base pair that it separates. Protein-protein interaction between a *t* subunit of the DNA polymerase III holoenzyme and DnaB is essential for rapid replication fork progression. PriA protein is, like DnaB, a helicase associated with the primosome, but PriA moves 3' → 5' along the lagging strand. The binding of SSB protein to the single-stranded regions created behind these proteins prevents re-annealing.

Figure depicts a model for the replication of E. *coli* DNA as it occurs at one of the replication forks. Two "core" DNA polymerase III units are present. One is synthesizing the leading strand, using the parental 3' → 5' strand as template.

The other synthesizes the lagging strand using the 5' → 3' parental strand as template. The lagging strand template must be looped out so that it can be copied in the 3' → 5' direction. The*"core" pol III* on this lagging strand has completed the synthesis of an RNA-primedOkazaki fragment when it encounters the 5'-end of the previous fragment. Synthesis of the next RNA primer by the primosome triggers the γ complex to disassemble the β_2-sliding clamp, releasing the lagging strand from the DNA polymerase III holoenzyme. The γ complex assembles a new β_2-sliding clamp at the 3'-OH end of the next RNA primer, re-attaching the lagging strand "core" polymerase III to begin a new round of lagging strand synthesis.

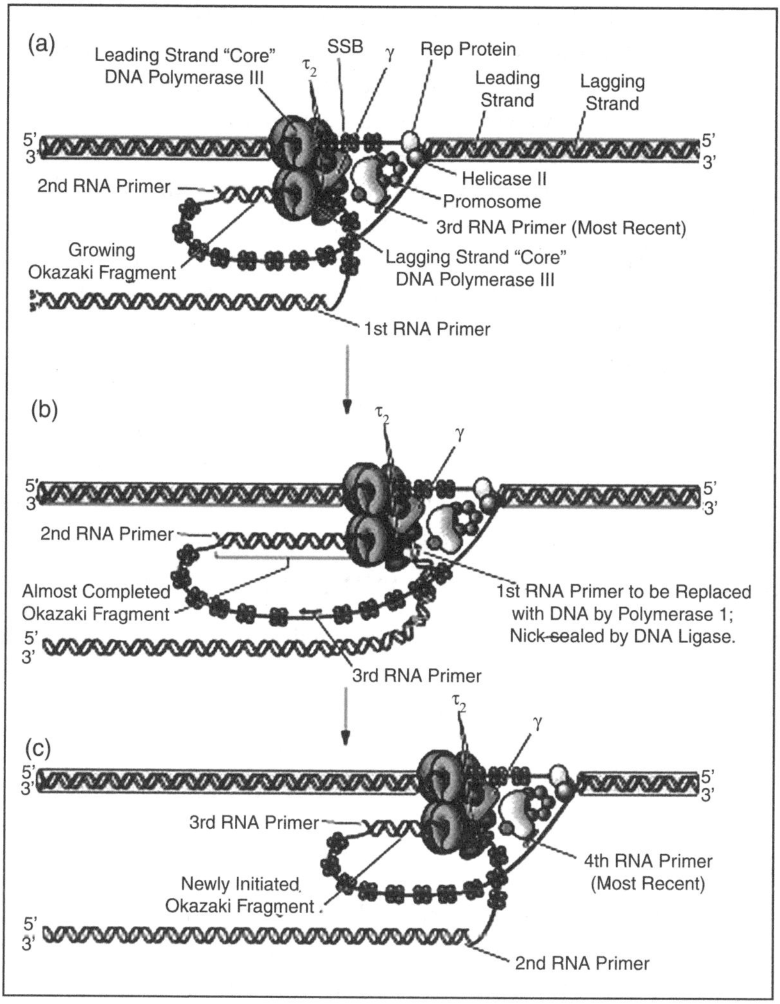

Fig. The Asymmetric Dimer Model for E. *coli* DNA Replication. Two Helicases (Rep protein and Helicase II) Act in Concert to Unwind the DNA. The Replisome Contains Two DNA "Core" Polymerase III Complexes that Remain Physically Associated even though their Tasks are Different: One Performs Leading Strand Synthesis while the Other Synthesizes Okazaki Fragments, the Precursors to the Lagging Strand.

TERMINATION

Located diametrically opposite from *oriC* on the E. *coli* circular map is a terminus region, the Ter, or τ, locus. The bidirectionally moving replication forks meet here and replication is terminated. The *Ter* region contains a number of short DNA sequences containing a consensus core element 5'-GTGTGTTGT. These *Ter* sequences act as terminators; clusters of three or four *Ter* sequences are organized into two sets inversely oriented with respect to one another. One set blocks the clockwise-moving replication fork, and its inverted counterpart blocks the counterclockwise-moving replication fork. A *Ter* sequence element will impede replication fork progression *only if oriented in the proper direction* with respect to the approaching replication fork and then only if a specific 36-kD replication termination protein, Tus protein, is bound to it. Tus protein is a contrahelicase.

That is, Tus protein prevents the DNA duplex from unwinding by blocking progression of the replication fork and inhibiting the ATP-dependent DnaB helicase activity. Mutations in either the *Ter* locus or the gene encoding Tus protein do not grossly affect DNA replication, demonstrating that this termination mechanism is not essential.

Final synthesis of both duplexes is completed. Replication usually leaves the circular progeny chromosomes intertwined by 20 to 30 coils about each other, a so-called catenated state. In order to disengage the individual duplexes from each other prior to their distribution to daughter cells, double-stranded cuts must be made so that the double helices can pass through one another. Topoisomerase II (DNA gyrase) can catalyze this process.

REPAIR OF DNA DAMAGE AND REPLICATION ERRORS

One of the benefits of the double stranded DNA structure is that it lends itself to repair, when structural damage or replication errors occur.

Several kinds of chemical change may cause damage to DNA:

- Spontaneous hydrolysis of a nucleoside removes the heterocyclic base component.
- Spontaneous hydrolysis of cytosine changes it to a uracil.
- Various toxic metabolites may oxidise or methylate heterocyclic base components.
- Ultraviolet light may dimerise adjacent cytosine or thymine bases.

All these transformations disrupt base pairing at the site of the change, and this produces a structural deformation in the double helix.. Inspection-repair enzymes detect such deformations, and use the undamaged nucleotide at that site as a template for replacing the damaged unit. These repairs reduce errors in DNA structure from about one in ten million to one per trillion.

RNA AND PROTEIN SYNTHESIS

The genetic information stored in DNA molecules is used as a blueprint for making proteins. Why proteins? Because these macromolecules have diverse primary, secondary and tertiary structures that equip them to carry out the numerous functions necessary to maintain a living organism.

As noted in the protein, these functions include:

- Structural integrity (hair, horn, eye lenses etc.).

- Molecular recognition and signaling (antibodies and hormones).
- Catalysis of reactions (enzymes)..
- Molecular transport (hemoglobin transports oxygen).
- Movement (pumps and motors).

The critical importance of proteins in life processes is demonstrated by numerous genetic diseases, in which small modifications in primary structure produce debilitating and often disasterous consequences.

Such genetic diseases incluse Tay-Sachs, phenylketonuria (PKU), sickel cell anemia, achondroplasia, and Parkinson disease. The unavoidable conclusion is that proteins are of central importance in living cells, and that proteins must therefore be continuously prepared with high structural fidelity by appropriate cellular chemistry.

Early geneticists identifed genes as hereditary units that determined the appearance and/ or function of an organism (*i.e.* its phenotype). We now define genes as sequences of DNA that occupy specific locations on a chromosome. The original proposal that each gene controlled the formation of a single enzyme has since been modified as: one gene = one polypeptide. The intriguing question of how the information encoded in DNA is converted to the actual construction of a specific polypeptide has been the subject of numerous studies, which have created the modern field of Molecular Biology.

CENTRAL DOGMA AND TRANSCRIPTION

It is proposed that information flows from DNA to RNA in a process called transcription, and is then used to synthesise polypeptides by a process called translation. Transcription takes place in a manner similar to DNA replication. A characteristic sequence of nucleotides marks the beginning of a gene on the DNA strand, and this region binds to a promoter protein that initiates RNA synthesis. The double stranded structure unwinds at the promoter site., and one of the strands serves as a template for RNA formation, as depicted in the following diagram.

The RNA molecule thus formed is single stranded, and serves to carry information from DNA to the protein synthesis machinery called ribosomes. These RNA molecules are therefore called messenger-RNA (mRNA).

An important distinction must be made here. One of the DNA strands in the double helix holds the genetic information used for protein synthesis. This is called the sense strand, or information strand. The complementary strand that binds to the sense strand is called the anti-sense strand and it serves as a template for generating a mRNA molecule that delivers a copy of the sense strand information to a ribosome.

The promoter protein binds to a specific nucleotide sequence that identifies the sense strand, relative to the anti-sense strand. RNA synthesis is then initiated in the 3' direction, as nucleotide triphosphates bind to complementary bases on the template strand, and are joined by phosphate diester linkages. An animation of this process for DNA replication was presented earlier.

A characteristic "stop sequence" of nucleotides terminates the RNA synthesis. The messenger molecule (coloured orange above) is released into the cytoplasm to find a ribosome, and the DNA then rewinds to its double helix structure.

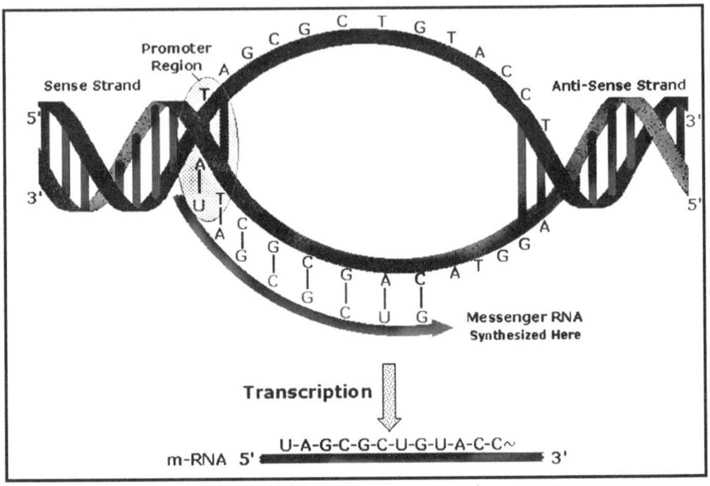

In eucaryotic cells the initially transcribed m-RNA molecule is usually modified and shortened by an "editing" process that removes irrelevent material. The DNA of such organisms is often thousands of times larger and more complex than that composing the single chromosome of a procaryotic bacterial cell.

This difference is due in part to repetitive nucleotide sequences (ca. 25 per cent iu the human genome). Furthermore, over 95 per cent of human DNA is found in intervening sequences that separate genes and parts of genes. The informational DNA segments that make up genes are called exons, and the non-coding segments are called introns.

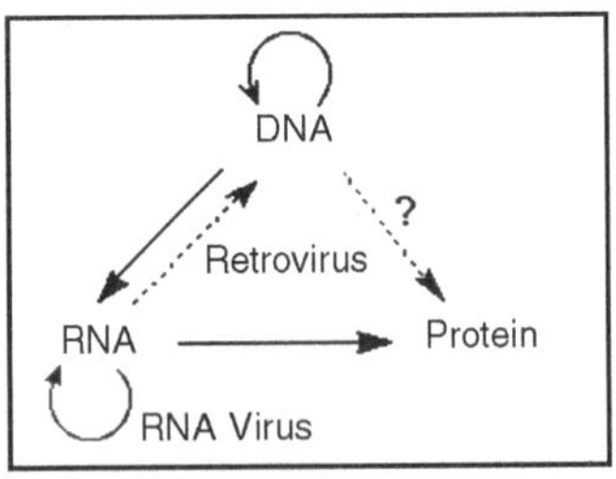

Before the mRNA molecule leaves the nucleus, the non-sense bases that make up the introns are cut out, and the informationally useful exons are joined together in a step known as RNA splicing. In this fashion shorter mRNA molecules carrying the blueprint for a specific protein are sent on their way to the ribosome factories.

The Central Dogma of molecular biology, which at first was formulated as a simple linear progression of information from DNA to RNA to Protein, is summarised in the following illustration. The replication process on the left consists of passing information from a parent DNA molecule to daughter molecules. The middle transcription process copies this information to a mRNA molecule. Finally, this information is used by the chemical machinery of the ribosome to make polypeptides.

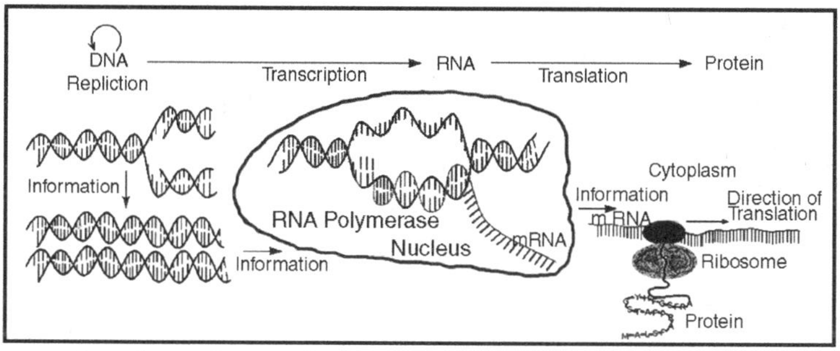

As more has been learned about these relationships, the central dogma has been refined to the representation displayed on the right. The dark blue arrows show the general, well demonstrated, information transfers noted above. It is now known that an RNA-dependent DNA polymerase enzyme, known as a reverse transcriptase, is able to transcribe a single-stranded RNA sequence into double-stranded DNA (magenta arrow).

Such enzymes are found in all cells and are an essential component of retroviruses (*e.g.* HIV), which require RNA replication of their genomes (green arrow). Direct translation of DNA information into protein synthesis (orange arrow) has not yet been observed in a living organism. Finally, proteins appear to be an informational dead end, and do not provide a structural blueprint for either RNA or DNA. In the following section the last fundamental relationship, that of structural information translation from mRNA to protein, will be described

TRANSLATION

Translation is a more complex process than transcription. This would, of course, be expected. After all, the coded messages produced by the Enigma machine could be copied easily, but required a considerable decoding effort before they could be read with understanding.

In a similar sense, DNA replication is simply a complementary base pairing exercise, but the translation of the four letter (bases) alphabet code of RNA to the twenty letter (amino acids) alphabet of protein literature is far from trivial. Clearly, there could not be a direct one-to-one correlation of bases to amino acids, so the nucleotide letters must form short words or codons that define specific amino acids.

Many questions pertaining to this genetic code were posed:

- How many RNA nucleotide bases designate a specific amino acid?
  If separate groups of nucleotides, called codons, serve this purpose, at least three are needed. There are $4^3 = 64$ different nucleotide triplets, compared with $4^2 = 16$ possible pairs.
- Are the codons linked separately or do they overlap?
  Sequentially joined triplet codons will result in a nucleotide chain three times longer than the protein it describes. If overlapping codons are used then fewer total nucleotides would be required.
- If triplet segments of mRNA designate specific amino acids in the protein, how are the codons identified?
  For the sequence ~CUAGGU~ are the codons CUA & GGU or ~C, UAG & GU~ or ~CU, AGG & U~?
- Are all the codon words the same size?

In Morse code the most widely used letters are shorter than less common letters. Perhaps nature employs a similar scheme. Physicists and mathematicians, as well as chemists and microbiologists all contributed to unravelling the genetic code.

Although earlier proposals assumed efficient relationships that correlated the nucleotide codons uniquely with the twenty fundamental amino acids, it is now apparent that there is considerable redundancy in the code as it now operates. Furthermore, the code consists exclusively of non-overlapping triplet codons. Clever experiments provided some of the earliest breaks in deciphering the genetic code. It is found that RNA from many different organisms could initiate specific protein synthesis when combined with broken E.coli cells (the enzymes remain active). A synthetic polyuridine RNA induced synthesis of poly-phenylalanine, so the UUU codon designated phenylalanine. Likewise an alternating ~CACA~ RNA led to synthesis of a ~His-Thr-His-Thr~ polypeptide.

The following table presents the present day interpretation of the genetic code. Note that this is the RNA alphabet, and an equivalent DNA codon table would have all the U nucleotides replaced by T. Methionine and tryptophan are uniquely represented by a single codon. At the other extreme, leucine is represented by eight codons. The average redundancy for the twenty amino acids is about three. Also, there are three stop codons that terminate polypeptide synthesis.

RNA Codons for Protein Synthesis

The translation process is fundamentally straightforward. Thc mRNΛ strand bearing the transcribed code for synthesis of a protein interacts with relatively small RNA molecules (about 70-nucleotides) to which individual amino acids have been attached by an ester bond at the 3'-end. These transfer

RNA's(tRNA) have distinctive three-dimensional structures cosisting of loops of single-stranded RNA connected by double stranded segments.

First Position	Second Position: U	C	A	G	Third Position
U	UUU Phe [F]	UCU Ser [S]	UAU Tyr [Y]	UGU Cys [C]	U
	UUC Phe [F]	UCC Ser [S]	UAC Tyr [Y]	UGC Cys [C]	C
	UUA Leu [L]	UCA Ser [S]	UAA Stop	UGA Stop	A
	UUG Leu [L]	UCG Ser [S]	UAG Stop	UGG Trp [W]	G
C	CUU Leu [L]	CCU Pro [P]	CAU His [H]	CGU Arg [R]	U
	CUC Leu [L]	CCC Pro [P]	CAC His [H]	CGC Arg [R]	C
	CUA Leu [L]	CCA Pro [P]	CAA Gln [Q]	CGA Arg [R]	A
	CUG Leu [L]	CCG Pro [P]	CAG Gln [Q]	CGG Arg [R]	G
A	AUU Ile [I]	ACU Thr [T]	AAU Asn [N]	AGU Ser [S]	U
	AUC Ile [I]	ACC Thr [T]	AAC Asn [N]	AGC Ser [S]	C
	AUA Ile [I]	ACA Thr [T]	AAA Lys [K]	AGA Arg [R]	A
	AUG Met [M]	ACG Thr [T]	AAG Lys [K]	AGG Arg [R]	G
G	GUU Val [V]	GCU Ala [A]	GAU Asp [D]	GGU Gly [G]	U
	GUC Val [V]	GCC Ala [A]	GAC Asp [D]	GGC Gly [G]	C
	GUA Val [V]	GCA Ala [A]	GAA Glu [E]	GGA Gly [G]	A
	GUG Val [V]	GCG Ala [A]	GAG Glu [E]	GGG Gly [G]	G

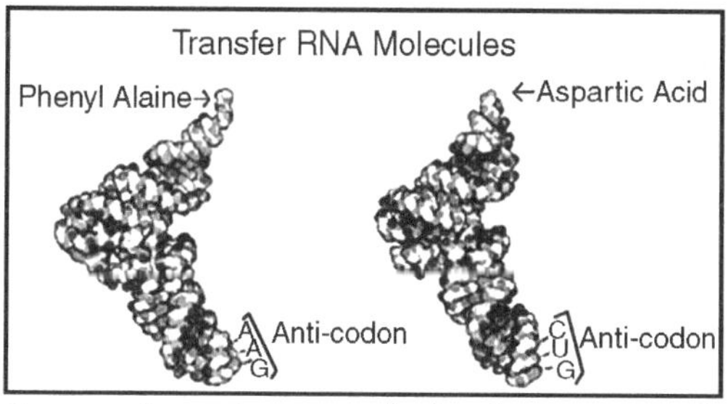

This cloverleaf secondary structure is further wrapped into an "L-shaped" assembly, having the amino acid at the end of one arm, and a characteristic anti-codon region at the other end. The anti-codon consists of a nucleotide triplet that is the complement of the amino acid's codon(s). Models of two such tRNA molecules are shown to the right.

When read from the top to the bottom, the anti-codons depicted here should complement a codon in the previous table. A cell's protein synthesis takes place in organelles called ribosomes. Ribosomes are complex structures made up of two distinct and separable subunits (one about twice the size of the other). Each subunit is composed of one or two RNA molecules (60-70 per cent) associated with 20 to 40 small proteins (30-40 per cent). The ribosome accepts a mRNA molecule, binding initially to a characteristic nucleotide sequence at the 5'-end (coloured light blue in the following diagram).

This unique binding assures that polypeptide synthesis starts at the right codon. A tRNA molecule with the appropriate ant-codon then attaches at the starting point and this is followed by a series of adjacent tRNA attachments, peptide bond formation and shifts of the ribosome along the mRNA chain to expose new codons to the ribosomal chemistry.

The following diagram is designed as a slide show illustrating these steps. The outcome is synthesis of a polypeptide chain corresponding to the mRNA blueprint.

A "stop codon" at a designated position on the mRNA terminates the synthesis by introduction of a "Release Factor".

MECHANISMS OF ALKYLATING AGENTS

Alkylating agents work by three different mechanisms all of which achieve the same end result - disruption of DNA function and cell death.In the first mechanism an alkylating agent (represented in the figure below as a pink star) attaches alkyl groups (small carbon compounds-depicted as pink triangles) to DNA bases.

This alteration results in the DNA being fragmented by repair enzymes in their attempts to replace the alkylated bases (frame 3 of the diagram below). Alkylated bases prevent DNA synthesis and RN transcription from the affected DNA.

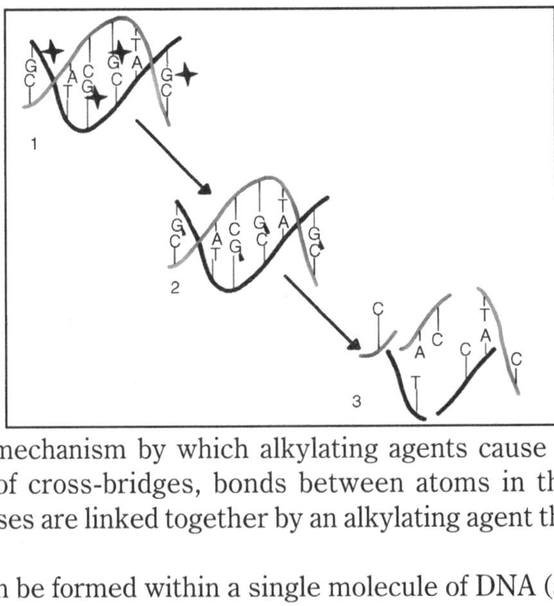

A second mechanism by which alkylating agents cause DNA damage is the formation of cross-bridges, bonds between atoms in the DNA. In this process, two bases are linked together by an alkylating agent that has two DNA binding sites.

Bridges can be formed within a single molecule of DNA (as shown below) or a cross-bridge may connect two different DNA molecules. Cross-linking prevents DNA from being separated for synthesis or transcription.

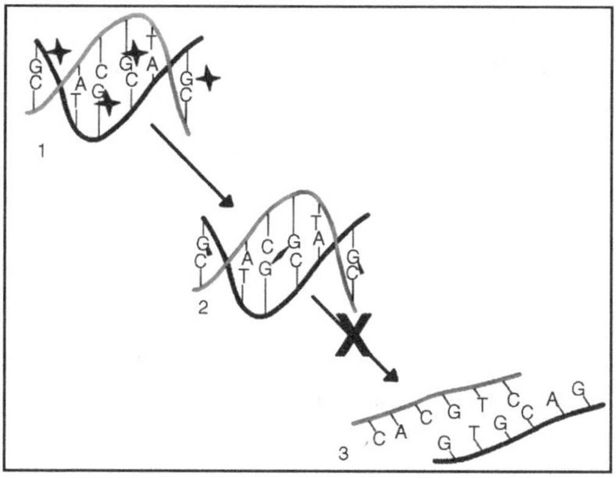

The third mechanism of action of alkylating agents is the induction of mispairing of the nucleotidesleading to mutations. In a normal DNA double helix, A always pairs with (is across from) T and G always pairs with C. As the figure below shows, alkylated G bases may erroneously pair with Ts. If this altered pairing is not corrected it may lead to a permanent mutation.

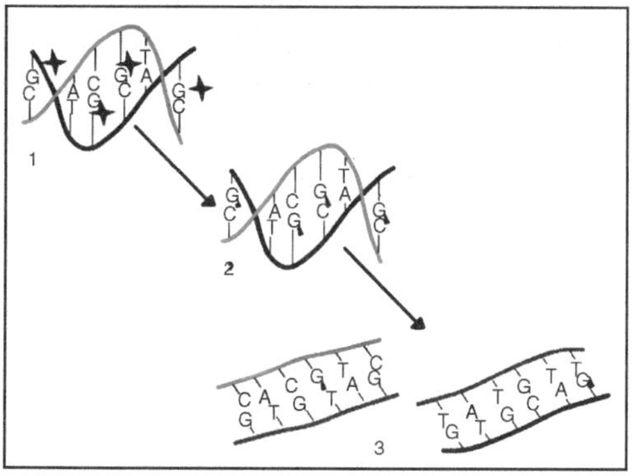

Alkylating agents are one of the earliest and most commonly used chemotherapy agents used for cancer treatments. Their use in cancer treatments started in early 1940s. Majority of alkaline agents are active or dormant nitrogen mustards, which are poisonous compound initially used for certain military purposes. Chlorambucil, Cyclophosphamide, CCNU, Melphalan, Procarbazine, Thiotepa, BCNU, and Busulfan are some of the commonly used alkylating agents.

Although they might differ in their clinical activity, action mechanism of all alkylating agents is the same. These agents work directly on the DNA and prevent the cell division process by cross-linking and breaking the DNA strands and causing abnormal base pairing. When a DNA is altered in this manner, undesired cellular activity comes to a halt and the cell dies eventually.

Alkylating chemotherapy drugs are effective during all phases of cell cycle. Therefore, they are used to treat a large number of cancers. However, they are more effective in treating slow-growing cancers such as solid tumors and leukemia. Long term use of alkylating agents can lead to permanent infertility by decreasing sperm production in males, and causing menstruation cessation in females. Many alkylating agents can also lead to secondary cancers such as Acute Myeloid Leukemia, years after the therapy.

ANTIMETABOLITES

Structure of antimetabolites (antineoplastic agents) is similar to certain compounds such as vitamins, amino acids, and precursors of DNA or RNA, found naturally in human body. Antimetabolites help in treatment cancer by inhibiting cell division thereby hindering the growth of tumor cells. These agents get incorporated in the DNA or RNA to interfere with the process of division of cancer cells.

Antimetabolites were first discovered and it is found that folic acid analog can reduce childhood leukemia. Out of 16 patients he tested, 10 displayed hematologic improvement. This discovery laid the foundation that enabled scientist to synthesise many new agents that could inhibit biological enzymatic reactions. Antimetabolites are found to be useful in treating chronic and acute cases of leukemia and various tumors. They are commonly used to treat gastrointestinal tract, breast, and ovary tumors.

Methotraxate, which is a commonly used antimetabolites chemotherapy agent, is effective in the S-phase of the cell cycle. It works by inhibiting an enzyme that is essential for DNA synthesis. 6-mercaptopurine and 5-fluorouracil (5FU) are two other commonly used antimetabolites. 5-Fluorouracil (5-FU) works by interfering with the DNA components, nucleotide, to stop DNA synthesis. This drug is used to treat many different types of cancers including breast, esophageal, head, neck, and gastric cancers. 6-mercaptopurine is an analogue of hypoxanthine and is commonly used to treat Acute Lymphoblastic Leukemia (ALL). Other popular antimetabolite chemotherapy drugs are Thioguanine, Cytarabine, Cladribine. Gemcitabine, and Fludarabine.

ANTHRACYCLINES

Anthracyclines were developed between 1970s and 1990s and are daunosamine and tetra-hydronaphthacenedione-based chemotherapy agents. These compounds are cell-cycle non-specific and are used to treat a large number of cancers including lymphomas, leukemia, and uterine, ovarian, lung and breast cancers.

Anthracyclines drugs are developed from natural resources. For instance, daunorubicin is developed by isolating it from soil-dwelling fungus Streptomyces. Similarly, Doxorubicin, which is another commonly used anthracycline chemotherapy agent, is isolated from mutated strain of Streptomyces. Although both the drugs have similar clinical action mechanisms, doxorubicin is more effective in treating solid tumors. Idarubicin, Epirubicin, and Mitoxantrone are few of the other commonly used anthracycline chemotherapy drugs. Anthracyclines work by forming free oxygen radicals that breaks DNA strands thereby inhibiting DNA synthesis and function. These chemotherapeutic agents form a complex with DNA and enzyme to inhibit the topoisomerase enzyme. Topoisomerase is an enzyme class that causes the supercoiling of DNA, allowing DNA repair, transcription, and replication.

One of the main side effects of anthracyclines is that it can damage cells of heart muscle along with the DNA of cancer cell leading to cardiac toxicity.

ANTITUMOR ANTIBIOTICS

Antitumor antibiotics are also developed from the soil fungus Streptomyces. These drugs are widely used to treat and suppress development of tumors in the body. Similar to anthracyclines, antitumor antibiotics drugs also form free oxygen radicals that result in DNA strand breaks, killing the growth of cancer cells. In most of the cases, these drugs are used in combination with other chemotherapy agents. Bleomycin is one of the commonly used antitumor antibiotic used to treat testicular cancer and hodgkin's lymphoma. The most serious side effect of this drug is lung toxicity that occurs when the oxygen radical formed by the antitumor antibiotics damages lung cells along with the cancer cells.

MONOCLONAL ANTIBODIE

Monoclonal antibodies are one of the newer chemotherapy agents approved for cancer treatment by the Food and Drug Administration (FDA) in 1997. Alemtuzumab (Campath), Bevacizumab (Avastin), Cetuximab (Erbitux), Gemtuzumab (Mylotarg), Ibritumomab (Zevalin), Panitumumab (Vectibix), Rituximab (Rituxan), Tositumomab (Bexxar), and Trastuzumab (Herceptin) are some of the FDA approved monoclonal drugs used in chemotherapeutic cancer treatments. The treatment is known to be useful in treating colon, lung, head, neck, and breast cancers. Some of the monoclonal drugs are used to treat chronic lymphocytic leukemia, acute myelogenous leukemia, and non-Hodgkin's lymphoma. Monoclonal antibodies work by attaching to certain parts of the tumor-specific antigens and make them easily recognisable by the host's immune system. They also prevent growth of cancer cells by blocking the cell receptors to which chemicals called 'growth factors' attach promoting cell growth. Monoclonal antibodies can be combined with radioactive particles and other powerful anticancer drugs to deliver them directly to cancer cells. Using this method, long term radioactive treatment and anticancer drugs can be given to patients without causing any serious harm to other healthy cells of the body.

PLATINUMS

Platinum-based natural metal derivatives were found to be useful for cancer treatments around 150 years ago with the synthesis of cisplatin. However, there clinical use did not commence until 30 years ago. Platinum-based chemotherapy agents work by cross-linking subunits of DNA. These agents act during any part of cell cycle and help in treating cancer by impairing DNA synthesis, transcription, and function.

Cisplatin, although found to be useful in treating testicular and lung cancer, is highly toxic and can severely damage the kidneys of the patient. Second generation platinum-complex carboplatin is found to be much less toxic in comparison to cisplatin and has fewer kidney-related side effects. Oxaliplatin, which is third generation platinum-based complex, is found to be helpful in treating colon cancer. Although, oxaliplatin does not cause any toxicity in kidney it can lead to severe neuropathies.

PLANT ALKALOIDS

Plant alkaloid chemotherapy agents, as the name suggests, are plant derivatives. They are cell-specific chemotherapy agents. However, the cycle affected is based on the drug used for the treatment. They are primarily categorised into four groups: topoisomerase inhibitors, vinca alkaloids, taxanes, and epipodophyllotoxins. Plant alkaloids are cell-cycle specific, but the cycle affected varies from drug to drug. Vincristine (Oncovin) is a plant alkaloid of interest in mesothelioma treatment.

TOPOISOMERASE INHIBITORS

Topoisomerase inhibitors are chemotherapy agents are categorised into Type I and Type II Topoisomerases inhibitors and they work by interfering with DNA transcription, replication, and function to prevent DNA supercoiling.

- *Type I Topoisomerase inhibitors:* These chemotherapy agents are extracted from the bark and wood of the Chinese tree Camptotheca accuminata. They work by forming a complex with topoisomerase DNA. This in turn suppresses the function of topoisomerase.
  Camptothecins which includes irinotecan and topotecan are commonly used type I topoisomerase inhibitors, first discovered in the late 1950s.
- *Type II Topoisomerase inhibitors:* These are extracted from the alkaloids found in the roots of May Apple plants. They work in the in the work in the late S and G2 phases of the cell cycle.
  Amsacrine, etoposide, etoposide phosphate, and teniposide are some of the examples of type II topoisomerase inhibitors.

VINCA ALKALOIDS

Vinca alkaloids are derived from the periwinkle plant, Vinca rosea (Catharanthus roseus) and are known to be used by the natives of Madagascar to treat diabetes. Although not useful in controlling diabetes, vinca alkaloids, are useful in treating leukemias. They are effective in the M phase of the cell cycle and work by inhibiting tubulin assembly in microtubules. Vincristine, Vinblastine, Vinorelbine, and Vindesine are some of the popularly used vinca alkaloid chemotherapy agents used today. Major side effect of vinca alkaloids is that they can cause neurotoxicity in patients.

TAXANES

Taxanes are plant alkaloids that were first developed in 1963 by isolating it from first isolated from the bark of the Pacific yew tree, Taxus brevifolia in 1963. Paclitaxel, which is the active components of taxanes was first discovered in 1971 and was made available for clinical use in the year 1993. Taxanes also work in the M-phase of the cell cycle and inhibit the function of microtubules by binding with them. Paclitaxel and docetaxel are commonly used taxanes. Taxanes chemotherapy agents are used to treat a large array of cancers including breast, ovarian, lung, head and neck, gastric, esophageal, prostrate and gastric cancers. The main side effect of taxanes is that they lower the blood counts in patients.

MECHANISM OF INTERCALATING

Intercalating agents wedge between bases along the DNA. The intercalated drug molecules affect the structure of the DNA, preventing polymerases and other DNA binding proteins from functioning properly. The result is prevention of DNA synthesis, inhibition of transcription and induction of mutations. The effect of principal alkaloids (sanguinarine, chelerythrine, coptisine, chelidonine) of greater celandine Chelidonium majus L., as well as the alkaloids from Colchicum autumnale L.(colchicine and colchamine) on calcium accumulation and oxidative phosphorylation in rat liver mitochondria has been studied. The obtained data were compared with DNA intercalating properties of alkaloids detected by the method of thermodenaturation (DNA melting curve plots). It was found that chelerythrine and sanguinarine blocked absorption and accumulation of calcium cations and inhibited oxidative phosphorylation, while the coptisine significantly diminished those indices. Chelidonine, colchicines and colchamine had no influence on the studied characteristics. The effect of alkaloids upon mitochondria functional state correlated tightly with their DNA intercalating properties: chelerythrine and sanguinarine were strong intercalators, while coptisine was a weak one, and chelidonine, colchicine and colchamine did not interact with DNA and caused no changes in its melting point.

Correlation coefficient between the intercalating properties of alkaloids and their inhibition of calcium accumulation was 0.89, and with their oxidative phosphorylation inhibition -0.93. It is suggested that the effect of studied alkaloids upon functional properties of mitochondria can be mediated by mtDNA. DNA topoisomerase I is essential for cellular metabolism and survival. It is also the target of a novel class of anticancer drugs active against previously refractory solid tumors, the camptothecins.

The present review describes the topoisomerase I catalytic mechanisms with particular emphasis on the cleavage complex that represents the enzyme's catalytic intermediate and the site of action for camptothecins. Roles of topoisomerase I in DNA replication, transcription and recombination are also reviewed. Because of the importance of topoisomerase I as a chemotherapeutic target, we review the mechanisms of action of camptothecins and the other topoisomerase I inhibitors identified to date.

The DNA intercalating agents 4'-(9-acridinyl-amino) methanesulfon-m-anisidide (m-AMSA) and adriamycin were studied by using filter elution methods to measure DNA single-strand breaks (SSB's), DNA-protein cross-links (DPC's), and double-stranded breaks (DSB's) in mouse leukemia L1210 cells. Both compounds produced SSB's and DPC's at nearly 1:1 ratios. The SSB's and DPC's were shown to be localised with respect to each other; this was inferred from the finding that filter assays based on protein adsorption completely prevented the elution of the DNA single-strand segments between SSB's. In the case of m-AMSA, which produces relatively high frequencies of DNA lesions, the possibility that a protein bridges across the SSB was excluded by alkaline sedimentation studies.

The o-AMSA isomer is much less cytotoxic than m-AMSA and did not produce protein-associated strand breaks. The simplest model to explain the results is that a protein becomes covalently bound to either the 3' or the 5' termini of the intercalator-induced strand breaks. At moderately cytotoxic doses, m-AMSA yielded much larger frequencies of protein-associated SSB's than did adriamycin. m-AMSA-induced protein-associated SSB's saturated at approximately 60000 per cell over a concentration range in which m-AMSA uptake by the cells was proportional to the drug concentration. m-AMSA was found to enter and exit from cells very rapidly at 37 degrees C; protein-associated SSB's and DSB's also appeared and disappeared rapidly. At reduced temperature, however, the appearance and disappearance of protein-associated SSB's could be blocked while m-AMSA entry and exit still occurred.

The saturation behaviour and temperature dependence suggest that the formation and disappearance of protein-associated strand breaks is enzymatic. The simplest hypothesis is that the linked protein is a nuclease, such as a topoisomerase, which becomes bound to one terminus of the strand break it produces. It is proposed that topoisomerases producing SSB's and DSB's are stimulated to different degrees by different intercalators. Topoisomerase II mediated DNA scission induced by both a non-intercalating agent [4'-demethylepipodophyllotoxin 4-(4,6-O-ethylidene-beta-D glucopyrano-side)(VP-16)] and an intercalator [4'-(9-acridinylamino) methanesulfon-m-anisidide (m-

AMSA)] wasstudied as a function of proliferation in Chinese hamster ovary (CHO), HeLa, and mouse leukemia L1210 cell lines. Log-phase CHO cells exhibited dose-dependent drug-induced DNA breaks, while plateau cells were found to be resistant to the effects of VP-16 and m-AMSA. Neither decreased viability nor altered drug uptake accounted for the drug resistance of these confluent cells. In contrast to CHO cells, plateau-phase HeLa and L1210 cells remained sensitive to VP-16 and m-AMSA. Recovery of drug sensitivity by plateau-phase CHO cells was found to reach a maximum approximately 18 h after these cells regained exponential growth and was independent of DNA synthesis.

DNA strand break frequency correlated with cytotoxicity in CHO cells; log cells demonstrated an inverse log linear relationship between drug dose (or DNA damage) and colony survival, whereas plateau-derived colony survival was virtually unaffected by increasing drug dose. Topoisomerase II activity, whether determined by decatenation of kinetoplast DNA, by cleavage of pBR322 DNA, or by precipitation of the DNA-topoisomerase II complex, was uniformly severalfold greater in log-phase CHO cells compared to plateau-phase cells.

The biochemical characteristics of the formation and disappearance of intercalator-induced DNA double-strand breaks (DSB) were studied in nuclei from mouse leukemia L1210 cells by using filter elution methodology. The three intercalators used were 4'-(9-acridinylamino)-methanesulfon-m-anisidide (m-AMSA), 5-iminodaunorubicin (5-ID), and ellipticine. These compounds differ in that they produced predominantly DNA single-strand breaks (SSB)(m-AMSA) or predominantly DNA double-strand breaks (ellipticine) or a mixture of both SSB and DSB (5-ID) in whole cells. In isolated nuclei, each intercalator produced DSB at a frequency comparable to that which is produced in whole cells. Moreover, these DNA breaks reversed within 30 min after drug removal. It thus appeared that neither ATP nor other nucleotides were necessary for intercalator-dependent DNA nicking-closing reactions. The formation of the intercalator-induced DSB was reduced at ice temperature. Break formation was also reduced in the absence of magnesium, at a pH above 6.4 and at NaCl concentrations above 200 mM. In the presence of ATP and ATP analogues, the intercalator-induced cleavage was enhanced. These results suggest that the intercalator-induced DSB are enzymatically mediated and that the enzymes involved in these reactions can catalyse DNA double-strand cleavage and rejoining in the absence of ATP, although the occupancy of an ATP binding site might convert the enzyme to a form more reactive to intercalators. Three inhibitors of DNA topoisomerase II—novobiocin, nalidixic acid, and norfloxacin—reduced the formation of DNA strand breaks.

DNA TECHNOLOGY

The most sweeping ethical argument against biotechnology would be one that finds the manipulation of genes or cells to be either categorically forbidden or presumptively wrong, so that compelling arguments would need to be adduced in its favour. Fable and myth provide a basis for the idea that certain forms of knowledge or technology may be subjected to such proscription.

It is not clear whether members of the lay public who express ethical reservations about gene technology have such a view in mind, but it is reasonable to presume that some do. There are many ways in which such a claim might be stated. Empirical research indicates that many members of the lay public who find food or agricultural biotechnology ethically objectionable base their judgement on the view that it is unnatural. Philosophers have called these objections to biotechnology "intrinsic objections," meaning that it is the activity of genetic manipulation itself that is wrong, not its consequences.

Statements to the effect that biotechnology is unnatural convey a judgement of disapproval, but do little to articulate the basis for that judgement. In one sense, all of agriculture is an unnatural activity, but we should not infer that all of agriculture is therefore of ethical concern. How would one spell out the belief that biotechnology is unnatural in a way that would form the basis for an argument against its use to develop agricultural crops or animals? How would one articulate an intrinsic objection to gene transfer that would cover its use in plants and animals, as well as human beings? A few strategies that have been attempted in the literature can be summarised.

1. *Genes and essences*: Since antiquity, people have thought of living things as having "essences" that constitute their essential being. Nelkin and Lindee note a general cultural tendency to interpret genes as bearers of the traditional notions of essence and purpose that would achieve moral significance in some teleological conceptions of nature. One view of biotechnology may see it as "tampering" with these "essences" suggest such a judgement, and it is particularly associated with those who have suggested that genetic engineering violates a species' telos. The term 'telos' is derived from the philosophy of Aristotle, where it was used to indicate a thing's guiding or final purpose, realised in the case of living organisms through the processes of growth, development and reproduction that are characteristic of their species. It is associated with teleology, a philosophy of nature that seeks to explain biological processes in terms of function, purpose and design. Although teleology does not necessarily prescribe particular ethical norms, versions of teleology that find a predetermined design in nature, often the work of a supernatural intelligence, move quickly to the ethical judgement that humans deviate from the preordained purposes of this plan at their physical and spiritual peril.
2. *Species boundaries and natural kinds*: Human cultures display a

remarkable constancy with respect to the way that species boundaries are taken to reflect a kind of natural order, reflected in the linguistic tendency to build the system of meanings around natural kinds. Plants and animals visible to human senses and important for human purposes are described as kinds, rather than as particular things not amenable to classification. Although different cultures parse the world around them in different ways, human languages tend to have equivalent kind-terms for 'dog' 'cat' 'tree' or 'flower'. Verhoog suggests that this tendency is evidence for an underlying system of purposes such as those discussed immediately above. He also makes the separate argument that biologists lack any special authority to redefine these terms to more faithfully reflect the scientific construal of kinds as interbreeding populations. The force of this second argument is that modern biology is challenging the most basic way in which human beings have made sense of the world since antiquity—and so much the worse for modern biology.

3. *Emotional repugnance*: Genetic modification of foods causes an immediate reaction of repugnance among many. The most sophisticated philosophical statement of the ethical significance that should be associated with that reaction was made in brief article by Kass, commenting on the announcement of Dolly, the sheep cloned by the Roslyn Institute in 1997. Kass's central argument is that mammalian cloning elicits a repulsive reaction from many, and that this repugnance is sufficient ground to regard cloning as intrinsically wrong. In making this case, Kass relies on a conservative tradition in ethics that harks back to the philosophical writings of David Hume, Adam Smith and Edmund Burke. These philosophers believed that morality was based on sentiments of sympathy with others, and that emotional attachments were a key component in any moral judgement. Although they lived and wrote in a pre-Darwinian culture, they also believed that emotional reactions like repugnance reflect a deep-seated and culturally ingrained wisdom. Societal stability is the result of respecting these emotional reactions, and departure from them entails the risk of upheaval and dissolution. Kass's argument has since formed the basis for a similar argument against applications of recombinant technology to foods.
4. *Religious arguments*: Many people clearly attach religious significance to species boundaries and question the wisdom of genetic engineering. Furthermore, many of the world's religions endorse specific injunctions against crossing species boundaries, interfering in reproductive processes, and consuming proscribed foods. As noted already, some of the most plausible ways of understanding the view

that biotechnology is unnatural or that it tampers with the natural order against the demands of morality involve appeals to divine authority. Furthermore, worldviews that construe nature as bearing specific forms of moral significance may also be considered as resting on religious foundations, especially when they involve beliefs that are not amenable to scientific characterisation and measurement.

EVALUATING SPECIAL ARGUMENTS

For the most part, professional philosophers have not been kind to the objection that biotechnology is unnatural. Roger Straughan and Gary Comstock review a series of ways to extend the claim that gene technology is unnatural into a more substantive ethical argument for regulating or restricting crop biotechnology. In each case, they find either that the substantive issues do not pertain specifically to the use of rDNA techniques for gene transfer, or that the characterisation of naturalness is too vague and fails to exclude many well-accepted uses of technology.

Bernard Rollin offers a similar analysis, and characterises arguments that appeal to the unnaturalness of gene transfer as "bad ethics." Mark Sagoff has replied to the suggestion that biotechnology is unnatural by reviewing the four ways in which John Stuart Mill found that something could be said to be natural, arguing that for the most part, no judgement against biotechnology can be maintained without also tarnishing ordinary plant breeding, if not agriculture itself. Philosopher Fred Gifford has shown how conceptions of the gene as a carrier of human essence fail to correspond with the conception of genes that is operative in contemporary molecular biology.

Scientific authors do not characterise the processes of cloning or genetic transformation in terms that would support the judgement that essences and telos are being affected. As such, there is a gap between the ethical understanding of nature implicit in philosophies that attribute essential or teleological significance to genes or gene processes, and the dominant scientific interpretation of the practices that constitute food and agricultural biotechnology. It is not clear who bears the burden of proof with respect to further development of this line of ethical concern. On the one hand, those who believe that genes have the ethical status of essence or telos have not shown how the idea of genes as sequences of DNA can be made compatible with traditional notions of essence or telos. One might argue that this line of criticism has reached a dead end until such an argument is forthcoming. On the other hand, one might argue that until scientists and practitioners of biotechnology bear the burden of defending biology against traditional notions of purpose and essence that may still be very active in the worldview of non-scientists, it is entirely appropriate to oppose biotechnology on the ground that it is intrinsically wrong.

The argument from natural kinds does not have widespread appeal, though it is one way of making sense out of the claims made by some of biotechnology's most vehement opponents. It deserves consideration if only as a possible way of explaining why biotechnology and molecular biology seem to cause such a profound sense of anxiety. It is not clear whether the next move should be a stronger statement of the reason why the need to preserve the basic categories of human language (and perhaps, by extension, of humanity's collective intelligence) entails any specific proscriptions or norms with respect to food and agricultural biotechnology. Alternatively, a need for better public education in biology might follow, on the assumption that the real problem is the underlying anxiety and disorder associated with shifting worldviews. Better ethical discourse on biotechnology might even be a means to resolving the tension felt by those who feel that modern molecular biology threatens the most basic categories that human beings use to make sense of the world.

Mark Sagoff's evaluation of the "naturalness" of biotechnology is relevant to repugnance arguments offered by Kass, Midgeley and Chadwick. Sagoff writes that Mill in fact offers us one sense of what it means to be natural that allows us to sort GM crops and animals into the unnatural basket while leaving traditional foods in the natural one. This is that things can be "unnatural" in the sense of being inauthentic, not true to themselves. Here, Sagoff admits that we (meaning our culture) might find biotechnology to be unnatural in the sense of being inconsistent with our aesthetic sensibilities.

He argues that we should allow ourselves free reign to indulge our aesthetic tastes, but only under the condition that we recognise the full implications of doing so. Sagoff's view on this point is that the human and environmental costs of rejecting biotechnology would be significant. Nevertheless, if a public informed about the technology and its likely benefits still found it repugnant, such a result it would strengthen the repugnance argument, and presumably Sagoff would be forced to concede that biotechnology is "out" on aesthetic grounds. Streiffer and Hedemann suggest that opinion research supporting the demand for labelling suggests that a majority of people have already found biotechnology to be intrinsically unacceptable, and on this basis argue that political decision makers can no longer reject this sentiment in good conscience. The ethical significance of religious views can be pursued in two ways. First, one may examine the theological or doctrinal basis for this judgement, given the sacred texts, sectarian juridical processes and doctrinal traditions of specific religions. Clearly, religious deliberations represent an important source of insight with respect to the application of cloning, genetic engineering and other forms of gene technology to human beings.

Second, one may simply acknowledge that the principle of religious tolerance affords people with wide latitude for deriving faith-based opinions on food and agricultural biotechnology, and enquire how these intrinsically personal ethical judgements entail social norms. Worldviews and normative beliefs about nature and natural order must be regarded as protected by principles of religious tolerance even if they do not derive from recognised religious traditions, churches or theological traditions, and even if they do not involve belief in a supernatural power.

Arguably, the second approach converts the significance of religious beliefs about gene technology into a problem of consumer and social policy. The norms that guide action are based on a secular principle of religious tolerance, rather than (or in addition to) norms that make specific appeal to religious inspiration or doctrine.

Tolerance implies that religious believers should be able to act on their beliefs. If those who find biotechnology unnatural are working from conceptions of nature that are so inconsistent with those of contemporary biology that we must regard them as "faith-based" (even if they make no specific appeal to God or recognised religion), then one of the main implications of calling these views faith-based is that the individuals who hold these views are regarded as having a right to hold and act on these views irrespective of modern science. But this line of reasoning may convert the argument from a "special concern" to an ordinary principle of technological ethics.

The fact that people have faith-based views prohibiting a practice does not ordinarily provide a public basis for constraining or regulating that practice. Rather this fact establishes a prima facie obligation to respect these beliefs and to accommodate a believer's desire to act on faith-based beliefs in their daily life. Any form of technology that compromised people's ability to hold and act on faith-based beliefs would raise ethical concern, so the ethical issue that is raised here is a general concern of technological ethics, rather than a special concern associated with gene technology. Streiffer and Hedemann resist this turn in the argument, suggesting that if a sufficient number of people hold faith-based beliefs, it becomes appropriate to take whatever public action such beliefs dictate, subject to the qualification that the full range of political values must be taken into consideration when doing. On this ground, they argue that, at a minimum, intrinsic arguments provide a powerful argument for segregating and labelling products of biotechnology, and could conceivably provide an argument for banning them altogether.

RIBONUCLEIC ACID

Ribonucleic acid (RNA) is a biologically important type of molecule that consists of a long chain of nucleotide units. Each nucleotide consists of a nitrogenous base, a ribose sugar, and a phosphate. RNA is very similar to DNA, but differs in a few important structural details: in the cell, RNA is usually single-stranded, while DNA is usually double-stranded; RNA nucleotides contain ribose while DNA contains deoxyribose (a type of ribose that lacks one oxygen atom); and RNA has the base uracil rather than thymine that is present in DNA. Ribonucleic acid (RNA) is a linear molecule composed of four types of smaller molecules called ribonucleotide bases: adenine (A), cytosine (C), guanine (G), and uracil (U). RNA is often compared to a copy from a reference book, or a template, because it carries the same information as its DNA template but is not used for long-term storage. Each ribonucleotide base consists of a ribose sugar, a phosphate group, and a nitrogenous base. Adjacent ribose nucleotide bases are chemically attached to one another in a chain via chemical bonds called phosphodiester bonds. Unlike DNA, RNA is usually single-stranded. Additionally, RNA contains ribose sugars rather than deoxyribose sugars, which makes RNA more unstable and more prone to degradation. RNA is synthesised from DNA by an enzyme known as RNA polymerase during a process called transcription. The new RNA sequences are complementary to their DNA template, rather than being identical copies of the template. RNA is then translated into proteins by structures called ribosomes. There are three types of RNA involved in the translation process: messenger RNA (mRNA), transfer RNA (tRNA), and ribosomal RNA (rRNA). Although some RNA molecules are passive copies of DNA, many play crucial, active roles in the cell. For example, some RNA molecules are involved in switching genes on and off, and other RNA molecules make up the critical protein synthesis machinery in ribosomes.

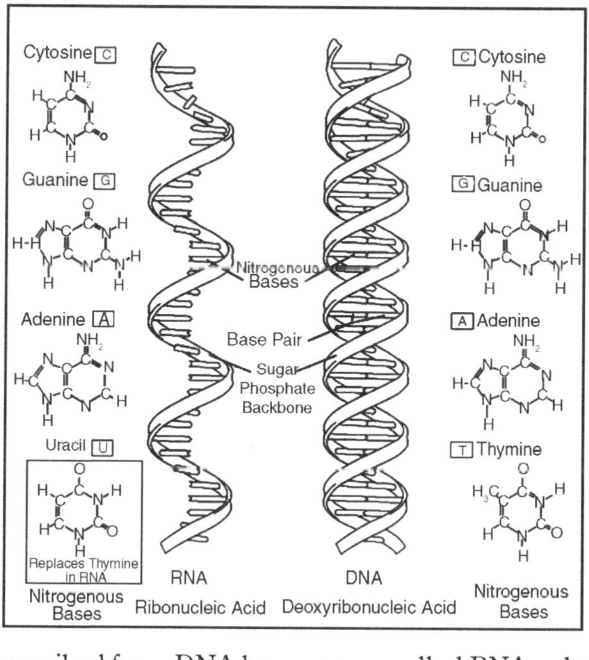

RNA is transcribed from DNA by enzymes called RNA polymerases and is generally further processed by other enzymes. RNA is central to the synthesis of proteins. Here, a type of RNA called messenger RNA carries information from DNA to structures called ribosomes.

These ribosomes are made from proteins and ribosomal RNAs, which come together to form a molecular machine that can read messenger RNAs and translate the information they carry into proteins. There are many RNAs with other roles – in particular regulating which genes are expressed, but also as the genomes of most viruses.

MESSENGER RNA (MRNA) AND ITS PRECURSORS

The primary RNA transcripts are produced by copying the DNA of a protein-encoding gene. Subsequent processing steps include addition, removal, and modification of nucleotides as well as splicing events that excise internal segments.

The mature mRNA is translated to produce a protein whose amino acid sequence is determined by the sequence of the coding region in the gene. The messenger RNA, as the name implies, is the molecule that carries the message from the gene to the protein synthesis machinery.

RIBOSOMAL RNA (RRNA)

The ribosomes are the most important part of the translation machinery and it has long been known that much of the mass of ribosomes is due to the presence several types of ribosomal RNA. These are non-coding RNAs produced by transcription of ribosomal RNA genes. One of the key steps in translation—formation of the peptide bond—is catalysed by the rRNA component of the ribosome. It is the major catalytic RNA in cells.

TRANSFER RNA (TRNA)

tRNAs are intermediates in protein synthesis. There are many different tRNA molecules in every cell and each one binds a specific amino acid, yielding an aminoacylated-tRNA (aa-tRNA). Each different aminoacylated tRNA interacts with a particular codon in mRNA thus delivering the correct amino acid to to the site of protein synthesis.

SMALL RNAS

The small RNAs represent a heterogeneous category of RNAs covering a wide ranges of functions. Some of them have catalytic functions—RNAse P is the classic example. Some of them are structural components of ribonucleoprotein complexes (*e.g.* signal recognition particle). Some of them are guide RNAs involved in various processing events. The best known examples of guide RNAs are the small RNAs of the spliceosome complexes that mediate the splicing of mRNA precursors. Other small RNAs were known to be involved in the regulation of gene expression.

GENOMIC RNA

Some viruses, notably retroviruses, have an RNA genome instead of a DNA genome. In addition, the mobility of various transposons is due to an intermediate RNA copy of the transposon sequence (retrotransposons). This was the state of knowledge 25 years ago. Since then, the study of RNA has made remarkable progress. Our knowledge of all the fundamental processes—transcription, processing, and catalysis—has expanded enormously.

The biggest change is in the area of small RNAs. Today there are several categories of small RNAs—siRNA, microRNA, piRNA—that were only discovered in the past 10-15 years. The functions of these small RNA molecules are still being worked out. There's little doubt that some of them have important biological roles but there's considerable controversy over what percentage might be artifacts of one sort or another.

RNA replaces DNA as the genetic material. As with DNA, RNA follows specific base pairing rules, except that in RNA the base uracil replaces the base thymine (*i.e.*, instead of an adenine-thymine or A-T pairing, there is an adenine-uracil or A-U pairing). Accordingly, when RNA acts as a carrier of genetic information, uracil replaces thymine in the genetic code.

In humans, messenger RNA (mRNA) is the product of transcription and acts to convey genetic information from the nucleus to the protein assembly complex at the ribosome. The ribsome is composed of ribosomal RNA (rRNA) and other proteins. Transfer RNAs (tRNA) act to catalyse the translation process by acting as carriers of specific amino acids. Because tRNAs bind to specific sites on the strand of mRNA, the sequence of amino acids subsequently inserted into the synthesised protein is both specific and genetically determined by the nucleotide sequence in DNA from which the mRNA strand was originally transcribed.

Other forms of RNA perform important roles in other biochemical reactions. Regardless of function, RNA is a biopolymer made up of ribonucleotide units and is present in all living cells and some viruses. The chemical units of RNA are ribonucleotide monomers consisting of a ribose sugar ($C_5H_{10}O_5$) phosphorylated at the third carbon (C3) and linked to one of four bases through a type of chemical linkage formed between a sugar and a base by a condensation reaction (glycosidic bond). The four bases found in RNA are adenine (A), guanine (G), cytosine (C), and uracil (U). Other bases may also be found, although they are generally modified versions of these four (*e.g.*, methylated bases are found in parts of tRNA).

The single nucleotides (monomers) of RNA form a linear chain by linking their phosphate groups and sugars in phosphodiester bonds. RNA does not form a double stranded alpha-helix as does DNA. In some parts of the RNA molecule, there is folding into alpha-helical-like regions. Corresponding to their unique functions, messenger RNA (mRNA), ribosomal RNA (rRNA), and transfer RNA (tRNA) all have different three-dimensional structures. In higher eukaryotic organisms, different RNAs are found distributed throughout the cell—in the nucleus, cytoplasm, and also in cytoplasmic organelles such as mitochondria and, in plants, chloroplasts.

The nucleus is the chief site of RNA synthesis and the source of all cytoplasmic RNA, while mitochondria and chloroplasts synthesise their RNA from their own DNA. rRNA is synthesised by the nucleoli within the nucleus, while the high molecular weight precursor to cytoplasmic mRNA, sometimes termed heterogeneous nuclear or hnRNA, is transcribed on the DNA chromatin. Low molecular weight RNA also occurs in the nucleus and consists partly of tRNA and partly of RNA, which has a regulatory function in geneactivation. The cytoplasm contains tRNA and rRNA in the ribosomes and mRNA in polysomes, or polyribosomes. The latter are the structural units of protein biosynthesis, consisting of several ribosomes attached to a strand of mRNA.

The function of mRNA is to transcribe the information held in DNA. In the cells of eukaryotic organisms, the first transcriptional product is the long, heterogenous nuclear RNA, or hnRNA. This contains both the nucleotide sequences eventually transcribed into polypeptides and large tracts of sequences not translated. Non-translated sequences are termed introns (or intervening sequences). Removal of introns, and other untranslated portions of the molecule, edits hnRNA into mRNA molecules. After editing removes as much as 90 per cent of hnRNA, the resulting mRNA molecules are transported into the cytoplasm.

rRNA is located within ribosomes, the sites of protein biosynthesis. Ribosomes are large ellipsoid cytoplasmic organelles consisting of RNA and protein.

tRNA, the smallest known functional RNA, is essential for protein biosynthesis. Its purpose is to transfer a specific amino acid from the cytoplasm and incorporate it into the growing polypeptide chain on the polysome. Different tRNAs contain between 70 and 85 nucleotides. The most characteristic feature of tRNA is that it contains the anticodon, a sequence of three nucleotides specific for the mRNA codon sequence. There is at least one tRNA per cell bearing the anticodon for each of the 20 amino acids. The aminoacyl-tRNA (the tRNA carrying the amino acid) binds to the large subunit of a ribosome, where antiparallel basepairing occurs between the anticodon of the tRNA and the complementary codon of the associated mRNA. The specificity of this base pairing ensures that the amino acid inserts into the correct position in the growing protein polypeptide chain. During translation, the deacylated tRNA (*i.e.*, with its amino acid removed) is released from the ribosome and becomes available once again for recharging with its amino acid.

DNA-dependent RNA synthesis is the process of RNA sythesis on a template of DNA. According to the rules of base pairing, the base sequence of

DNA determines the synthesis of a complementary base sequence in RNA. Assisted (catalysed) by the enzyme RNA polymerase, the growing RNA chain releases from the template so that the process can start again, even before the previous molecule is complete. Termination codons and a termination factor known as rho-factor end the synthesis process. In certain viruses, RNA-dependent RNA synthesis occurs, with the viral RNA acting as a template for the synthesis of new RNA. Each nucleotide in RNA contains a ribose sugar, with carbons numbered 1' through 5'. A base is attached to the 1' position, generally adenine (A), cytosine (C), guanine (G) or uracil (U). Adenine and guanine are purines, cytosine and uracil are pyrimidines. A phosphate group is attached to the 3' position of one ribose and the 5' position of the next. The phosphate groups have a negative charge each at physiological pH, making RNA a charged molecule (polyanion). The bases may form hydrogen bonds between cytosine and guanine, between adenine and uracil and between guanine and uracil.However other interactions are possible, such as a group of adenine bases binding to each other in a bulge, or the GNRA tetraloop that has a guanine–adenine base-pair.

An important structural feature of RNA that distinguishes it from DNA is the presence of a hydroxyl group at the 2' position of the ribose sugar. The presence of this functional group causes the helix to adopt the A-form geometry rather than the B-form most commonly observed in DNA. This results in a very deep and narrow major groove and a shallow and wide minor groove. A second consequence of the presence of the 2'-hydroxyl group is that in conformationally flexible regions of an RNA molecule (that is, not involved in formation of a double helix), it can chemically attack the adjacent phosphodiester bond to cleave the backbone. RNA is transcribed with only four bases (adenine, cytosine, guanine and uracil), but there are numerous modified bases and sugars in mature RNAs. Pseudouridine (Ø), in which the linkage between uracil and ribose is changed from a C–N bond to a C–C bond, and ribothymidine, are found in various places (most notably in the TØC loop of tRNA).Another notable modified base is hypoxanthine, a deaminated adenine base whose nucleoside is called inosine.

Inosine plays a key role in the wobble hypothesis of the genetic code.There are nearly 100 other naturally occurring modified nucleosides,of which pseudouridine and nucleosides with 2'-O-methylribose are the most common.The specific roles of many of these modifications in RNA are not fully understood. However, it is notable that in ribosomal RNA, many of the post-transcriptional modifications occur in highly functional regions, such as the peptidyl transferase center and the subunit interfacc, implying that thcy arc important for normal function.

RNA and DNA are both nucleic acids, but differ in three main ways. First, unlike DNA which is double-stranded, RNA is a single-stranded molecule in most of its biological roles and has a much shorter chain of nucleotides. Second, while DNA contains deoxyribose, RNA contains ribose, (there is no hydroxyl group attached to the pentose ring in the 2' position in DNA). These hydroxyl groups make RNA less stable than DNA because it is more prone to hydrolysis. Third, the complementary base to adenine is not thymine, as it is in DNA, but rather uracil, which is an unmethylated form of thymine.Like DNA, most biologically active RNAs, including mRNA, tRNA, rRNA, snRNAs and other non-coding RNAs, contain self-complementary sequences that allow parts of the RNA to fold and pair with itself to form double helices. Structural analysis of these RNAs have revealed that they are highly structured.

9

Isolation of Genomic DNA Technology

INTRODUCTION

PCR allows isolation of DNA fragments from genomic DNA by selective amplification of a specific region of DNA. This use of PCR augments many methods, such as generating hybridization probes for Southern or northern hybridization and DNA cloning, which require larger amounts of DNA, representing a specific DNA region. PCR supplies these techniques with high amounts of pure DNA, enabling analysis of DNA samples even from very small amounts of starting material.

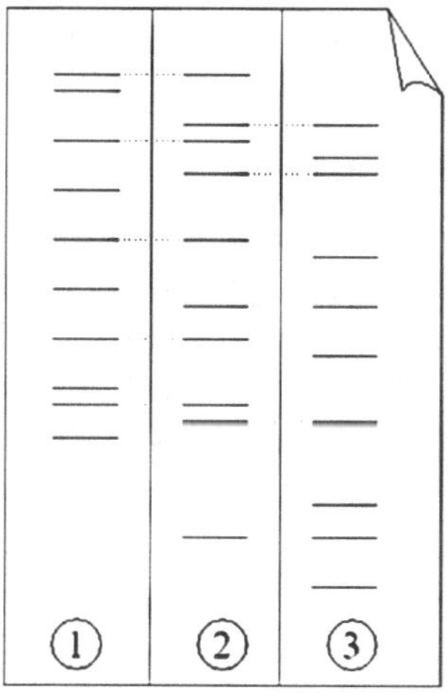

Fig: Electrophoresis of PCR-Amplified DNA Fragments. (1) Father. (2) Child. (3) Mother. The Child has Inherited some, but not all of the Fingerprint of Each of its Parents, Giving it a New, Unique Fingerprint.

Other applications of PCR include DNA sequencing to determine unknown PCR-amplified sequences in which one of the amplification primers may be used in Sanger sequencing, isolation of a DNA sequence to expedite recombinant DNA technologies involving the insertion of a DNA sequence into a plasmid or the genetic material of another organism. Bacterial colonies (E.coli) can be rapidly screened by PCR for correct DNA vector constructs. PCR may also be used for genetic fingerprinting; a forensic technique used to identify a person or organism by comparing experimental DNAs through different PCR-based methods. Some PCR 'fingerprints' methods have high discriminative power and can be used to identify genetic relationships between individuals, such as parent-child or between siblings, and are used in paternity testing. This technique may also be used to determine evolutionary relationships among organisms.

Amplification and Quantitation of DNA

Because PCR amplifies the regions of DNA that it targets, PCR can be used to analyze extremely small amounts of sample. This is often critical for forensic analysis, when only a trace amount of DNA is available as evidence. PCR may also be used in the analysis of ancient DNA that is thousands of years old. These PCR-based techniques have been successfully used on animals, such as a forty-thousand-year-old mammoth, and also on human DNA, in applications ranging from the analysis of Egyptian mummies to the identification of a Russian Tsar.

Quantitative PCR methods allow the estimation of the amount of a given sequence present in a sample – a technique often applied to quantitatively determine levels of gene expression. Real-time PCR is an established tool for DNA quantification that measures the accumulation of DNA product after each round of PCR amplification.

PCR in Diagnosis of Diseases

PCR allows early diagnosis of malignant diseases such as leukemia and lymphomas, which is currently the highest developed in cancer research and is already being used routinely. PCR assays can be performed directly on genomic DNA samples to detect translocation-specific malignant cells at a sensitivity which is at least 10,000 fold higher than other methods.PCR also permits identification of non-cultivatable or slow-growing microorganisms such as mycobacteria, anaerobic bacteria, or viruses from tissue culture assays and animal models. The basis for PCR diagnostic applications in microbiology is the detection of infectious agents and the discrimination of non-pathogenic from pathogenic strains by virtue of specific genes.Viral DNA can likewise be detected by PCR. The primers used need to be specific to the targeted sequences in the DNA of a virus, and the PCR can be used for diagnostic analyses or DNA sequencing of the viral genome. The high sensitivity of PCR permits virus detection soon after infection and even before the onset of disease. Such early detection may give physicians a significant lead in treatment. The amount of virus ("viral load") in a patient can also be quantified by PCR-based DNA quantitation techniques.

VARIATIONS ON THE BASIC PCR TECHNIQUE

- Allele-specific PCR: This diagnostic or cloning technique is used to identify or utilize single-nucleotide polymorphisms (SNPs) (single base differences in DNA). It requires prior knowledge of a DNA sequence, including differences between alleles, and uses primers whose 3' ends encompass the SNP. PCR amplification under stringent conditions is much less efficient in the presence of a mismatch between template and primer, so successful amplification with an SNP-specific primer signals presence of the specific SNP in a sequence.
- Assembly PCR or Polymerase Cycling Assembly (PCA): Assembly PCR is the artificial synthesis of long DNA sequences by performing PCR on a pool of long oligonucleotides with short overlapping segments. The oligonucleotides alternate between sense and antisense directions, and the overlapping segments determine the order of the PCR fragments thereby selectively producing the final long DNA product.
- Asymmetric PCR: Asymmetric PCR is used to preferentially amplify one strand of the original DNA more than the other. It finds use in some types of sequencing and hybridization probing where having only one of the two complementary stands is required. PCR is carried out as usual, but with a great excess of the primers for the chosen strand. Due to the slow (arithmetic) amplification later in the reaction after the limiting primer has been used up, extra cycles of PCR are required. A recent modification on this process, known as Linear-After-The-Exponential-PCR (LATE-PCR), uses a limiting primer with a higher melting temperature (Melting temperature|Tm) than the excess primer to maintain reaction efficiency as the limiting primer concentration decreases mid-reaction.
- Helicase-dependent amplification: This technique is similar to traditional PCR, but uses a constant temperature rather than cycling through denaturation and annealing/extension cycles. DNA Helicase, an enzyme that unwinds DNA, is used in place of thermal denaturation.
- Hot-start PCR: This is a technique that reduces non-specific amplification during the initial set up stages of the PCR. The technique may be performed manually by heating the reaction components to the melting temperature (*e.g.*, 95ÚC) before adding the polymerase. Specialized enzyme systems have been developed that inhibit the polymerase's activity at ambient temperature, either by the binding of an antibody or by the presence of covalently bound inhibitors that only dissociate after a high-temperature activation step. Hot-start/

cold-finish PCR is achieved with new hybrid polymerases that are inactive at ambient temperature and are instantly activated at elongation temperature.

- Intersequence-specific (ISSR) PCR: a PCR method for DNA fingerprinting that amplifies regions between some simple sequence repeats to produce a unique fingerprint of amplified fragment lengths.
- Inverse PCR: a method used to allow PCR when only one internal sequence is known. This is especially useful in identifying flanking sequences to various genomic inserts. This involves a series of DNA digestions and self ligation, resulting in known sequences at either end of the unknown sequence.
- Ligation-mediated PCR: This method uses small DNA linkers ligated to the DNA of interest and multiple primers annealing to the DNA linkers; it has been used for DNA sequencing, genome walking, and DNA footprinting.
- Methylation-specific PCR (MSP): The MSP method was developed by Stephen Baylin and Jim Herman at the Johns Hopkins School of Medicine, and is used to detect methylation of CpG islands in genomic DNA. DNA is first treated with sodium bisulfite, which converts unmethylated cytosine bases to uracil, which is recognized by PCR primers as thymine. Two PCRs are then carried out on the modified DNA, using primer sets identical except at any CpG islands within the primer sequences. At these points, one primer set recognizes DNA with cytosines to amplify methylated DNA, and one set recognizes DNA with uracil or thymine to amplify unmethylated DNA. MSP using qPCR can also be performed to obtain quantitative rather than qualitative information about methylation.
- Miniprimer PCR: Miniprimer PCR uses a novel thermostable polymerase (S-Tbr) that can extend from short primers ("smalligos") as short as 9 or 10 nucleotides, instead of the approximately 20 nucleotides required by Taq. This method permits PCR targeting smaller primer binding regions, and is particularly useful to amplify unknown, but conserved, DNA sequences, such as the 16S (or eukaryotic 18S) rRNA gene. 16S rRNA miniprimer PCR was used to characterize a microbial mat community growing in an extreme environment, a hypersaline pond in Puerto Rico. In that study, deeply divergent sequences were discovered with high frequency and included representatives that deûned two new division-level taxa, suggesting that miniprimer PCR may reveal new dimensions of microbial diversity. By enlarging the "sequence space" that may be queried by PCR primers, this technique may enable novel PCR strategies that are not possible within the limits of primer design imposed by Taq and other commonly used enzymes.

- Multiplex Ligation-dependent Probe Amplification (MLPA): permits multiple targets to be amplified with only a single primer pair, thus avoiding the resolution limitations of multiplex PCR.
- Multiplex-PCR: The use of multiple, unique primer sets within a single PCR mixture to produce amplicons of varying sizes specific to different DNA sequences. By targeting multiple genes at once, additional information may be gained from a single test run that otherwise would require several times the reagents and more time to perform. Annealing temperatures for each of the primer sets must be optimized to work correctly within a single reaction, and amplicon sizes, *i.e.*, their base pair length, should be different enough to form distinct bands when visualized by gel electrophoresis.
- Nested PCR: increases the specificity of DNA amplification, by reducing background due to non-specific amplification of DNA. Two sets of primers are being used in two successive PCRs. In the first reaction, one pair of primers is used to generate DNA products, which besides the intended target, may still consist of non-specifically amplified DNA fragments. The product(s) are then used in a second PCR with a set of primers whose binding sites are completely or partially different from and located 3' of each of the primers used in the first reaction. Nested PCR is often more successful in specifically amplifying long DNA fragments than conventional PCR, but it requires more detailed knowledge of the target sequences.
- Overlap-extension PCR: is a genetic engineering technique allowing the construction of a DNA sequence with an alteration inserted beyond the limit of the longest practical primer length.
- Quantitative PCR (Q-PCR): is used to measure the quantity of a PCR product (preferably real-time). It is the method of choice to quantitatively measure starting amounts of DNA, cDNA or RNA. Q-PCR is commonly used to determine whether a DNA sequence is present in a sample and the number of its copies in the sample. The method with currently the highest level of accuracy is Quantitative real-time PCR. It is often confusingly known as RT-PCR (Real Time PCR) or RQ-PCR. QRT-PCR or RTQ-PCR are more appropriate contractions. RT-PCR commonly refers to reverse transcription PCR, which is often used in conjunction with Q-PCR. QRT-PCR methods use fluorescent dyes, such as Sybr Green, or fluorophore-containing DNA probes, such as TaqMan, to measure the amount of amplified product in real time.
- RT-PCR: (Reverse Transcription PCR) is a method used to amplify, isolate or identify a known sequence from a cellular or tissue RNA. The PCR is preceded by a reaction using reverse transcriptase to

convert RNA to cDNA. RT-PCR is widely used in expression profiling, to determine the expression of a gene or to identify the sequence of an RNA transcript, including transcription start and termination sites and, if the genomic DNA sequence of a gene is known, to map the location of exons and introns in the gene. The 5' end of a gene (corresponding to the transcription start site) is typically identified by an RT-PCR method, named RACE-PCR, short for *Rapid Amplification of cDNA Ends*.

- Solid Phase PCR: encompasses multiple meanings, including Polony Amplification (where PCR colonies are derived in a gel matrix, for example), 'Bridge PCR' (the only primers present are covalently linked to solid support surface), conventional Solid Phase PCR (where Asymmetric PCR is applied in the presence of solid support bearing primer with sequence matching one of the aqueous primers) and Enhanced Solid Phase PCR (where conventional Solid Phase PCR can be improved by employing high Tm solid support primer with application of a thermal 'step' to favour solid support priming).
- TAIL-PCR: Thermal asymmetric interlaced PCR is used to isolate unknown sequence flanking a known sequence. Within the known sequence TAIL-PCR uses a nested pair of primers with differing annealing temperatures; a degenerate primer is used to amplify in the other direction from the unknown sequence.
- Touchdown PCR: a variant of PCR that aims to reduce non-specific background by gradually lowering the annealing temperature as PCR cycling progresses. The annealing temperature at the initial cycles is usually a few degrees (3-5ÚC) above the T_m of the primers used, while at the later cycles, it is a few degrees (3-5ÚC) below the primer T_m. The higher temperatures give greater specificity for primer binding, and the lower temperatures permit more efficient amplification from the specific products formed during the initial cycles.
- PAN-AC: This method uses isothermal conditions for amplification, and may be used in living cells.
- Universal Fast Walking: this method allows genome walking and genetic fingerprinting using a more specific 'two-sided' PCR than conventional 'one-sided' approaches (using only one gene-specific primer and one general primer - which can lead to artefactual 'noise') by virtue of a mechanism involving lariat structure formation. Streamlined derivatives of UFW are LaNe RAGE (lariat-dependent nested PCR for rapid amplification of genomic DNA ends), 5'RACE LaNe and 3'RACE LaNe.

HISTORY

A 1971 paper in the Journal of Molecular Biology by Kleppe and co-workers first described a method using an enzymatic assay to replicate a short DNA template with primers *in vitro*. However, this early manifestation of the basic PCR principle did not receive much attention, and the invention of the polymerase chain reaction in 1983 is generally credited to Kary Mullis. At the core of the PCR method is the use of a suitable DNA polymerase able to withstand the high temperatures of >90°C (>195°F) required for separation of the two DNA strands in the DNA double helix after each replication cycle. The DNA polymerases initially employed for in vitro experiments presaging PCR were unable to withstand these high temperatures. So the early procedures for DNA replication were very inefficient, time consuming, and required large amounts of DNA polymerase and continual handling throughout the process.

A 1976 discovery of Taq polymerase a DNA polymerase purified from the thermophilic bacterium, *Thermus aquaticus*, which naturally occurs in hot (50 to 80 °C (120 to 175 °F)) environments paved the way for dramatic improvements of the PCR method. The DNA polymerase isolated from *T. aquaticus* is stable at high temperatures remaining active even after DNA denaturation, thus obviating the need to add new DNA polymerase after each cycle. This allowed an automated thermocycler-based process for DNA amplification.

At the time he developed PCR in 1983, Mullis was working in Emeryville, California for Cetus Corporation, one of the first biotechnology companies. There, he was responsible for synthesizing short chains of DNA. Mullis has written that he conceived of PCR while cruising along the Pacific Coast Highway one night in his car. He was playing in his mind with a new way of analyzing changes (mutations) in DNA when he realized that he had instead invented a method of amplifying any DNA region through repeated cycles of duplication driven by DNA polymerase.

In *Scientific American*, Mullis summarized the procedure: "Beginning with a single molecule of the genetic material DNA, the PCR can generate 100 billion similar molecules in an afternoon. The reaction is easy to execute. It requires no more than a test tube, a few simple reagents, and a source of heat." He was awarded the Nobel Prize in Chemistry in 1993 for his invention, seven years after he and his colleagues at Cetus first put his proposal to practice. However, some controversies have remained about the intellectual and practical contributions of other scientists to Mullis' work, and whether he had been the sole inventor of the PCR principle.

Patent Wars

The PCR technique was patented by Cetus Corporation, where Mullis worked when he invented the technique in 1983. The *Taq* polymerase enzyme was also covered by patents. There have been several high-profile lawsuits related to the technique, including an unsuccessful lawsuit brought by DuPont. The pharmaceutical company Hoffmann-La Roche purchased the rights to the patents in 1992 and currently holds those that are still protected. A related patent battle over the Taq polymerase enzyme is still ongoing in several jurisdictions around the world between Roche and Promega. The legal arguments have extended beyond the life of the original PCR and Taq polymerase patents.

GENOMICS

The term "Genome" was coined by H. Winkler to describe 'the complete set of chromosomal and extrachrmosomal genes of:

- *Definition*: Genomics is the scientific discipline of mapping, sequencing, and analyzing genomes. This is the field in which basically one uses the genomic information in light of the new experimental data obtained, to answer the related biological, medical, or industrial questions and challenges faced by the human population. Genome is the complete set of chromosomal genes of an organism, a cell, an organelle or a virus.

The term was coined by Thomas Roderick in 1986. There are different branches of this field now. Genomics can now be broadly divided into:

Structural Genomics

Structural genomics involves the construction of high- resolution genetic, physical, or transcript maps of the organism usually during the initial phase of genome analysis which results in the ultimate physical map of an organism with its complete genome sequencing.

In recent years, due to the completion of the various genome-sequencing projects, the term structural genomics now also includes the high-throughput determination of three-dimensional structures of all proteins. Hence the structural genomics, establishes the relationship between the sequence and structure of a genome.

Functional Genomics

Functional Genomics deals with the reconstruction of the genome to determine the biological function of genes and gene interactions. The information provided by the structural genomics is used to design experiments to identify the function of the genes.

The investigations range from studying the single genes or proteins to studying all genes or proteins at once in a systematic manner which also involves large scale experimental methodology with statistical and computational analysis of the results. Hence, the functional genomics helps to co-relate the structure and sequence of the genome to its function.

GENOME SEQUENCING PROJECTS

To understand the functional and structural aspects of a gene, one needs to know the sequence of a gene. Sequencing can be mainly by using two methodologies:

- *Directed Sequencing of BAC Contigs*: BAC (Bacterial Artificial Chromosome) vectors are used to make genomic libraries in which the insert size is 80-100 kb. After screening and locating the common restriction fragments, the BAC clones are then mapped to find the overlapping arrays of contiguous clones called contigs. The large DNA fragments are then broken into small pieces to complete the sequencing of mapped contigs. Hence, using this directed strategy, DNA pieces from adjacent stretches of a chromosome are sequenced.
- *Random Shotgun Sequencing*: In this method, the genome is randomly broken up into sizes of 2-10 kb range and then inserted into a vector (plasmids). These are then sequenced with the help of overlapping or common sequences. A computer programme identifies the overlapping sequences and joins them into one continuous stretch.

The Advantages of Sequencing the Genome

- It provides a basis for the discovery of all genes and thus one can construct an inventory of genes.
- The sequence shows the relationships between the genes which helps in future experimentation.
- The sequencing provides an index to draw and organise all genetic information about the organism.
- For times to come, the whole genome sequence becomes an archive for the future which contains all the genetic information required to make an organism.

After the sequencing of the genes, we can investigate the functional properties such as the rate of transcription, the level of expression under different environmental conditions of genes. Two techniques have been used to understand the working of the genes

MICROARRAY

A microarray is typically a glass slide, on to which DNA molecules are attached at fixed locations or spots. This technology works on the base pairing or hybridisation of single-stranded nucleic acids. The most commonly used molecule is cDNA or complementary DNA which is derived from messenger RNA.

The cDNA is labeled with a reporter molecule (generally fluorescent molecules) to detect it when bound to microarray. A particular genome generally has a large number of genes which have their own RNA transcripts. Under *in vitro* conditions RNA is unstable (prone to attack by RNases) therefore it is converted into cDNA by using RT-PCR technique. Since, each cDNA is derived from a distinct messenger RNA, each feature represents an expressed gene. The reporter fluorescent molecule emits fluorescence which indicates high expression levels of a particular gene. Thus by using this technique the expression pattern and functioning of thousands of genes can be analysed on a small slide only.

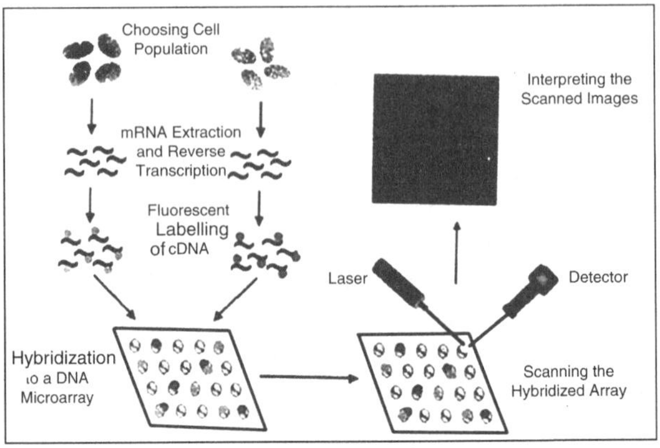

Fig. Major Steps Involved in Comparative Microarray Hybridisation Experiments

STRUCTURE OF DNA AND ENCODES

The structure of DNA encodes all the information every cell needs to function and thrive. In addition, DNA carries hereditary information in a form that can be copied and passed intact from generation to generation. A gene is a segment of DNA.

The biochemical instructions found within most genes, known as the genetic code, specify the chemical structure of a particular protein. Proteins are composed of long chains of amino acids, and the specific sequence of these amino acids dictates the function of each protein. The DNA structure of a gene determines the arrangement of amino acids in a protein, ultimately determining the type and function of the protein manufactured.

DNA Structure

DNA molecules form from chains of building blocks called nucleotides. Each nucleotide consists of a sugar molecule called deoxyribose that bonds to a phosphate molecule and to a nitrogen-containing compound, known as a base. DNA uses four bases in its structure: adenine (A), cytosine (C), guanine (G), and thymine (T). The order of the bases in a DNA molecule—the genetic code—determines the amino acid sequence of a protein.

In the cells of most organisms, two long strands of DNA join in a single molecule that resembles a spiraling ladder, commonly called a double helix. Alternating phosphate and sugar molecules form each side of this ladder. Bases from one DNA strand join with bases from another strand to form the rungs of the ladder, holding the double helix together.

The pairing of bases in the DNA double helix is highly specific—adenine always joins with thymine, and guanine always links to cytosine. These base combinations, known as complementary base pairing, play a fundamental role in DNA's function by aiding in the replication and storage of genetic information. Complementary base pairing also enables scientists to predict the sequence of bases on one strand of a DNA molecule if they know the order on the corresponding, or complementary, DNA strand. Scientists use complementary base pairing to help identify the genes on a particular chromosome and to develop methods used in genetic engineering.

Genes line up in a row along the length of a DNA molecule. In humans a single gene can vary in length from 100 to over 1,000,000 bases. Genes make up less than 2 per cent of the length of a DNA molecule. The rest of the DNA molecule is made up of long, highly repetitive nucleotide sequences. Once dismissed as "junk" DNA, scientists now believe these nucleotide sequences may play a role in the survival of cells. Identifying the function of these sequences is a thriving field of genetics research.

DNA Replication

In order for inherited traits to be transmitted from parent to child, the genetic information encoded in DNA must be copied with great precision during cell division. The accuracy of DNA replication depends upon the complementary pairing of bases. During replication, the DNA double helix unwinds and bonds joining the base pairs break, separating the DNA molecule into two separate strands. Each strand of DNA directs the synthesis of another complementary strand. The unpaired bases of each DNA strand attach to bases floating within the cell. But the DNA strand's unpaired bases bond only with specific, complementary bases—for example, an adenine base will bond only with a thymine base and a cytosine bases will pair only with a guanine base.

Once all of the bases of a DNA strand bond to complementary bases, the complementary bases then link to each other, forming a new DNA double-helix molecule. Thus the original DNA molecule replicates into two DNA molecules that are exact duplicates.

STRUCTURAL ANALYSIS OF GENOME RELATIONSHIPS

Cytogenetic analysis of somatic or meiotic preparations by conventional staining, chromosome banding and *in situ* hybridization ('chromosome painting') provides information on the overall organization of plant genomes. Meiotic chromosome pairing behaviour in natural or artificial hybrids is also an important guide to genome relationships, and until recently was the only method for assaying genome similarities. However, pairing can be strongly affected by single genes such as *Ph1* in wheat, and the degree of pairing between related genomes may not directly reflect overall genome homology.

Nevertheless, chromosome pairing has been shown to be in good agreement with the conventional taxonomy of the *Triticeae*. Analysis of hybrids is important because allo- and autopolyploidy are common in plant evolution.

Allopolyploids include crop species such as oilseed rape (canola), oat, wheat, groundnut, upland and sea island cotton, some bananas, strawberry, arabica coffee and tobacco. The cultivated species derived from polyploid hybrids are generally thought to have combined agronomically important traits from their diploid ancestors.

GENOME SIZE AND COMPOSITION

The amount of nuclear DNA varies greatly between plant species. For example, the haploid genome of hexaploid wheat contains about 1.7 3 1010 bp, each of the three genomes being roughly similar in size to that of barley. The genome of rice, in contrast, contains only about 4 3 108 bp.

Variation in DNA amount, other than variation due to ploidy level, is probably due almost entirely to variation in the copy number of repeated DNA sequences. Large genome species contain higher proportions of short tandem repeats, often present as blocks of heterochromatin, and a higher proportion of dispersed repeats.

A major component of the latter in wheat and barley are retroelement-like sequences such as Wis-1, Wis-2 and Bis-1. Long-range analysis of repeat structure in large cereal genomes suggests that chromosome evolution has involved duplication of blocks of repeats.

Nuclear DNA amount can also vary within plant species. This variation, and variation between species, can often be correlated with ecogeographical variables suggesting that DNA amount is of adaptive significance. In maize, for example, DNA amount declines with increasing altitude or increasing latitude.

This variation can be as much as 30 per cent of the genome size and involves variation in the amount of C-band heterochromatin as well as variation in dispersed repeats. Proposed mechanisms producing changes in repetitive sequences include unequal crossing over, transposition, retrotransposition, gene conversion and replication slippage. They are rapid in evolutionary terms and consequently the copy number of specific sequences can vary greatly in different species or lineages. New sequence variants also arise frequently, probably because there is little or no selection constraining the DNA sequence of many classes of repeats.

As a result, there are families of repeated sequences which are specific to individual grass species, or genomes, or which differ greatly in copy number between closely related species, and in general the divergence of repetitive sequences can be used to indicate the degree of relatedness of whole genomes in diploid species and their polyploid derivatives.

Repeat sequence plasticity can be exploited for Southern blot analysis and for chromosome painting using dispersed repeats or total genomic DNA. For example, wheat chromosomes can be clearly differentiated from those of rye or barley in hybrids and the three genomes can be differentiated in hexaploid wheat.

Fluorescent *in situ* hybridization (FISH) is emerging as the method of choice for such work because several probes, each labelled to give a different colour when illuminated by specific wavelengths, can be hybridized to the same preparation. *In situ* hybridization is extremely valuable for identifying hybrids and introgression lines in crop species and for studies of plant evolution in general.

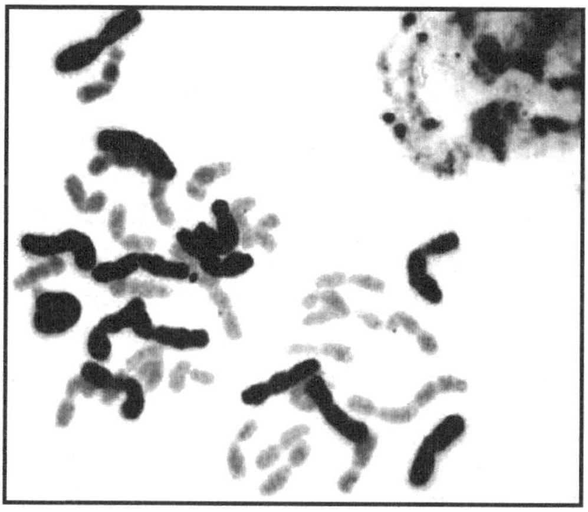

Fig. Genomic in Situ Hybridization Using Digoxigenin-labelled rye DNA on root-tip Metaphase Chromosomes of Hexaploid Triticale. The 14 rye Chromosomes are Brightly Stained.

DNA FINGERPRINTING

DNA Fingerprinting, method of identification that compares fragments of deoxyribonucleic acid (DNA) It is sometimes called DNA typing. DNA is the genetic material found within the cell nuclei of all living things. In mammals the strands of DNA are grouped into structures called chromosomes. With the exception of identical twins, the complete DNA of each individual is unique.

A DNA fingerprint is constructed by first extracting a DNA sample from body tissue or fluid such as hair, blood, or saliva. The sample is then segmented using enzymes, and the segments are arranged by size using a process called electrophoresis. The segments are marked with probes and exposed on X-ray film, where they form a characteristic pattern of black bars—the DNA fingerprint. If the DNA fingerprints produced from two different samples match, the two samples probably came from the same person.

DNA fingerprinting was first developed as an identification technique in 1985. Originally used to detect the presence of genetic diseases, DNA fingerprinting soon came to be used in criminal investigations and forensic science. The first criminal conviction based on DNA evidence in the United States occurred in 1988. In criminal investigations, DNA fingerprints derived from evidence collected at the crime scene are compared to the DNA fingerprints of suspects. The DNA evidence can implicate or exonerate a suspect.

Generally, courts have accepted the reliability of DNA testing and admitted DNA test results into evidence. However, DNA fingerprinting is controversial in a number of areas: the accuracy of the results, the cost of testing, and the possible misuse of the technique.

The accuracy of DNA fingerprinting has been challenged for several reasons. First, because DNA segments rather than complete DNA strands are "fingerprinted," a DNA fingerprint may not be unique; large-scale research to confirm the uniqueness of DNA fingerprinting test results has not been conducted. In addition, DNA fingerprinting is often performed in private laboratories that may not follow uniform testing standards and quality controls. Also, since human beings must interpret the test, human error could lead to false results. DNA fingerprinting is expensive. Suspects who are unable to provide their own DNA experts may not be able to adequately defend themselves against charges based on DNA evidence.

In the United States, the Federal Bureau of Investigation (FBI) has created a national database of genetic information called the National DNA Index System. The database contains DNA obtained from convicted criminals and from evidence found at crime scenes. Some experts fear that this database might be used for unauthorized purposes, such as identifying individuals with stigmatizing illnesses such as acquired immunodeficiency syndrome (AIDS).

DNA SEQUENCING

The term DNA sequencing encompasses biochemical methods for determining the order of the nucleotide bases, adenine, guanine, cytosine, and thymine, in a DNA oligonucleotide. The sequence of DNA constitutes the heritable genetic information in nuclei, plasmids, mitochondria, and chloroplasts that forms the basis for the developmental programmes of all living organisms.

Determining the DNA sequence is therefore useful in basic research studying fundamental biological processes, as well as in applied fields such as diagnostic or forensic research. The advent of DNA sequencing has significantly accelerated biological research and discovery.

The rapid speed of sequencing attained with modern DNA sequencing technology has been instrumental in the large-scale sequencing of the human genome, in the Human Genome Project. Related projects, often by scientific collaboration across continents, have generated the complete DNA sequences of many animal, plant, and microbial genomes.

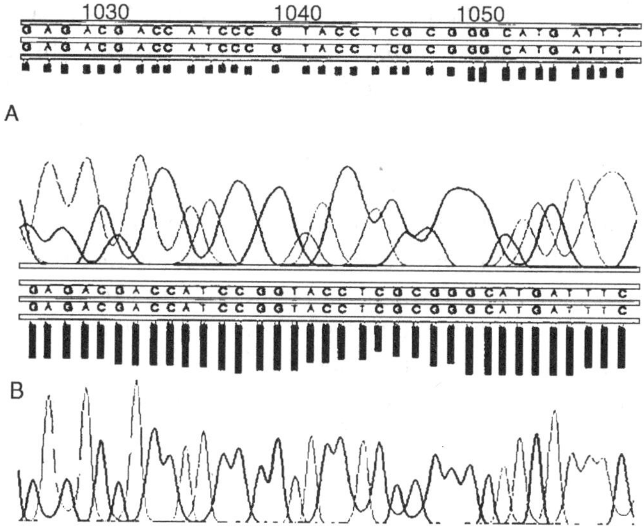

Fig. DNA Sequence Trace

Early Methods

For thirty years, a large proportion of DNA sequencing has been carried out with the chain-termination method developed by Frederick Sanger and coworkers in 1975. Prior to the development of rapid DNA sequencing methods in the early 1970s by Sanger in England and Walter Gilbert and Allan Maxam at Harvard, a number of laborious methods were used. For instance, in 1973 Gilbert and Maxam reported the sequence of 24 basepairs using a method known as wandering-spot analysis. RNA sequencing, which for technical reasons is easier to perform than DNA sequencing, was one of the earliest forms of nucleotide sequencing. The major landmark of RNA sequencing, dating from the pre-recombinant DNA era, is the sequence of the first complete gene and then the complete genome of Bacteriophage MS2..

Maxam-Gilbert Sequencing

In 1976-1977, Allan Maxam and Walter Gilbert developed a DNA sequencing method based on chemical modification of DNA and subsequent cleavage at specific bases. Chemical sequencing method two years after the ground-breaking paper of Sanger and Coulson on plus-minus sequencing, Maxam-Gilbert sequencing rapidly became more popular, since purified DNA could be used directly, while the initial Sanger method required that each read start be cloned for production of single-stranded DNA. However, with the development and improvement of the chain-termination method Maxam-Gilbert sequencing has fallen out of favour due to its technical complexity, extensive use of hazardous chemicals, and difficulties with scale-up. In addition, unlike the chain-termination method, chemicals used in the Maxam-Gilbert method cannot easily be customized for use in a standard molecular biology kit.

In brief, the method requires radioactive labelling at one end and purification of the DNA fragment to be sequenced. Chemical treatment generates breaks at a small proportion of one or two of the four nucleotide bases in each of four reactions (G, A+G, C, C+T). Thus a series of labelled fragments is generated, from the radiolabelled end to the first 'cut' site in each molecule. The fragments are then size-separated by gel electrophoresis, with the four reactions arranged side by side. To visualize the fragments generated in each reaction, the gel is exposed to X-ray film for autoradiography, yielding an image of a series of dark 'bands' corresponding to the radiolabelled DNA fragments, from which the sequence may be inferred. Also sometimes known as 'chemical sequencing', this method originated in the study of DNA-protein interactions (footprinting), nucleic acid structure and epigenetic modifications to DNA, and within these it still has important applications.

CHAIN-TERMINATION METHODS

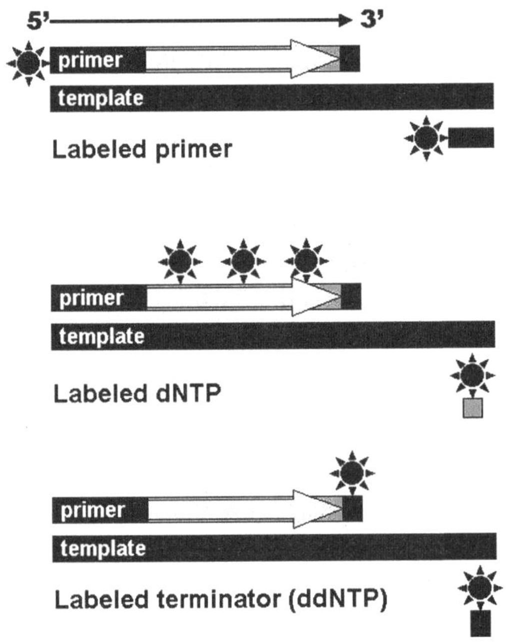

Fig: Part of a radioactively labelled sequencing gel

While the chemical sequencing method of Maxam and Gilbert, and the plus-minus method of Sanger and Coulson were orders of magnitude faster than previous methods, the chain-terminator method developed by Sanger was even more efficient, and rapidly became the method of choice. The Maxam-Gilbert technique requires the use of highly toxic chemicals, and large amounts of radiolabeled DNA, whereas the chain-terminator method uses fewer toxic chemicals and lower amounts of radioactivity. The key principle of the Sanger method was the use of dideoxynucleotides triphosphates (ddNTPs) as DNA chain terminators.

The classical chain-termination or Sanger method requires a single-stranded DNA template, a DNA primer, a DNA polymerase, radioactively or fluorescently labeled nucleotides, and modified nucleotides that terminate DNA strand elongation. The DNA sample is divided into four separate sequencing reactions, containing the four standard deoxynucleotides (dATP, dGTP, dCTP and dTTP) and the DNA polymerase.

To each reaction is added only one of the four dideoxynucleotides (ddATP, ddGTP, ddCTP, or ddTTP). These dideoxynucleotides are the chain-terminating nucleotides, lacking a 3'-OH group required for the formation of a phosphodiester bond between two nucleotides during DNA strand elongation. Incorporation of a dideoxynucleotide into the nascent (elongating) DNA strand therefore terminates DNA strand extension, resulting in various DNA fragments of varying length. The dideoxynucleotides are added at lower concentration than the standard deoxynucleotides to allow strand elongation sufficient for sequence analysis.

The newly synthesized and labeled DNA fragments are heat denatured, and separated by size (with a resolution of just one nucleotide) by gel electrophoresis on a denaturing polyacrylamide-urea gel. Each of the four DNA synthesis reactions is run in one of four individual lanes (lanes A, T, G, C); the DNA bands are then visualized by autoradiography or UV light, and the DNA sequence can be directly read off the X-ray film or gel image. In the image on the right, X-ray film was exposed to the gel, and the dark bands correspond to DNA fragments of different lengths. A dark band in a lane indicates a DNA fragment that is the result of chain termination after incorporation of a dideoxynucleotide (ddATP, ddGTP, ddCTP, or ddTTP). The terminal nucleotide base can be identified according to which dideoxynucleotide was added in the reaction giving that band. The relative positions of the different bands among the four lanes are then used to read (from bottom to top) the DNA sequence as indicated.There are some technical variations of chain-termination sequencing.

In one method, the DNA fragments are tagged with nucleotides containing radioactive phosphorus for radiolabelling. Alternatively, a primer labeled at the 5' end with a fluorescent dye is used for the tagging. Four separate reactions are still required, but DNA fragments with dye labels can be read using an optical system, facilitating faster and more economical analysis and automation. This approach is known as 'dye-primer sequencing'.

The later development by L Hood and coworkers of fluorescently labeled ddNTPs and primers set the stage for automated, high-throughput DNA sequencing. The different chain-termination methods have greatly simplified the amount of work and planning needed for DNA sequencing. For example, the chain-termination-based "Sequenase" kit from USB Biochemicals contains most of the reagents needed for sequencing, prealiquoted and ready to use. Some sequencing problems can occur with the Sanger method, such as non-specific binding of the primer to the DNA, affecting accurate read-out of the DNA sequence. In addition, secondary structures within the DNA template, or contaminating RNA randomly priming at the DNA template can also affect the fidelity of the obtained sequence. Other contaminants affecting the reaction may consist of extraneous DNA or inhibitors of the DNA polymerase.

SOME INTERESTING USES OF DNA FORENSIC IDENTIFICATION

IDENTIFYING SEPTEMBER 11TH VICTIMS

Identifying the victims of the September 11, 2001, World Trade Centre attack presented a unique forensic challenge because the number and identity of the victims were unknown and many victims were represented only by bone and tissue fragments. At the time of the attack, no systems were in place for rapidly identifying victims in disasters with more than 500 fatalities. The National Institutes of Justice assembled a panel of experts from the National Institutes of Health and other institutions to develop processes to identify victims using DNA collected at the site. Panel members produced forms and kits needed to enable the medical examiner's office to collect reference DNA from victims' previously stored medical specimens.

These specimens were collected and entered into a database. The medical examiner's office also received about 20,000 pieces of human remains from the World Trade Centre site, and a database of the victims' DNA profiles was created. New information technology infrastructure was developed for data transfer between the state police and medical examiner's office and to interconnect the databases and analytical tools used by panel members. In 2005 the search was declared at an end because many of the unidentified remains were too small or too damaged to be identified by the DNA extraction methods available at that time. Remains of only 1585, of the 2792 people known to have died had been identified. In 2007, the medical examiner's office reopened the search after the Bode Technology Group developed a new methodology of DNA extraction that required much less sample material than previously necessary. The victim DNA database and the new methods have allowed more victims to be identified, and further identifications will be possible as forensic DNA technology improves.

The DNA Shoah Project

The DNA Shoah Project is a genetic database of people who lost family during the Holocaust. The database will serve to reunite families separated during wartime and aid in identifying victims who remain buried anonymously throughout Europe.

Disappeared Children in Argentina

Numerous people (known as "the Disappeared") were kidnapped and murdered in Argentina in the 1970s. Many were pregnant. Their children were taken at birth and, along with other kidnapped children, were raised by their kidnappers. The grandparents of these children have been looking for them for many years. Read an article about a DNA researcher who has been helping them.

- Tomb of the Unknowns
- Son of Louis XVI and Marie Antionette PARIS, Apr 19, 2000 (Reuters) — Scientists cracked one of the great mysteries of European history by using DNA tests to prove that the son of executed French King Louis XVI and Marie-Antoinette died in prison as a child. Royalists have argued for 205 years over whether Louis-Charles de France perished in 1795 in a grim Paris prison or managed to escape the clutches of the French Revolution. In December 1999, the presumed heart of the child king was removed from its resting place to enable scientists to compare its DNA makeup with samples from living and dead members of the royal family — including a lock of his mother Marie-Antoinette's hair.
- The Murdered Nicholas Romanov, the Last Czar of Russia, and His Family
- Peruvian Ice Maiden The Ice Maiden was a 12-to-14-year old girl sacrificed by Inca priests 500 years ago to satisfy the mountain gods of the Inca people. She was discovered in 1995 by climbers on Mt. Ampato in the Peruvian Andes. She is perhaps the best preserved mummy found in the Andes because she was in a frozen state. Analysis of the Ice Maiden's DNA offers a wonderful opportunity for understanding her genetic origin. If we could extract mitochondrial DNA from the Ice Maiden's tissue and successfully amplify and sequence it, then we could begin to trace her maternal line of descent and possibly locate past and current relatives.
- African Lemba Tribesmen In southern Africa, a people known as the Lemba heed the call of the shofar. They have believed for generations that they are Jews, direct descendants of the biblical patriarchs Abraham, Isaac, and Jacob. However unlikely the Lemba's claims may seem, modern science is finding ways to test them. The ever-growing understanding of human genetics is revealing connections between peoples that have never been seen before.

- Super Bowl XXXIV Footballs and 2000 Summer Olympic Souvenirs The NFL used DNA technology to tag all the Super Bowl XXXIV balls, ensuring their authenticity for years to come and helping to combat the growing epidemic of sports memorabilia fraud. The footballs were marked with an invisible, yet permanent, strand of synthetic DNA. The DNA strand is unique and is verifiable any time in the future using a specially calibrated laser.
  A section of human genetic code taken from several unnamed Australian athletes was added to ink used to mark all official goods — everything from caps to socks — from the 2000 Summer Olympic Games. The technology is used as a way to mark artwork or one-of-a-kind sports souvenirs.
- Migration Patterns Evolutionarily stable mitochondrial DNA and Y chromosomes have allowed bioanthropologists to begin to trace human migration patterns around the world and identify family lineage
- Wine Heritage Using DNA fingerprinting techniques akin to those used to solve crimes and settle paternity suits, scientists at the University of California, Davis, have discovered that 18 of the world's most renowned grapevine varieties, or cultivars are close relatives. These include varieties long grown in northeastern France such as Chardonnay, the "king of whites," and reds such as Pinot and Gamay noir, are close relatives.
- DNA Banks for Endangered Animal Species
- Poached Animals
- Declining Grizzly Bear Population
- Snowball the Cat A woman was murdered in Prince Edward Island, Canada. Her estranged husband was implicated because a snowy white cat hair was found in a jacket near the scene of the crime, and DNA fragments from the hair matched DNA fragments from Snowball, the cat belonging to the husband's parents. See M. Menotti-Raymond *et al.*, "Pet cat hair implicates murder suspect," Nature, 386, 774, 1997. Also see Holmes, Judy,Feline Forensics, Syracuse University Magazine, Summer 2001.
- Angiosperm Witness for the Prosecution The first case in which a murderer was convicted on plant DNA evidence was described in the PBS TV series, "Scientific American Frontiers." A young woman was murdered in Phoenix, Arizona, and a pager found at the scene of the crime led the police to a prime suspect. He admitted picking up the victim but claimed she had robbed him of his wallet and pager. The forensic squad examined the suspect's pickup truck and collected pods later identified as the fruits of the palo verde tree (*Cercidium spp.*). One detective went back to the murder scene and found several

Palo Verde trees, one of which showed damage that could have been caused by a vehicle. The detective's superior officer innocently suggested the possibility of linking the fruits and the tree by using DNA comparison, not realising that this had never been done before. Several researchers were contacted before a geneticist at the University of Arizona in Tucson agreed to take on the case. Of course, it was crucial to establish evidence that would stand up in court on whether individual plants (especially Palo Verde trees) have unique patterns of DNA. A preliminary study on samples from different trees at the murder scene and elsewhere quickly established that each Palo Verde tree is unique in its DNA pattern. It was then a simple matter to link the pods from the suspect's truck to the damaged tree at the murder scene and obtain a conviction. [WNED-TV (PBS - Buffalo, N.Y.)]

DNA FORENSICS DATABASES

NATIONAL DNA DATABANK: CODIS

The COmbined DNA Index System, CODIS, blends computer and DNA technologies into a tool for fighting violent crime. The current version of CODIS uses two indexes to generate investigative leads in crimes where biological evidence is recovered from the crime scene. The Convicted Offender Index contains DNA profiles of individuals convicted of felony sex offences (and other violent crimes). The Forensic Index contains DNA profiles developed from crime scene evidence. All DNA profiles stored in CODIS are generated using STR (short tandem repeat) analysis.

CODIS utilises computer software to automatically search its two indexes for matching DNA profiles. Law enforcement agencies at federal, state, and local levels take DNA from biological evidence (*e.g.*, blood and saliva) gathered in crimes that have no suspect and compare it to the DNA in the profiles stored in the CODIS systems. If a match is made between a sample and a stored profile, CODIS can identify the perpetrator.

This technology is authorised by the DNA Identification Act of 1994. All 50 states have laws requiring that DNA profiles of certain offenders be sent to CODIS. As of August 2007, the database contained over 5 million DNA profiles in its Convicted Offender Index and about 188,000 DNA profiles collected from crime scenes but not connected to a particular offender.

As more offender DNA samples are collected and law enforcement officers become better trained and equipped to collect DNA samples at crime scenes, the backlog of samples awaiting testing throughout the criminal justice system is increasing dramatically. In March 2003 President Bush proposed $1 billion in funding over 5 years to reduce the DNA testing backlog, build crime lab capacity, stimulate research and development, support training, protect the innocent, and identify missing persons. For more information, see the U.S. Department of Justice's Advancing Justice Through DNA Technology.

MORE ON CODIS

- CODIS: Combined DNA Index System - Information from the FBI.
- The FBI Laboratory's Combined DNA Index System Programme - Enter regional information to learn more about CODIS in your area. From Promega Corporation, a major supplier of reagents and other materials to support molecular biology research.
- National Commission on the Future of DNA Evidence.
- Postconviction DNA Testing: Recommendations for Handling Requests - Report from the National Commission on the Future of DNA Evidence.
- What Every Law Enforcement Officer Should Know About DNA Evidence (September 1999) - Report from the National Commission on the Future of DNA Evidence.
- Slide Show: Forensic DNA Legislation 2002 - A look at states' CODIS legislation.
- Ethics of State DNA Collection (2004 meeting presentations and handouts from National Conference of State Legislatures' Criminal Justice Programme, Genetic Technologies Project, and Centre for Ethics in Government)

Bibliography

A.G. Ponniah and George John: *Fish Chromosome Atlas*, National Bureau of Fish Genetic Resources, 1998.

A.K. Sharma and Ramavtar Sharma: *Crop Improvement And Mutation Breeding*, Scientific Publication, Jaipur, 2014.

Biological Control of Plant Pathogens, Weeds and Phytoparasitic Nematodes : *M. Reddi Kumar, M. John Sudheer, B. Chenchu Reddy and B. Srinivasulu*, B.S. Publication, Delhi, 2008.

Dashavantha Reddy, Nagaraja Rao. P and Venkateswara Rao. K: *Pests and Pathogens : Management Strategies*, BS Publication, Delhi, 2010.

J.D. Murray, G.B. Anderson, A.M. Oberbauer and M.M. McGloughlin: *Transgenic Animals in Agriculture*, New India Publication, Delhi, 2005.

K. Manibhushan Rao and A. Mahadevan: *Recent Developments in Biocontrol of Plant Pathogens*, Today & Tomorrow Publication, Delhi, 1996.

Malcolm B Willis: *Dalton's Introduction to Practical Animal Breeding*, Blackwell Science, 2004.

Mousumi Debnath: *Cell and Molecular Biology*, Pointer Publication, Delhi, 2008.

Ramesh Bhandari: *Basic Concepts of Molecular Biology : Problems and Solutions*, Cyber Tech Publications, New Delhi, 2011.

Rehana Abidi: *Fish Pathogens and Diseases in India : A Bibliography (1898-2001)*, National Bureau of Fish Genetic Resources, 2002.

S Kannaiyan and A Gopalam: *Agrobiodiversity, Vol. III. Diversity in Pests and Pathogens*, Associated Publishing Company, Delhi, 2008.

Shilpa Pandey: *Basics of Molecular Biology*, Anmol Publication, Delhi, 2008.

Sohrab Ali: *Elements of Veterinary Genetics*, Arise Publication, Delhi, 2008.

Sohrab Ali: *Practical Animal Breeding*, Arise Publication, Delhi, 2008.

Surzycki: *Basic Techniques in Molecular Biology*, Springer Publication, London, 2003.

Yashodhara Dalmia and Salima Hashmi: *Memory Metaphor Mutations: Contemporary Art of India and Pakistan*, Oxford University Press, Delhi, 2007.

Abha Bhardwaj and Samresh Choudhuri: *Molecular Cell Biology*, Wisdom Press, Delhi, 2012.

Amar Tyagi: *Cell Biology*, Anmol Publications, Delhi, 2006.

Arunima Mukherjee: *Cell Biology*, Oxford Book Company, Jaipur, 2009.

B.D. Sharma and J.C. Rana.: *Plant Genetic Resources of Western Himalayas: Status and Prospects*, Bishen Singh Mahendra Pal Singh, Dehradun, 2005.

B.P. Singh and Umesh Srivastava.: *Plant Genetic Resources in Indian Perspective Theory and Practices*, Indian Council of Agricultural Research, 2004.

E.B. Babcock.: *Genetics and Plant Breeding*, Agrobios Publications, Jodhpur, 2007.

I.D. Pandian.: *Botanical Analysis of Plant Cells*, A.K. Publications, Delhi, 2008.

I.D. Tyagi.: *Plant Breeding and Genetics at a Glance*, South Asian Publications, Delhi, 2005.

K.V. Mohanan.: *Essentials of Plant Breeding*, PHI Learning, Delhi, 2007.

P K Shivaraj Kumar: *Genetic Analysis in Cell Biology*, RBSA Publications, Delhi, 2007.

Prabhakar Sharma, Samresh Choudhuri and Abhijit Paintal: *Handbook of Cell Biology*, Wisdom Press, Delhi, 2011.

S.K. Singh: *Manual of Cell Biology*, Campus Books, Delhi, 2012.

Sachin Padgaonkar: *Basics of Cell Biology*, Anmol Publications, Delhi, 2008.

Index